Physical Geology

Physical Geology

Lily Jones
Emma Byrne
Amelia Byrne

Kruger Brentt
Publishers
2023

Kruger Brentt Publishers UK. LTD.
Company Number 9728962

Regd. Office: 68 St Margarets Road, Edgware, Middlesex HA8 9UU

ISBN: 978-1-78715-104-8

For information on all our publications visit our website at http://krugerbrentt.com/

PREFACE

In its broadest sense, geology is the study of Earth—its interior and its exterior surface, the minerals, rocks and other materials that are around us, the processes that have resulted in the formation of those materials, the water that flows over the surface and through the ground, the changes that have taken place over the vastness of geological time, and the changes that we can anticipate will take place in the near future. Geology is a science, meaning that we use deductive reasoning and scientific methods to understand geological problems. It is, arguably, the most integrated of all of the sciences because it involves the understanding and application of all of the other sciences: physics, chemistry, biology, mathematics, astronomy, and others. But unlike most of the other sciences, geology has an extra dimension, that of time—deep time—billions of years of it. Geologists study the evidence that they see around them, but in most cases, they are observing the results of processes that happened thousands, millions, and even billions of years in the past. Those were processes that took place at incredibly slow rates—millimetres per year to centimetres per year—but because of the amount of time available, they produced massive results. The sedimentary rock that these mountains are made of formed in ocean water over 500 million years ago. A few hundred million years later, these beds were pushed east for tens to hundreds of kilometres by tectonic plate convergence and also pushed up to thousands of metres above sea level. The distinctive line on the slope on the left side of both photos shows the elevation of the edge of the glacier a few hundred years ago. Like almost all other glaciers in the world, it receded after the 18th century because of natural climate change, is now receding even more rapidly because of human-caused climate change.

The term Physical Geology was coined by an English mathematician and ageologist, William Hopkins in 1883. The subject deals with the physical forces andprocesses that bring about changes in the earth's crust or to the surface of the earthon account of their prolonged existence and action. Physical Geology isbroadly divided into two branches—those dealing with the internal dynamics of the earth (Endogenous Geology) and those with external dynamics (ExogenousGeology).

The present book contains twenty-two chapters covering all related disciplines. These chapters include Geology: an introduction, Minerals, Igneous rocks that are intrusive, Magma and volcanism, Soil formation and weathering, Sedimentary rocks and sediments, Metamorphic rocks and metamorphism, Geological time scale, Interior of the earth, Plate tectonics theory, Earthquakes, Understanding of geological structures, Streams and floods, Groundwater, Mass wasting, Glaciers and glaciation, Shorelines, Geology of the oceans, Climate change and its impacts, Geological resources, Western canada's geological history, The solar system's and earth's history. Related terminology is given at the end for ready reference.

We are grateful to all those persons as well as various books, manuals, periodicals, magazines, journals etc. that helped in the preparation of this book. In spite of the best efforts, it is possible that some errors may have occurred into the compilation and editing of the book. Further queries, constructive suggestions and criticisms for the improvement of the book are always welcome and shall be thankfully acknowledged.

Lily Jones
Emma Byrne
Amelia Byrne

CONTENTS

CHAPTER

1 GEOLOGY: AN INTRODUCTION

1.1 GEOLOGY: AN OVERVIEW

In its broadest sense, geology is the study of Earth — its interior and its exterior surface, the rocks and other materials that are around us, the processes that have resulted in the formation of those materials, the water that flows over the surface and lies underground, the changes that have taken place over the vastness of geological time, and the changes that we can anticipate will take place in the near future. Geology is a science, meaning that we use deductive reasoning and scientific methods (see Box 1.1) to understand geological problems. It is, arguably, the most integrated of all of the sciences because it involves the understanding and application of all of the other sciences: physics, chemistry, biology, mathematics, astronomy, and others. But unlike most of the other sciences, geology has an extra dimension, that of time — deep time — billions of years of it. Geologists study the evidence that they see around them, but in most cases, they are observing the results of processes that happened thousands, millions, and even billions of years in the past. Those were processes that took place at incredibly slow rates — millimetres per year to centimetres per year — but because of the amount of time available, they produced massive results.

Geology is displayed on a grand scale in mountainous regions, perhaps nowhere better than the Rocky Mountains in Canada (Figure 1). The peak on the right is Rearguard Mountain, which is a few kilometres northeast of Mount Robson, the tallest peak in the Canadian Rockies (3,954 m). The large glacier in the middle of the photo is the Robson Glacier. The river flowing from Robson Glacier drains into Berg Lake in the bottom right. There are many geological features portrayed here. The sedimentary rock that these mountains are made of formed in ocean water over 500 million years ago. A few hundred million years later, these beds were pushed east for tens to hundreds of kilometres by tectonic plate convergence and also pushed up to thousands of metres above sea level. Over the past two million years this area — like most of the rest of Canada — has been repeatedly glaciated, and the erosional effects of those glaciations are obvious. The Robson Glacier is now only a small remnant of its size during the Little Ice Age of the 15th to 18th centuries, as shown by the distinctive line on the slope on the left. Like almost all other glaciers in the world, it is now receding even more rapidly because of human-caused climate change.

Figure 1: Rearguard Mountain and Robson Glacier in the Rocky Mountains of British Columbia [SE]

Geology is also about understanding the evolution of life on Earth; about discovering resources such as metals and energy; about recognizing and minimizing the environmental implications of our use of those resources; and about learning how to mitigate the hazards related to earthquakes, volcanic eruptions, and slope failures. All of these aspects of geology, and many more, are covered in this textbook.

Box 1 What are scientific methods?

There is no single method of inquiry that is specifically the "scientific method"; furthermore, scientific inquiry is not necessarily different from serious research in other disciplines. The key feature of serious inquiry is the creation of a hypothesis (a tentative explanation) that could explain the observations that have been made, and then the formulation and testing (by experimentation) of one or more predictions that follow from that hypothesis. For example, we might observe that most of the cobbles in a stream bed are well rounded (see photo in this box), and then derive the hypothesis that the rocks become rounded during transportation along the stream bed. A prediction that follows from this hypothesis is that cobbles present in a stream will become increasingly rounded over time as they are transported downstream. An experiment to test this prediction would be to place some angular cobbles in a stream, label them so that we can be sure to find them again later, and then return at various time intervals (over a period of months or years) to carefully measure their locations and roundness. A critical feature of a good hypothesis and any resulting predictions is that they must be testable. For example, an alternative hypothesis to the one above is that an extraterrestrial organization creates rounded cobbles and places them in streams when nobody is looking. This may indeed be the case, but there is no practical way to test this hypothesis. Most importantly, there is no way to prove that it is false, because if we aren't able to catch the aliens at work, we still won't know if they did it!

1.2 STUDY THE EARTH?

The simple answer to this question is that Earth is our home — our only home for the foreseeable future — and in order to ensure that it continues to be a great place to live, we need to understand how it works. Another answer is that some of us can't help but study it because it's fascinating. But there is more to it than that:

- We rely on Earth for valuable resources such as soil, water, metals, industrial minerals, and energy, and we need to know how to find these resources and exploit them sustainably.
- We can study rocks and the fossils they contain to understand the evolution of our environment and the life within it.
- We can learn to minimize our risks from earthquakes, volcanoes, slope failures, and damaging storms.
- We can learn how and why Earth's climate has changed in the past, and use that knowledge to understand both natural and human-caused climate change.
- We can recognize how our activities have altered the environment in many ways and the climate in increasingly serious ways, and how to avoid more severe changes in the future.
- We can use our knowledge of Earth to understand other planets in our solar system, as well as those around distant stars.

An example of the importance of geological studies for minimizing risks to the public is illustrated in Figure 2. This is a slope failure that took place in January 2005 in the Riverside Drive area of North Vancouver. The steep bank beneath the house shown gave way, and a slurry of mud and sand flowed down, destroying another house below and killing one person. This event took place following a heavy rainfall, which is a common occurrence in southwestern B.C. in the winter.

Figure 2: The aftermath of a deadly debris flow in the Riverside Drive area of North Vancouver in January, 2005

The irony of the 2005 slope failure is that the District of North Vancouver had been warned in a geological report written in 1980 that this area was prone to slope failure and that steps should be taken to minimize the risk to residents. Very little was done in the intervening 25 years, and the results were deadly.

1.3 HOW DO GEOLOGISTS WORK?

Geologists are involved in a range of widely varying occupations with one thing in common: the privilege of studying this fascinating planet. In Canada, many geologists work in the resource industries, including mineral exploration and mining and energy exploration and extraction. Other major areas where geologists work include hazard assessment and mitigation (e.g., assessment of risks from slope failures, earthquakes, and volcanic eruptions); water supply planning, development, and management; waste management; and assessment of geological issues on construction projects such as highways, tunnels, and bridges. Most geologists are employed in the private sector, but many work for government-funded geological organizations, such as the Geological Survey of Canada or one of the provincial geological surveys. And of course, many geologists are involved in education at the secondary and the postsecondary levels.

Some people are attracted to geology because they like to be outdoors, and it is true that many geological opportunities involve fieldwork in places that are as amazing to see as they are interesting to study. But a lot of geological work is also done in offices or laboratories. Geological work tends to be varied and challenging, and for these reasons and many others, geologists are among those who are the most satisfied with their employment.

Figure 3: Geologists examining ash-layer deposits at Kilauea Volcano, Hawaii [SE photo]

In Canada, most working geologists are required to be registered with an association of professional geoscientists. This typically involves meeting specific postsecondary educational standards and gaining several years of relevant professional experience under the supervision of a registered geoscientist.

1.4 ROCKS AND MINERALS

The rest of this chapter is devoted to a brief overview of a few of the important aspects of physical geology, starting with minerals and rocks. This is followed by a review of Earth's internal structure and the processes of plate tectonics, and an explanation of geological time.

Like everything else in the universe, Earth is made up of varying proportions of the 90 naturally occurring elements — hydrogen, carbon, oxygen, magnesium, silicon, iron, and so on. In most geological materials, these combine in various ways to make minerals. Minerals will be covered in some detail in Chapter 2, but here we will briefly touch on what minerals are, and how they are related to rocks.

A mineral is a naturally occurring combination of specific elements that are arranged in a particular repeating three-dimensional structure or lattice. Terms in bold are defined in the glossary at the end of the book. The mineral halite is shown as an example in Figure 4. In this case, atoms of sodium (Na: purple) alternate with atoms of chlorine (Cl: green) in all three dimensions, and the angles between the bonds are all 90°. Even in a tiny crystal, like the ones in your salt shaker, the lattices extend in all three directions for thousands of repetitions. Halite always has this composition and this structure.

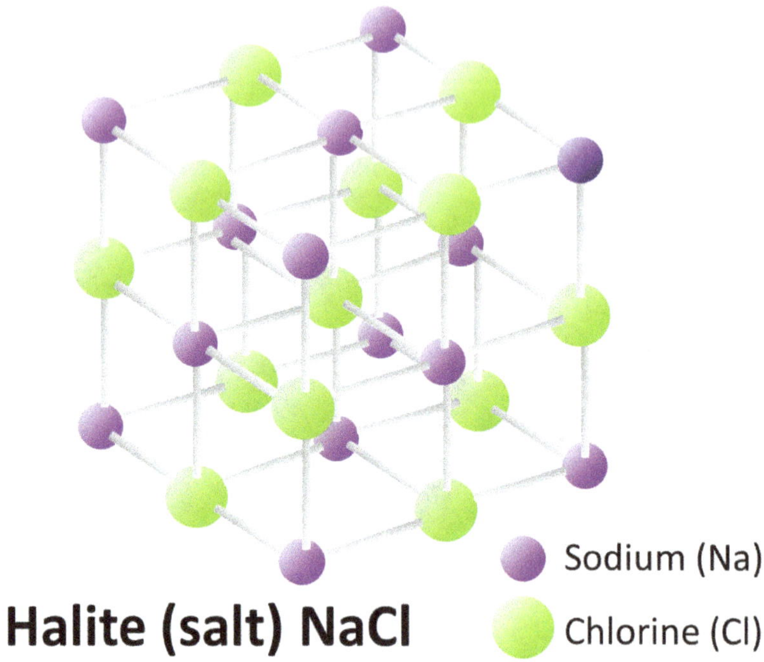

Figure 4: The lattice structure and composition of the mineral halite (common table salt) [SE]

Note: Element symbols (e.g., Na and Cl) are used extensively in this book. In Appendix 1, you will find a list of the symbols and names of the elements common in minerals and a copy of the periodic table. Please use those resources if you are not familiar with the element symbols.

There are thousands of minerals, although only a few dozen are mentioned in this book. In nature, minerals are found in rocks, and the vast majority of rocks are composed of at least a few different minerals. A close-up view of granite, a common rock, is shown in Figure 5. Although a hand-sized piece of granite may have thousands of individual mineral crystals in it, there are typically only a few different minerals, as shown here.

Rocks can form in a variety of ways. Igneous rocks form from magma (molten rock) that has either cooled slowly underground (e.g., to produce granite) or cooled quickly at the surface after a volcanic eruption (e.g., basalt). Sedimentary rocks, such as sandstone, form when the weathered products of other rocks accumulate at the surface and are then buried by other sediments. Metamorphic rocks form when either igneous or sedimentary rocks are heated and squeezed to the point where some of their minerals are unstable and new minerals form to create a different type of rock. An example is schist.

Figure 5: A close-up view of the rock granite and some of the minerals that it typically contains (H = hornblende (amphibole), Q = quartz and F = feldspar). The crystals range from about 0.1 to 3 mm in diameter. Most are irregular in outline, but some are rectangular. [SE]

A key point to remember is the difference between a mineral and a rock. A mineral is a pure substance with a specific composition and structure, while a rock is typically a mixture of several different minerals (although a few types of rock may include only one type of mineral). Examples of minerals are feldspar, quartz, mica, halite, calcite, and amphibole. Examples of rocks are granite, basalt, sandstone, limestone, and schist.

1.5 BASICS OF PLATE TECTONICS

Plate tectonics is the model or theory that has been used for the past 60 years to understand Earth's development and structure — more specifically the origins of continents and oceans, of folded rocks and mountain ranges, of earthquakes and volcanoes, and of continental drift. It is explained in some detail in Chapter 10, but is introduced here because it includes concepts that are important to many of the topics covered in the next few chapters.

Key to understanding plate tectonics is an understanding of Earth's internal structure, which is illustrated in Figure 6. Earth's core consists mostly of iron. The outer core is hot enough for the iron to be liquid. The inner core, although even hotter, is under so much pressure that it is solid. The mantle is made up of iron and magnesium silicate minerals. The bulk of the mantle, surrounding the outer core, is solid rock, but is plastic enough to be able to flow slowly. Surrounding that part of the mantle is a partially molten layer (theasthenosphere), and the outermost part of the mantle is rigid. The crust — composed mostly of granite on the continents and mostly of basalt beneath the oceans — is also rigid. The crust and outermost rigid mantle together make up the lithosphere. The lithosphere is divided into about 20 tectonic plates that move in different directions on Earth's surface. (For a more accurate depiction of the components of the Earth's interior, please see Figure 9.2.)

An important property of Earth (and other planets) is that the temperature increases with depth, from close to 0°C at the surface to about 7000°C at the centre of the core. In the crust, the rate of temperature increase is about 30°C/km. This is known as the **geothermal gradient**.

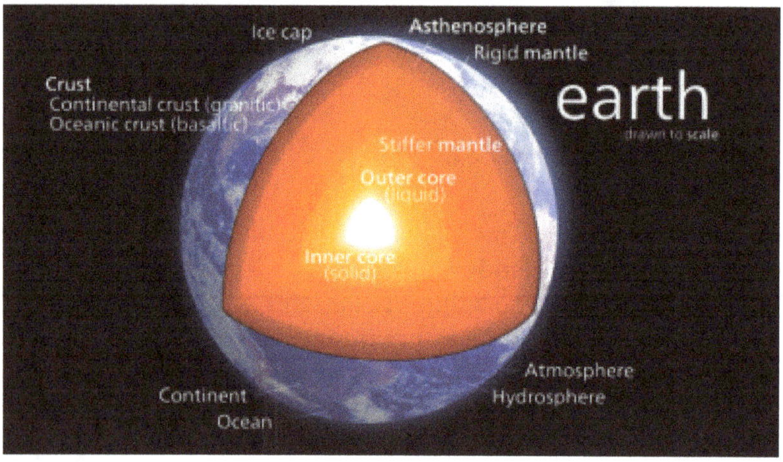

Figure 6: The structure of Earth's interior showing the inner and outer core, the different layers of the mantle, and the crust [Wikipedia]

Heat is continuously flowing outward from Earth's interior, and the transfer of heat from the core to the mantle causes convection in the mantle (Figure 7). This convection is the primary driving force for the movement of tectonic plates. At places where convection currents in the mantle are moving upward, new lithosphere forms (at ocean ridges), and the plates move apart (diverge). Where two plates are converging (and the convective flow is downward), one plate will be **subducted** (pushed down) into the mantle beneath the other. Many of Earth's major earthquakes and volcanoes are associated with convergent boundaries.

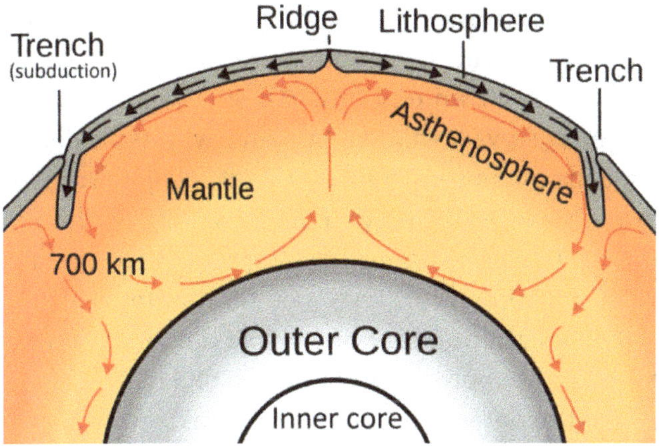

Figure 7: A model of convection within Earth's mantle

Earth's major tectonic plates and the directions and rates at which they are diverging at sea-floor ridges, are shown in Figure 1.8.

Exercises

Exercise 1 Plate Motion During Your Lifetime

Using either a map of the tectonic plates from the Internet or Figure 8, determine which tectonic plate you are on right now, approximately how fast it is moving, and in what direction. How far has that plate moved relative to Earth's core since you were born?

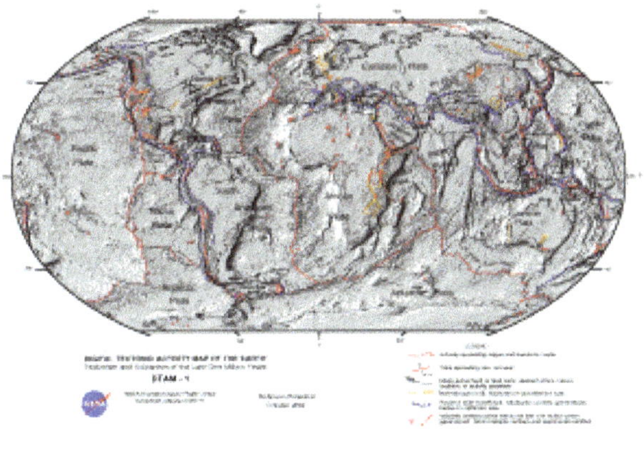

Figure 8: Earth's tectonic plates and tectonic features that have been active over the past 1 million years

1.6 GEOLOGICAL TIME

In 1788, after many years of geological study, James Hutton, one of the great pioneers of geology, wrote the following about the age of Earth: The result, therefore, of our present enquiry is, that we find no vestige of a beginning — no prospect of an end. Hutton, J, 1788. Theory of the Earth; or an investigation of the laws observable in the composition, dissolution, and restoration of land upon the Globe. Transactions of the Royal Society of Edinburgh. Of course he wasn't exactly correct, there was a beginning and there will be an end to Earth, but what he was trying to express is that geological time is so vast that we humans, who typically live for less than a century, have no means of appreciating how much geological time there is. Hutton didn't even try to assign an age to Earth, but we now know that it is approximately 4,570 million years old. Using the scientific notation for geological time, that is 4,570 Ma (for mega annum or "millions of years") or 4.57 Ga (for giga annum or billions of years). More recent dates can be expressed in ka (kilo annum); for example, the last cycle of glaciation ended at approximately 11.7 ka or 11,700 years ago. This notation will be used for geological dates throughout this book.

Exercises

Exercise 2 Using Geological Time Notation

To help you understand the scientific notation for geological time, write the following out in numbers (for example, 3.23 Ma = 3,230,000 years).

2.75 ka
0.93 Ga
14.2 Ma

We use this notation to describe times from the present, but not to express time differences in the past. For example, we could say that the dinosaurs lived from about 225 Ma to 65 Ma, which is 160 million years, but we would not say that they lived for 160 Ma.

Unfortunately, knowing how to express geological time doesn't really help us understand or appreciate its extent. A version of the geological time scale is included as Figure 9. Unlike time scales you'll see in other places, or even later in this book, this time scale is linear throughout its length, meaning that 50 Ma during the Cenozoic is the same thickness as 50 Ma during the Hadean—in each case about the height of the "M" in Ma. The Pleistocene glacial epoch began at about 2.6 Ma, which is equivalent to half the thickness of the thin grey line at the top of the yellow bar marked "Cenozoic." Most other time scales have earlier parts of Earth's history compressed so that more detail can be shown for the more recent parts. That makes it difficult to appreciate the extent of geological time.

To create some context, the **Phanerozoic** Eon (the last 542 million years) is named for the time during which visible (*phaneros*) life (*zoi*) is present in the geological record. In fact, large organisms — those that leave fossils visible to the naked eye — have existed for a little longer than that, first appearing around 600 Ma, or a span of just over 13% of geological time. Animals have been on land for 360 million years, or 8% of geological time. Mammals have dominated since the demise of the dinosaurs around 65 Ma, or 1.5% of geological time, and the genus *Homo* has existed since approximately 2.2 Ma, or 0.05% (1/2,000th) of geological time.

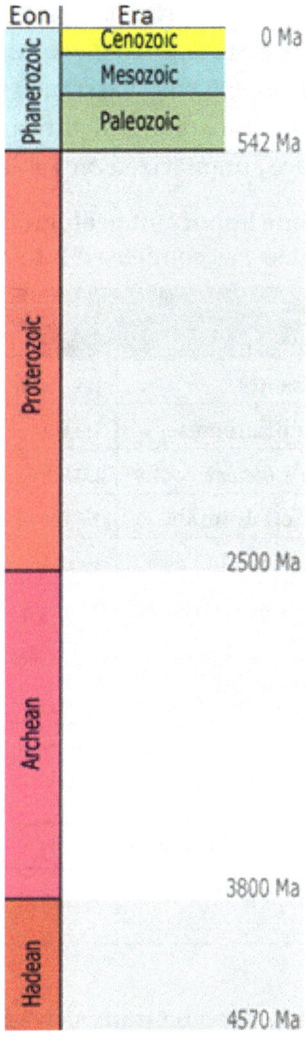

Figure 9: The geological time scale [SE]

Geologists (and geology students) need to understand geological time. That doesn't mean simply memorizing the geological time scale; instead, it means getting your mind around the concept that although most geological processes are extremely slow, very large and important things can happen if such processes continue for enough time.

For example, the Atlantic Ocean between Nova Scotia and northwestern Africa has been getting wider at a rate of about 2.5 cm per year. Imagine yourself taking a journey at that rate — it would be impossibly and ridiculously slow. And yet, since it started to form around 200 Ma (just 4% of geological time), the Atlantic Ocean has grown to a width of over 5,000 km!

A useful mechanism for understanding geological time is to scale it all down into one year. The origin of the solar system and Earth at 4.57 Ga would be represented by January 1, and the present year would be represented by the last tiny fraction of a second on New Year's Eve. At this scale, each day of the year represents 12.5 million years; each hour represents about 500,000 years; each minute represents 8,694 years; and each second represents 145 years. Some significant events in Earth's history, as expressed on this time scale, are summarized on Table 1.

Table 1: A summary of some important geological dates expressed as if all of geological time was condensed into one year [SE]

Event	Approximate Date	Calendar Equivalent
Formation of oceans and continents	4.5 – 4.4 Ga	January
Evolution of the first primitive life forms	3.8 Ga	early March
Formation of British Columbia's oldest rocks	2.0 Ga	July
Evolution of the first multi-celled animals	0.6 Ga or 600 Ma	November 15
Animals first crawled onto land	360 Ma	December 1
Vancouver Island reached North America and the Rocky Mountains were formed	90 Ma	December 25
Extinction of the non-avian dinosaurs	65 Ma	December 26
Beginning of the Pleistocene ice age	2 Ma or 2000 ka	8 p.m., December, 31
Retreat of the most recent glacial ice from southern Canada	14 ka	11:58 p.m., December 31
Arrival of the first people in British Columbia	10 ka	11:59 p.m., December 31
Arrival of the first Europeans on the west coast of what is now Canada	250 years ago	2 seconds before midnight, December 31

AT A GLANCE

The topics covered in this chapter can be summarized as follows:

1.1	What Is Geology?	Geology is the study of Earth. It is an integrated science that involves the application of many of the other sciences, but geologists also have to consider geological time because most of the geological features that we see today formed thousands, millions, or even billions of years ago.
1.2	Why Study Earth?	Geologists study Earth out of curiosity and for other, more practical reasons, including understanding the evolution of life on Earth; searching for resources; understanding risks from geological events such as earthquakes, volcanoes, and slope failures; and documenting past environmental and climate changes so that we can understand how human activities are affecting Earth.

1.3	What Do Geologists Do?	Geologists work in the resource industries and in efforts to protect our resources and the environment in general. They are involved in ensuring that risks from geological events (e.g., earthquakes) are minimized and that the public understands what the risks are. Geologists are also engaged in fundamental research about Earth and in teaching.
1.4	Minerals and Rocks	Minerals are naturally occurring, specific combinations of elements that have particular three-dimensional structures. Rocks are made up of mixtures of minerals and can form though igneous, sedimentary, or metamorphic processes.
1.5	Fundamentals of Plate Tectonics	Convection currents move through Earth's mantle because the mantle is being heated from below by the hot core. Those convection currents cause the movement of tectonic plates (which are composed of the crust and the uppermost rigid mantle). Plates are formed at divergent boundaries and consumed (subducted) at convergent boundaries. Many important geological processes take place at plate boundaries.
1.6	Geological Time	Earth is approximately 4,570,000,000 years old; that is, 4.57 billion years or 4.57 Ga or 4,570 Ma. It's such a huge amount of time that even extremely slow geological processes can have an enormous impact.

CHAPTER 2
MINERALS

Minerals are all around us: the graphite in your pencil, the salt on your table, the plaster on your walls, and the trace amounts of gold in your computer. Minerals can be found in a wide variety of consumer products including paper, medicine, processed foods, cosmetics, and many more. And of course, everything made of metal is also derived from minerals.

As defined , a mineral is *a naturally occurring combination of specific elements arranged in a particular repeating three-dimensional structure*(Figure 1).

"Naturally occurring" implies that minerals are not artificially made, although many naturally occurring minerals (e.g., diamond) are also made in laboratories. That doesn't disqualify them from being minerals.

"Specific elements" means that most minerals have a specific chemical formula or composition. The mineral pyrite, for example, is FeS_2 (two atoms of sulphur for each atom of iron), and any significant departure from that would make it a different mineral. But many minerals have variable compositions within a specific range. The mineral olivine, for example, can range all the way from Fe_2SiO_4 to Mg_2SiO_4. Intervening compositions are written as $(Fe,Mg)_2SiO_4$ meaning that Fe and Mg can be present in any proportion. This type of substitution is known as **solid solution**.

Figure 1: Crystals of native sulphur at an outlet of volcanic gases, Kilauea volcano, Hawaii

Most important of all, a mineral has a specific "repeating three-dimensional structure" or "lattice," which is the way in which the atoms are arranged. We've already seen in Chapter 1 how sodium and chlorine atoms in halite alternate in a regular pattern. That happens to be about the simplest mineral lattice of all; most mineral lattices are much more complicated, as we'll see.

2.1 ELECTRONS, PROTONS, NEUTRONS AND ATOMS

All matter, including mineral crystals, is made up of atoms, and all atoms are made up of three main particles: protons, neutrons, and electrons. As summarized in Table 1, protons are positively charged, neutrons are uncharged and electrons are negatively charged. The negative charge of one electron balances the positive charge of one proton. Both protons and neutrons have a mass of 1, while electrons have almost no mass.

Table 1: Charges and masses of the particles within atoms

Elementary Particle	Charge	Mass
Proton	+1	1
Neutron	0	1
Electron	-1	~0

The element hydrogen has the simplest atoms, each with just one proton and one electron. The proton forms the nucleus, while the electron orbits around it. All other elements have neutrons as well as protons in their nucleus, such as helium, which is depicted in Figure 2. The positively charged protons tend to repel each other, and the neutrons help to hold the nucleus together. The number of protons is the atomic number, and the number of protons plus neutrons is the atomic mass. For hydrogen, the atomic mass is 1 because there is one proton and no neutrons. For helium, it is 4: two protons and two neutrons.

For most of the 16 lightest elements (up to oxygen) the number of neutrons is equal to the number of protons. For most of the remaining elements, there are more neutrons than protons, because extra neutrons are needed to keep the nucleus together by overcoming the mutual repulsion of the increasing numbers of protons concentrated in a very small space. For example, silicon has 14 protons and 14 neutrons. Its atomic number is 14 and its atomic mass is 28. The most common isotope of uranium has 92 protons and 146 neutrons. Its atomic number is 92 and its atomic mass is 238 (92 + 146).

The dot in the middle is the nucleus, and the surrounding cloud represents where the two electrons might be at any time. The darker the shade, the more likely that an electron will be there. An angstrom (Å) is 10^{-10}m . A femtometre (fm) is 10^{-15}m. In other words, a helium atom's electron cloud is about 100,000 times bigger than its nucleus.

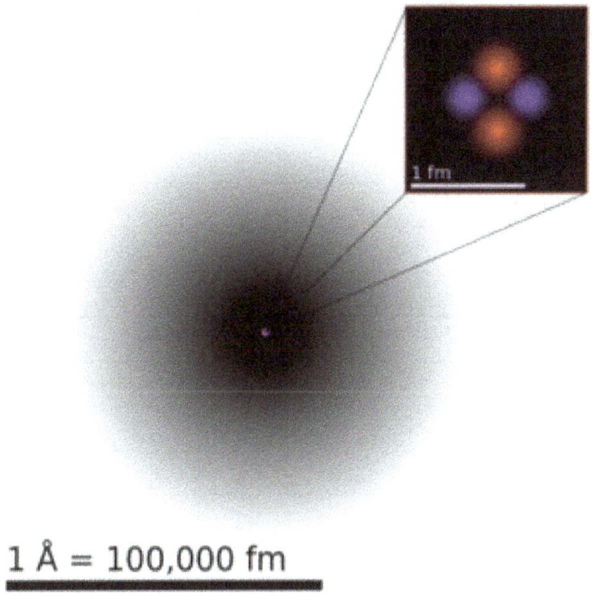

Figure 2: A depiction of a helium atom.

Electrons orbiting around the nucleus of an atom are arranged in shells — also known as "energy levels." The first shell can hold only two electrons, while the next shell holds up to eight electrons. Subsequent shells can hold more electrons, but the outermost shell of any atom holds no more than eight electrons. The electrons in the outermost shell play an important role in bonding between atoms. Elements that have a full outer shell are inert in that they do not react with other elements to form compounds. They all appear in the far-right column of the periodic table: helium, neon, argon, etc. For elements that do not have a full outer shell, the outermost electrons can interact with the outermost electrons of nearby atoms to create chemical bonds. The electron shell configurations for 29 of the first 36 elements are listed in Table 2.

Table 2: Electron shell configurations of some of the elements up to element 36. (The inert elements, with filled outer shells, are bolded.)

Number of Electrons in Each Shell						
Element	**Symbol**	**Atomic No.**	**First**	**Second**	**Third**	**Fourth**
Hydrogen	H	1	1			
Helium	**He**	**2**	**2**			
Lithium	Li	3	2	1		
Beryllium	Be	4	2	2		
Boron	B	5	2	3		
Carbon	C	6	2	4		

Number of Electrons in Each Shell						
Nitrogen	N	7	2	5		
Oxygen	O	8	2	6		
Fluorine	F	9	2	7		
Neon	**Ne**	**10**	**2**	**8**		
Sodium	Na	11	2	8	1	
Magnesium	Mg	12	2	8	2	
Aluminum	Al	13	2	8	3	
Silicon	Si	14	2	8	4	
Phosphorus	P	15	2	8	5	
Sulphur	S	16	2	8	6	
Chlorine	Cl	17	2	8	7	
Argon	**Ar**	**18**	**2**	**8**	**8**	
Potassium	K	19	2	8	8	1
Calcium	Ca	20	2	8	8	2
Scandium	Sc	21	2	8	9	2
Titanium	Ti	22	2	8	10	2
Vanadium	V	23	2	8	11	2
Chromium	Cr	24	2	8	13	1
Manganese	Mn	25	2	8	13	2
Iron	Fe	26	2	8	14	2
.	.	.	.	.	.	.
Selenium	Se	34	2	8	18	6
Bromine	Br	35	2	8	18	7
Krypton	**Kr**	**36**	**2**	**8**	**18**	**8**

2.2 BONDING AND LATTICES

As we've just seen, an atom seeks to have a full outer shell (i.e., eight electrons for most elements, or two electrons for hydrogen and helium) to be atomically stable. This is accomplished by transferring or sharing electrons with other atoms. Elements that already have their outer orbits filled are considered to be inert; they do not readily take part in chemical reactions.

Sodium has 11 electrons: two in the first shell, eight in the second, and one in the third (Figure 3). Sodium readily gives up the third shell electron; when it loses this one negative charge, it becomes positively charged. By giving up its lone third shell electron, sodium ends up with a full outer second shell. Chlorine, on the other hand, has 17 electrons: two in the first shell, eight in the second, and seven in the third. Chlorine readily accepts an eighth electron to fill its third shell, and therefore becomes negatively charged because of an imbalance between the number of protons (17) and electrons (18). In changing their number of electrons, these atoms become ions — the sodium loses an electron to become a positive ion or cation, and the chlorine gains an electron to become a negative ion or anion (Figure 2.3). Since negative and positive charges attract, sodium and chlorine ions stick together, creating an ionic bond. Electrons can be thought of as being transferred from one atom to another in an ionic bond. Common table salt (NaCl) is a mineral composed of chlorine and sodium linked together by ionic bonds. The mineral name for NaCl is halite.

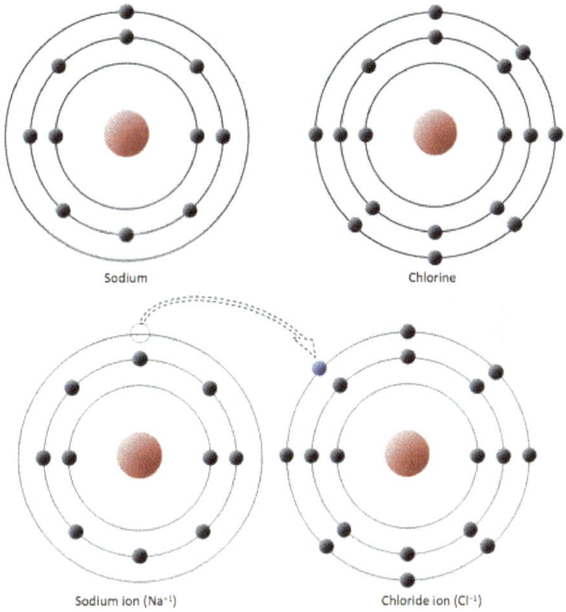

Figure 3: A very simplified electron configuration of sodium and chlorine atoms (top). Sodium gives up an electron to become a cation (bottom left) and chlorine accepts an electron to become an anion (bottom right).

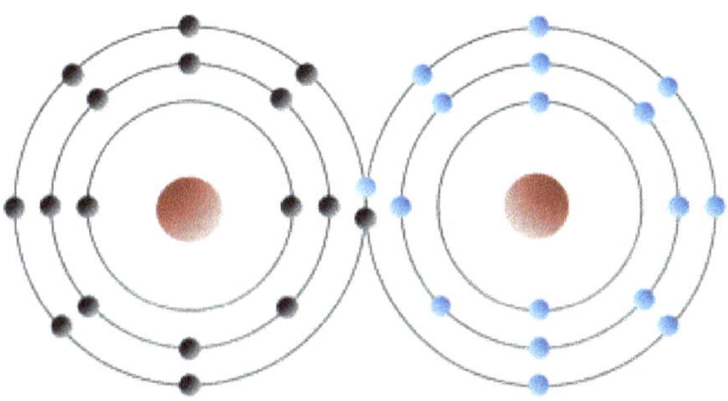

Figure 4: Depiction of a covalent bond between two chlorine atoms. The electrons are black, in the left atom and blue in the right atom. Two electrons are shared

(one black and one blue) so that each atom "appears" to have a full outer shell. [SE]

An element like chlorine can also form bonds without forming ions. For example, two chlorine atoms, which each seek an eighth electron in their outer shell, can share an electron in what is known as acovalent bond, to form chlorine gas (Cl2) (Figure 4). Electrons are shared in a covalent bond.

What's with all of these "sili" names?

The element silicon is one of the most important geological elements and is the second-most abundant element in Earth's crust (after oxygen). *Silicon* bonds readily with oxygen to form a silica tetrahedron (Figure 6). Pure *silicon* crystals (created in a lab) are used to make semiconductive media in electronic devices. A silicate mineral is one in which silicon and oxygen are present as *silica* tetrahedra. *Silica* also refers to a chemical component of a rock and is expressed as % SiO_2. The mineral quartz is made up entirely of *silica* tetrahedra, and some forms of quartz are known as *silica*. Silicone is a synthetic product (e.g., *silicone* rubber, resin, or caulking) made from *silicon*-oxygen chains and various organic molecules. To help you keep the "sili" names straight, here is a summary table:

"Sili" Names Summary Table	
Silicon	The 14th element
Silicon wafer	A crystal of pure silicon sliced very thinly and used for electronics
Silica tetrahedron	A combination of one silicon atom and four oxygen atoms that form a tetrahedron
% silica	The proportion of a rock that is composed of the components Si + O_2
Silica	A form of the mineral quartz (SiO_2)
Silicate	A mineral that contains silica tetrahedra (e.g., quartz, feldspar, mica, olivine)
Silicone	A flexible material made up of Si–O chains with attached organic molecules

As described in Chapter 1, all minerals are characterized by a specific three-dimensional pattern known as a lattice or crystal structure. These structures range from the simple cubic pattern of halite (NaCl) , to the very complex patterns of some silicate minerals. Two minerals may have the same composition, but very different crystal structures and properties. Graphite and diamond, for example, are both composed only of carbon, but while diamond is the hardest substance known, graphite is softer than paper. Their lattice structures are compared in Figure 5.

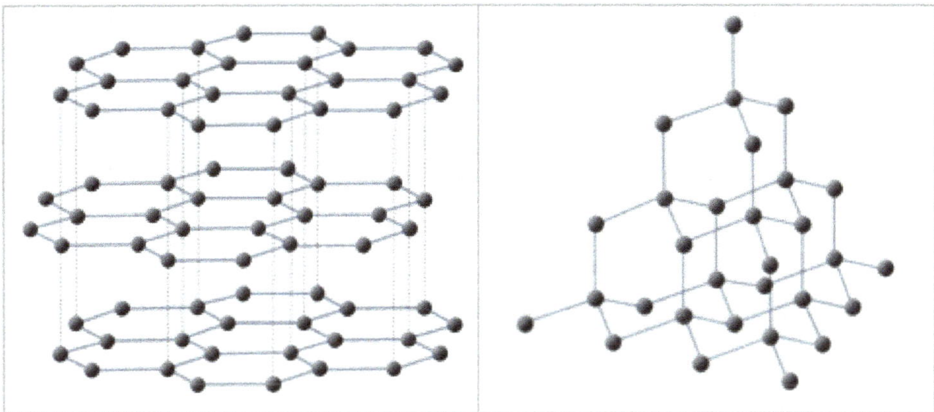

Graphite (solid lines are strong covalent bonds, dotted lines are weak inter-layer bonds) | Diamond (all bonds are strong covalent bonds)

Figure 5: A depiction of the lattices of graphite and diamond.

Figure 6: Cubic crystals (left) and right-angle cleavage planes (right) of the mineral halite. If you look closely at the cleavage fragment in the middle, you can see where it would break again (cleave) along a plane parallel to the existing surface.

Mineral lattices have important implications for mineral properties, as exemplified by the relative hardnesses of diamond and graphite. Lattices also determine the shape that mineral crystals grow in and how they break. For example, the right angles in the lattice of the mineral halite influence both the shape of its crystals (typically cubic), and the way those crystals break (Figure 2.6).

2.3 GROUPS OF MINERALS

Most minerals are made up of a cation (a positively charged ion) or several cations and an anion (a negatively charged ion (e.g., S^{2-})) or an anion complex (e.g., SO_4^{2-}). For example, in the mineral hematite (Fe_2O_3), the cation is Fe_3^+ (iron) and the anion is O^{2-} (oxygen). We group minerals into classes on the basis of their predominant anion or anion group. These include oxides, sulphides, carbonates, silicates, and others. Silicates are by far the predominant group in terms of their abundance within the crust and mantle. Some examples of minerals from the different mineral groups are given in Table 3.

Table 3: The Mineral Groups and Examples

Group	Examples
Oxides	Hematite (iron oxide Fe_2O_3), corundum (aluminum oxide Al_2O_3), water ice (H_2O)
Sulphides	Galena (lead sulphide PbS), pyrite (iron sulphide FeS_2), chalcopyrite (copper-iron sulphide $CuFeS_2$)
Sulphates	Gypsum (calcium sulphate $CaSO_4{\cdot}H_2O$), barite (barium sulphate $BaSO_4$) *(Note that sulphates are different from sulphides. Sulphates have the SO_4^{-2} ion while sulphides have the S^{-2} ion)*
Halides	Fluorite (calcium flouride CaF_2), halite (sodium chloride NaCl) *(Halide minerals have halogen elements as their anion — the minerals in the second last column on the right side of the periodic table, including F, Cl, Br, etc.)*
Carbonates	Calcite (calcium carbonate $CaCO_3$), dolomite (calcium-magnesium carbonate $(Ca,Mg)CO_3$)
Phosphates	Apatite ($Ca_5(PO_4)3(OH)$), Turquoise ($CuAl_6(PO_4)4(OH)_8{\cdot}5H_2O$)
Silicates	Quartz (SiO_2), feldspar (sodium-aluminum silicate $NaAlSi_3O_8$), olivine (iron or magnesium silicate $(Mg,Fe)_2SiO_4$) *(Note that in quartz the anion is oxygen, and while it could be argued, therefore, that quartz is an oxide, it is always classed with the silicates.)*
Native minerals	Gold (Au), diamond (C), graphite (C), sulphur (S), copper (Cu)

Oxide minerals have oxygen (O^{2-}) as their anion, but they exclude those with oxygen complexes such as carbonate (CO_3^{2-}), sulphate (SO_4^{2-}), and silicate (SiO_4^{4-}). The most important oxides are the iron oxides hematite and magnetite (Fe_2O_3 and Fe_3O_4, respectively). Both of these are important ores of iron. Corundum (Al_2O_3) is an abrasive, but can also be a gemstone in its ruby and sapphire varieties. If the oxygen is also combined with hydrogen to form the hydroxyl anion (OH^-) the mineral is known as a hydroxide. Some important hydroxides are limonite and bauxite, which are ores of iron and aluminium respectively. Frozen water (H_2O) is a mineral (an oxide), but liquid water is not because it doesn't have a regular lattice.

Sulphides are minerals with the S^{-2} anion, and they include galena (PbS), sphalerite (ZnS), chalcopyrite ($CuFeS_2$), and molybdenite (MoS_2), which are the most important ores of lead, zinc, copper, and molybdenum respectively. Some other sulphide minerals are pyrite (FeS_2), bornite (Cu_5FeS_4), stibnite (Sb_2S_3), and arsenopyrite (FeAsS).

Sulphates are minerals with the SO_4^{-2} anion, and these include anhydrite ($CaSO_4$) and its cousin gypsum ($CaSO_4.2H_2O$) and the sulphates of barium and strontium: barite ($BaSO_4$) and celestite ($SrSO_4$). In all of these minerals, the cation has a +2 charge, which balances the −2 charge on the sulphate ion.

The halides are so named because the anions include the halogen elements chlorine, fluorine, bromine, etc. Examples are halite (NaCl), cryolite (Na_3AlF_6), and fluorite (CaF_2).

The carbonates include minerals in which the anion is the CO_3^{-2} complex. The carbonate combines with +2 cations to form minerals such as calcite ($CaCO_3$), magnesite ($MgCO_3$), dolomite ($(Ca,Mg)CO_3$), and siderite ($FeCO_3$). The copper minerals malachite and azurite are also carbonates.

In phosphate minerals, the anion is the PO_4^{-3} complex. An important phosphate mineral is apatite ($Ca_5(PO_4)_3(OH)$), which is what your teeth are made of.

The silicate minerals include the elements silicon and oxygen in varying proportions ranging from Si : O_2 to Si : O_4.

Native minerals are single-element minerals, such as gold, copper, sulphur, and graphite.

2.4 SILICATE MINERALS

The vast majority of the minerals that make up the rocks of Earth's crust are silicate minerals. These include minerals such as quartz, feldspar, mica, amphibole, pyroxene, olivine, and a great variety of clay minerals. The building block of all of these minerals is the silica tetrahedron, a combination of four oxygen atoms and one silicon atom. These are arranged such that planes drawn through the oxygen atoms form a tetrahedron (Figure 6). Since the silicon ion has a charge of +4 and each of the four oxygen ions has a charge of −2, the silica tetrahedron has a net charge of −4.

In silicate minerals, these tetrahedra are arranged and linked together in a variety of ways, from single units to complex frameworks (Figure 9). The simplest silicate structure, that of the mineral olivine, is composed of isolated tetrahedra bonded to iron and/or magnesium ions. In olivine, the −4 charge of each silica tetrahedron is balanced by two divalent (i.e., +2) iron or magnesium cations. Olivine can be either Mg_2SiO_4 or Fe_2SiO_4, or some combination of the two $(Mg,Fe)_2SiO_4$. The divalent cations of magnesium and iron are quite close in radius (0.73 versus 0.62 angstromsAn angstrom is the unit commonly used for the expression of atomic-scale dimensions. One angstrom is 10−10 m or 0.0000000001 m. The symbol for an angstrom is Å.). Because of this size similarity, and because they are both divalent cations (both have a charge of +2), iron and magnesium can readily substitute for each other in olivine and in many other minerals.

Tetrahedron Configuration		Example Minerals
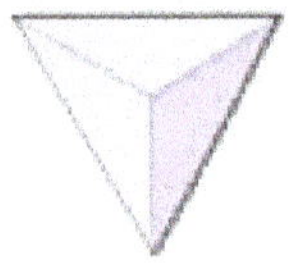	Isolated (nesosilicates)	Olivine, garnet, zircon, kyanite
	Pairs (sorosilicates)	Epidote, zoisite
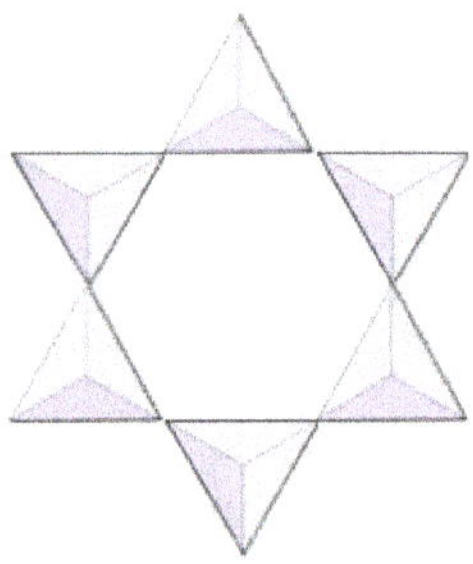	Rings (cyclosilicates)	Tourmaline
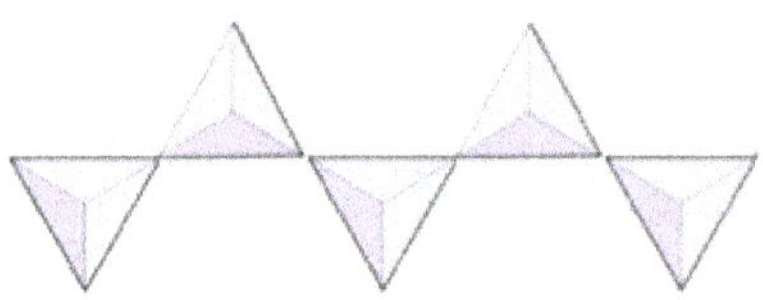	Single chains (inosilicates)	Pyroxenes, wollastonite
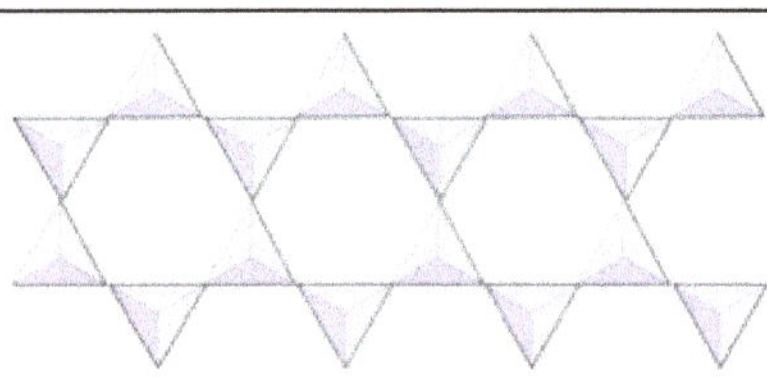	Double chains (inosilicates)	Amphiboles
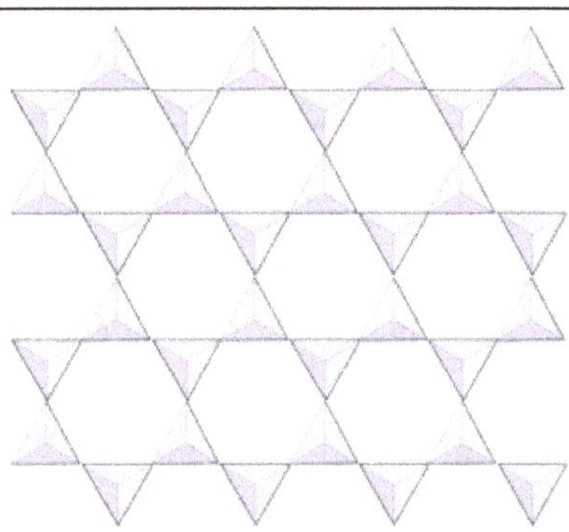	Sheets (phyllosilicates)	Micas, clay minerals, serpentine, chlorite
3-dimensional structure	Framework (tectosilicates)	Feldspars, quartz, zeolite

Figure 9: Silicate mineral configurations. The triangles represent silica tetrahedra.

2.5 MINERAL FORMATION

In order for a mineral crystal to grow, the elements needed to make it must be present in the appropriate proportions, the physical and chemical conditions must be favourable, and there must be sufficient time for the atoms to become arranged.

Physical and chemical conditions include factors such as temperature, pressure, presence of water, pH, and amount of oxygen available. Time is one of the most important factors because it takes time for atoms to become ordered. If time is limited, the mineral grains will remain very small. The presence of water enhances the mobility of ions and can lead to the formation of larger crystals over shorter time periods.

Most of the minerals that make up the rocks around us formed through the cooling of molten rock, known as magma. At the high temperatures that exist deep within Earth, some geological materials are liquid. As magma rises up through the crust, either by volcanic eruption or by more gradual processes, it cools and minerals crystallize. If the cooling process is rapid (minutes, hours, days, or years), the components of the minerals will not have time to become ordered and only small crystals can form before the rock becomes solid. The resulting rock will be fine-grained (i.e., crystals less than 1 mm). If the cooling is slow (from decades to millions of years), the degree of ordering will be higher and relatively large crystals will form. In some cases, the cooling will be so fast (seconds) that the texture will be glassy, which means that no crystals at all form. Volcanic glass is not composed of minerals because the magma has cooled too rapidly for crystals to grow, although over time (millions of years) the volcanic glass may crystallize into various silicate minerals.

Minerals can also form in several other ways:

1. Precipitation from aqueous solution (i.e., from hot water flowing underground, from evaporation of a lake or inland sea, or in some cases, directly from sea-water)
2. Precipitation from gaseous emanations (e.g., in volcanic regions as shown in Figure 1)
3. Metamorphism — formation of new minerals directly from the elements within existing minerals under conditions of elevated temperature and pressure
4. Weathering — during which minerals unstable at Earth's surface may be altered to other minerals
5. Organic formation — formation of minerals within shells (primarily calcite) and teeth and bones (primarily apatite) by organisms (these organically formed minerals are still called minerals because they can also form inorganically)

Opal is a mineraloid, because although it has all of the other properties of a mineral, it does not have a specific structure. Pearl is not a mineral because it can *only* be produced by organic processes.

2.6 PROPERTIES OF MINERALS

Minerals are universal. A crystal of hematite on Mars will have the same properties as one on Earth, and the same as one on a planet orbiting another star. That's good news for geology students who are planning interplanetary travel since we can use those properties to help us identify minerals anywhere. That doesn't mean that it's easy, however; identification of minerals takes a lot of practice. Some of the mineral properties that are useful for identification are as follows:

Colour	Streak	Lustre	Hardness
Habit	Cleavage/fracture	Density	Other

Colour

For most of us, colour is one of our key ways of identifying objects. While some minerals have particularly distinctive colours that make good diagnostic properties, many do not, and for many, colour is simply unreliable. The mineral sulphur (Figures 1 and 16) is always a distinctive and unique yellow. Hematite, on the other hand, is an example of a mineral for which colour is not diagnostic. In some forms hematite is deep dull red, but in others it is black and shiny metallic (Figure 7) Many other minerals can have a wide range of colours (e.g., quartz, feldspar, amphibole, fluorite, and calcite). In most cases, the variations in colours are a result of varying proportions of trace elements within the mineral. In the case of quartz, for example, yellow quartz (citrine) has trace amounts of ferric iron (Fe^{3+}), rose quartz has trace amounts of manganese, purple quartz (amethyst) has trace amounts of iron, and milky quartz, which is very common, has millions of fluid inclusions (tiny cavities, each filled with water).

Figure 7: Examples of the colours of the minerals sulphur and hematite

Streak

In the context of minerals, "colour" is what you see when light reflects off the surface of the sample. One reason that colour can be so variable is that the type of surface is variable. If we grind a small amount of the sample to a powder we get a much better indication of its actual colour. This can easily be done by scraping a corner of the sample across a streak plate (a piece of unglazed porcelain). The result is that some of the mineral gets ground to a powder and we can get a better impression of its "true" colour (Figure 17).

Figure 8: The streak colours of earthy hematite (left) and specular hematite (right). Although the specular hematite streak looks close to black, it does have red undertones that you can see if you look closely. [SE]

Lustre

Lustre is the way light reflects off the surface of a mineral, and the degree to which it penetrates into the interior. The key distinction is between metallic and non-metallic lustres. Light does not pass through metals, and that is the main reason they look "metallic." Even a thin sheet of metal – such as aluminum foil – will prevent light from passing through it. Many non-metallic minerals may look as if light will not pass through them, but if you take a closer look at a thin edge of the mineral you can see that it does. If a non-metallic mineral has a shiny, reflective surface, then it is called "glassy." If it is dull and non-reflective, it is "earthy." Other types of non-metallic lustres are "silky," "pearly," and "resinous." Lustre is a good diagnostic property, since most minerals will always appear either metallic or non-metallic. There are a few exceptions to this.

Hardness

One of the most important diagnostic properties of a mineral is its hardness. In 1812 German mineralogist Friedrich Mohs came up with a list of 10 reasonably common minerals that had a wide range of hardness. In fact, while each mineral on the list is harder than the one before it, the relative measured hardnesses (vertical axis) are not linear. For example apatite is about three times harder than fluorite and diamond is three times harder than corundum. Some commonly available reference materials are also shown on this diagram, including a typical fingernail , a piece of copper wire , a knife blade or a piece of window glass a hardened steel file , and a porcelain streak plate . These are tools that a geologist can use to measure the hardness of unknown minerals. For example, if you have a mineral that you can't scratch with your fingernail, but you can scratch with a copper wire, then its hardness is between 2.5 and 3.5. And of course the minerals themselves can be used to test other minerals.

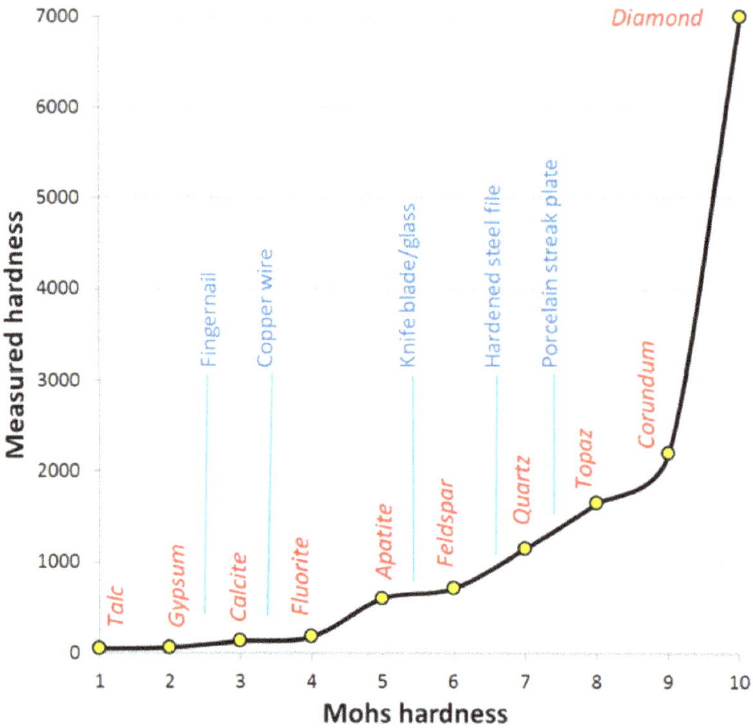

Figure 9: Minerals and reference materials in the Mohs scale of hardness. The "measured hardness" values are Vickers Hardness numbers.

Crystal Habit

When minerals form within rocks, there is a possibility that they will form in distinctive crystal shapes if they are not crowded out by other pre-existing minerals. Every mineral has one or more distinctive crystal habits, but it is not that common, in ordinary rocks, for the shapes to be obvious. Quartz, for example, will form six-sided prisms with pointed ends, but this typically happens only when it crystallizes from a hot water solution within a cavity in an existing rock (Figure 2.10). Pyrite can form cubic crystals, but can also form crystals with 12 faces, known as dodecahedra ("dodeca" means 12). The mineral garnet also forms dodecahedral crystals (Figure 10).

Figure 10: Hexagonal prisms of quartz (left), cubic crystals of pyrite (centre), and a dodecahedral crystal of garnet (right)

Because beautiful well-formed crystals are rare in ordinary rocks, habit isn't as useful a diagnostic feature as one might think. However, there are several minerals for which it is important. One is garnet, which is common in some metamorphic rocks and typically displays the dodecahedral shape. Another is amphibole, which forms long thin crystals, and is common in igneous rocks like granite (Figure 5).

Mineral habit is often related to the regular arrangement of the molecules that make up the mineral. Some of the terms that are used to describe habit include bladed, botryoidal (grape-like), dendritic (branched), drusy (an encrustation of minerals), equant (similar in all dimensions), fibrous, platy, prismatic (long and thin), and stubby.

Cleavage and fracture

Crystal habit is a reflection of how a mineral grows, while cleavage and fracture describe how it breaks. These characteristics are the most important diagnostic features of many minerals, and often the most difficult to understand and identify. Cleavage is what we see when a mineral breaks along a specific plane or planes, while fracture is an irregular break. Some minerals tend to cleave along planes at various fixed orientations, some do not cleave at all (they only fracture). Minerals that have cleavage can also fracture along surfaces that are not parallel to their cleavage planes.

As we've already discussed, the way that minerals break is determined by their atomic arrangement and specifically by the orientation of weaknesses within the lattice. Graphite and the micas, for example, have cleavage planes parallel to their sheets and halite has three cleavage planes parallel to the lattice directions.

Quartz has no cleavage because it has equally strong Si-O bonds in all directions, and feldspar has two cleavages at 90° to each other.

One of the main difficulties with recognizing and describing cleavage is that it is visible only in individual crystals. Most rocks have small crystals and it's very difficult to see the cleavage within a small crystal. Geology students have to work hard to understand and recognize cleavage, but it's worth the effort since it is a reliable diagnostic property for most minerals.

Density

Density is a measure of the mass of a mineral per unit volume, and it is a useful diagnostic tool in some cases. Most common minerals, such as quartz, feldspar, calcite, amphibole, and mica, have what we call "average density" (2.6 to 3.0 g/cm^3), and it would be difficult to tell them apart on the basis of their density. On the other hand, many of the metallic minerals, such as pyrite, hematite, and magnetite, have densities over 5 g/cm^3. They can easily be distinguished from the lighter minerals on the basis of density, but not necessarily from each other. A limitation of using density as a diagnostic tool is that one cannot assess it in minerals that are a small part of a rock with other minerals in it.

Other properties

Several other properties are also useful for identification of some minerals. For example, calcite is soluble in dilute acid and will give off bubbles of carbon dioxide. Magnetite is magnetic, so will affect a magnet. A few other minerals are weakly magnetic.

AT A GLANCE

The topics covered in this chapter can be summarized as follows.

2.1	Electrons, Protons, Neutrons, and Atoms	An atom is made up of protons and neutrons in the nucleus and electrons arranged in energy shells around the nucleus. The first shell holds two electrons, and outer shells hold more, but atoms strive to have eight electrons in their outermost shell (or two for H and He). They either gain or lose electrons (or share) to achieve this, and in so doing become either cations (if they lose electrons) or anions (if they gain them).
2.2	Bonding and Lattices	The main types of bonding in minerals are ionic bonding (electrons transferred) and covalent bonding (electrons shared). Some minerals have metallic bonding or other forms of weak bonding. Minerals form in specific three-dimensional lattices, and the nature of the lattices and the type of bonding within them have important implications for mineral properties.
2.3	Mineral Groups	Minerals are grouped according to the anion part of their formula, with some common types being oxides, sulphides, sulphates, halides, carbonates, phosphates, silicates, and native minerals.
2.4	Silicate Minerals	Silicate minerals are, by far, the most important minerals in Earth's crust. They all include silica tetrahedra (four oxygens surrounding a single silicon atom) arranged in different structures (chains, sheets, etc.). Some silicate minerals include iron or magnesium and are called ferromagnesian silicates.
2.5	Formation of Minerals	Most minerals in the crust form from the cooling and crystallization of magma. Some form from hot water solutions, during metamorphism or weathering, or through organic processes.
2.6	Mineral Properties	Some of the important properties for mineral identification include hardness, cleavage/fracture, density, lustre, colour, and streak colour.

CHAPTER

IGNEOUS ROCKS THAT ARE INTRUSIVE 3

A rock is a consolidated mixture of the same or different minerals. By consolidated, we mean hard and strong; real rocks don't fall apart in your hands! A mixture of minerals implies the presence of more than one mineral grain, but not necessarily more than one type of mineral (Figure 1). A rock can be composed of only one type of mineral (e.g., limestone is commonly made up of only calcite), but most rocks are composed of several different minerals. A rock can also include non-minerals, such as fossils or the organic matter within a coal bed or in some types of mudstone.

Figure 1: A fine-grained mafic dyke (dark green) intruded into a felsic dyke (pink) and into coarse diorite (grey), Quadra Island, B.C. All of these rocks are composed of more than one type of mineral. The mineral components are clearly visible in the diorite, but not in the other two rock types.

Rocks are grouped into three main categories based on how they form:

Igneous: formed from the cooling and crystallization of magma (molten rock)

Sedimentary: formed when weathered fragments of other rocks are buried, compressed, and cemented together, or when minerals precipitate directly from solution

Metamorphic: formed by alteration (due to heat, pressure, and/or chemical action) of a pre-existing igneous or sedimentary rock

3.1 ROCK CYCLE

The rock components of the crust are slowly but constantly being changed from one form to another and the processes involved are summarized in the rock cycle (Figure 2). The rock cycle is driven by two forces: (1) Earth's internal heat engine, which moves material around in the core and the mantle and leads to slow but significant changes within the crust, and (2) the hydrological cycle, which is the movement of water, ice, and air at the surface, and is powered by the sun.

The rock cycle is still active on Earth because our core is hot enough to keep the mantle moving, our atmosphere is relatively thick, and we have liquid water. On some other planets or their satellites, such as the Moon, the rock cycle is virtually dead because the core is no longer hot enough to drive mantle convection and there is no atmosphere or liquid water.

Figure 2: A schematic view of the rock cycle.

In describing the rock cycle, we can start anywhere we like, although it's convenient to start with magma. As we'll see in more detail below, magma is rock that is hot to the point of being entirely molten. This happens at between about 800° and 1300°C, depending on the composition and the pressure, onto the surface and cool quickly (within seconds to years) — forming extrusive igneous rock (Figure 3).

Figure 3: Magma forming pahoehoe basalt at Kilauea Volcano, Hawaii [SE]

Magma can either cool slowly within the crust (over centuries to millions of years) – forming intrusive igneous rock, or erupt onto the surface and cool quickly (within seconds to years) – forming extrusive igneous rock. Intrusive igneous rock typically crystallizes at depths of hundreds of metres to tens of kilometres below the surface. To change its position in the rock cycle, intrusive igneous rock has to be uplifted and exposed by the erosion of the overlying rocks.

Through the various plate-tectonics-related processes of mountain building, all types of rocks are uplifted and exposed at the surface. Once exposed, they are weathered, both physically (by mechanical breaking of the rock) and chemically (by weathering of the minerals), and the weathering products – mostly small rock and mineral fragments – are eroded, transported, and then deposited as sediments. Transportation and deposition occur through the action of glaciers, streams, waves, wind, and other agents, and sediments are deposited in rivers, lakes, deserts, and the ocean.

3.2 MAGMA AND ITS FORMATION

Magmas can vary widely in composition, but in general they are made up of only eight elements; in order of importance: oxygen, silicon, aluminum, iron, calcium, sodium, magnesium, and potassium (Figure 4). Oxygen, the most abundant element in magma, comprises a little less than half the total, followed by silicon at just over one-quarter. The remaining elements make up the other one-quarter. Magmas derived from crustal material are dominated by oxygen, silicon, aluminum, sodium, and potassium.

The composition of magma depends on the rock it was formed from (by melting), and the conditions of that melting. Magmas derived from the mantle have higher levels of iron, magnesium, and calcium, but they are still likely to be dominated by oxygen and silicon. All magmas have varying proportions of elements such as hydrogen, carbon, and sulphur, which are converted into gases like water vapour, carbon dioxide, and hydrogen sulphide as the magma cools.

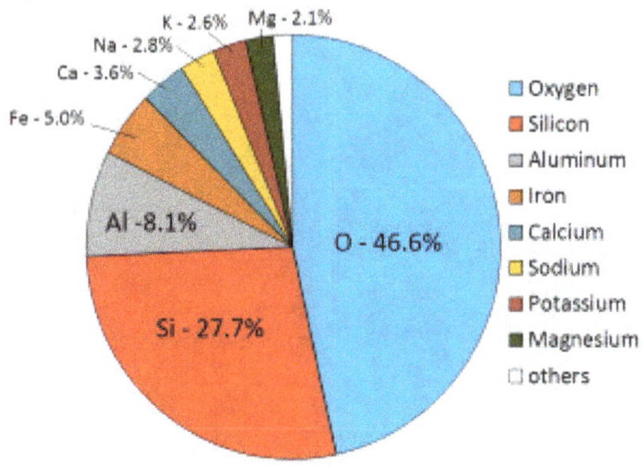

Figure 4: Average elemental proportions in Earth's crust, which is close to the average composition of magmas within the crust [SE]

Virtually all of the igneous rocks that we see on Earth are derived from magmas that formed from partial melting of existing rock, either in the upper mantle or the crust. Partial melting is what happens when only some parts of a rock melt; it takes place because rocks are not pure materials. Most rocks are made up of several minerals, each of which has a different melting temperature. The wax in a candle is a pure material. If you put some wax into a warm oven (50°C will do as the melting temperature of most wax is about 40°C) and leave it there for a while, it will soon start to melt. That's complete melting, not partial melting. If instead you took a mixture of wax, plastic, aluminum, and glass and put it into the same warm oven, the wax would soon start to melt, but the plastic, aluminum, and glass would not melt (Figure 5a). That's partial melting and the result would be solid plastic, aluminum, and glass surrounded by liquid wax (Figure 5b). If we heat the oven up to around 120°C, the plastic would melt too and mix with the liquid wax, but the aluminum and glass would remain solid (Figure 5c). Again this is partial melting. If we separated the wax/plastic "magma" from the other components and let it cool, it would eventually harden. As you can see from Figure 5d, the liquid wax and plastic have mixed, and on cooling, have formed what looks like a single solid substance. It is most likely that this is a very fine-grained mixture of solid wax and solid plastic, but it could also be some other substance that has formed from the combination of the two.

Figure 5: Partial melting of "pretend rock": (a) the original components of white candle wax, black plastic pipe, green beach glass, and aluminum wire, (b) after

heating to 50°C for 30 minutes only the wax has melted, (c) after heating to 120°C for 60 minutes much of the plastic has melted and the two liquids have mixed, (d) the liquid has been separated from the solids and allowed to cool to make a "pretend rock" with a different overall composition. [SE]

In this example, we partially melted some pretend rock to create some pretend magma. We then separated the magma from the source and allowed it to cool to make a new pretend rock with a composition quite different from the original material (it lacks glass and aluminum).

Of course partial melting in the real world isn't exactly the same as in our pretend-rock example. The main differences are that rocks are much more complex than the four-component system we used, and the mineral components of most rocks have more similar melting temperatures, so two or more minerals are likely to melt at the same time to varying degrees. Another important difference is that when rocks melt, the process takes thousands to millions of years, not the 90 minutes it took in the pretend-rock example.

Contrary to what one might expect, and contrary to what we did to make our pretend rock, most partial melting of real rock does not involve heating the rock up. The two main mechanisms through which rocks melt are decompression melting and flux melting. Decompression melting takes place within Earth when a body of rock is held at approximately the same temperature but the pressure is reduced. This happens because the rock is being moved toward the surface, either at a mantle plume (a.k.a., hot spot), or in the upwelling part of a mantle convection cell. Mantle plumes are described in Chapter 4 and mantle convection in Chapter 9. The mechanism of decompression melting is shown in Figure 6a. If a rock that is hot enough to be close to its melting point is moved toward the surface, the pressure is reduced, and the rock can pass to the liquid side of its melting curve. At this point, partial melting starts to take place. The process of flux melting is shown in Figure 6b. If a rock is close to its melting point and some water (a flux that promotes melting) is added to the rock, the melting temperature is reduced (solid line versus dotted line), and partial melting starts.

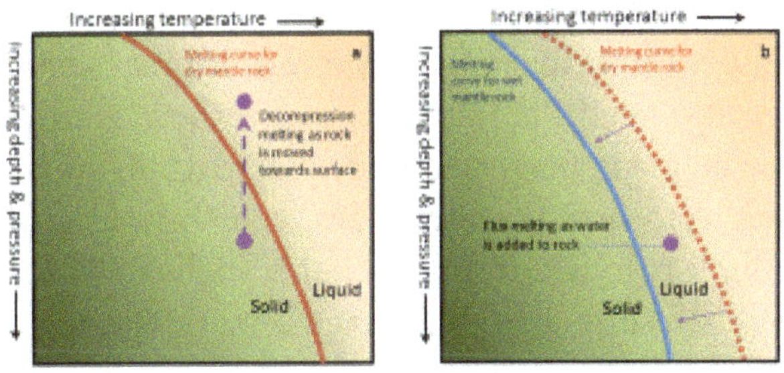

Figure 6: Mechanisms for (a) decompression melting (the rock is moved toward the surface) and (b) flux melting (water is added to the rock) and the melting curve is displaced. [SE]

The partial melting of rock happens in a wide range of situations, most of which are related to plate tectonics. The more important of these are shown in Figure 7. At both mantle plumes and in the upward parts of convection systems, rock is being moved toward the surface, the pressure is dropping, and at some point, the rock crosses to the liquid side of its melting curve. At subduction zones, water from the wet, subducting oceanic crust is transferred into the overlying hot mantle. This provides the flux needed to lower the melting temperature. In both of these cases, only partial melting takes place — typically only about 10% of the rock melts — and it is always the most silica-rich components of the rock that melt, creating a magma that is more silica-rich than the rock from which it is derived. (By analogy, the melt from our pretend rock is richer in wax and plastic than the "rock" from which it was derived.) The magma produced, being less dense than the surrounding rock, moves up through the mantle, and eventually into the crust.

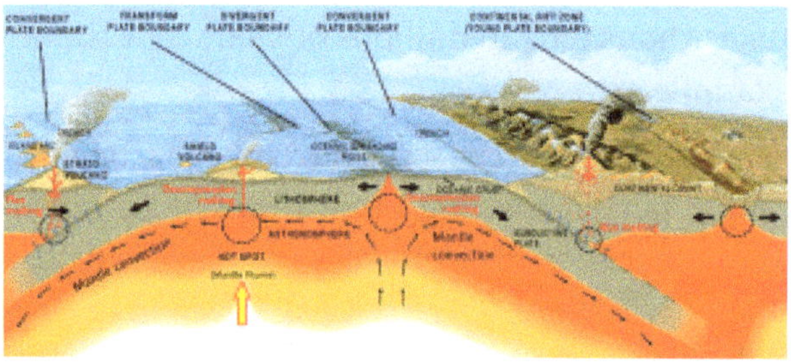

Figure 7: Common sites of magma formation in the upper mantle. The black circles are regions of partial melting. The blue arrows represent water being transferred from the subducting plates into the overlying mantle. [SE, after USGS (http://pubs.usgs.gov/gip/dynamic/Vigil.html)]

As it moves toward the surface, and especially when it moves from the mantle into the lower crust, the hot magma interacts with the surrounding rock. This typically leads to partial melting of the surrounding rock because most such magmas are hotter than the melting temperature of crustal rock. (In this case, melting is caused by an increase in temperature.) Again, the more silica-rich parts of the surrounding rock are preferentially melted, and this contributes to an increase in the silica content of the magma.

At very high temperatures (over 1300°C), most magma is entirely liquid because there is too much energy for the atoms to bond together. As the temperature drops, usually because the magma is slowly moving upward, things start to change. Silicon and oxygen combine to form silica tetrahedra, and then, as cooling continues, the tetrahedra start to link together to make chains (polymerize). These silica chains have the important effect of making the magma more viscous (less runny), and as we'll see in Chapter 4, magma viscosity has significant implications for volcanic eruptions. As the magma continues to cool, crystals start to form.

3.3 MAGMA CRYSTALLISATION

The minerals that make up igneous rocks crystallize at a range of different temperatures. This explains why a cooling magma can have some crystals within it and yet remain predominantly liquid. The sequence in which minerals crystallize from a magma is known as the Bowen reaction series (Figure 8 and Who was Bowen).

Of the common silicate minerals, olivine normally crystallizes first, at between 1200° and 1300°C. As the temperature drops, and assuming that some silica remains in the magma, the olivine crystals react (combine) with some of the silica in the magma (see Box 1) to form pyroxene. As long as there is silica remaining and the rate of cooling is slow, this process continues down the discontinuous branch: olivine to pyroxene, pyroxene to amphibole, and (under the right conditions) amphibole to biotite.

At about the point where pyroxene begins to crystallize, plagioclase feldspar also begins to crystallize. At that temperature, the plagioclase is calcium-rich (anorthite) (see Figure 8). As the temperature drops, and providing that there is sodium left in the magma, the plagioclase that forms is a more sodium-rich variety.

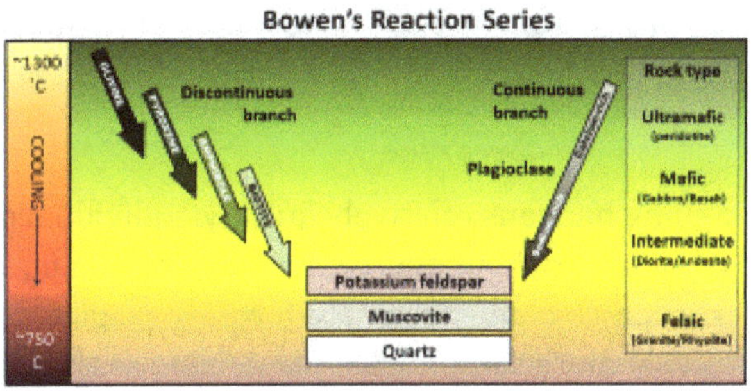

Figure 8: The Bowen reaction series describes the process of magma crystallization [SE]

Who was Bowen, and what's a reaction series?

Norman Levi Bowen, born in Kingston Ontario, studied geology at Queen's University and then at MIT in Boston. In 1912, he joined the Carnegie Institution in Washington, D.C., where he carried out groundbreaking experimental research into the processes of cooling magmas. Working mostly with basaltic magmas, he determined the order of crystallization of minerals as the temperature drops. The method, in brief, was to melt the rock to a magma in a specially made kiln, allow it to cool slowly to a specific temperature (allowing some minerals to form), and then quench it (cool it quickly) so that no new minerals form (only glass). The results were studied under the microscope and by chemical analysis. This was done over and over, each time allowing the magma to cool to a lower temperature before quenching.

The Bowen reaction series is one of the results of his work, and even a century later, it is an important basis for our understanding of igneous rocks. The word *reaction* is critical. In the discontinuous branch, olivine is typically the first mineral to form (at just below 1300°C). As the temperature continues to drop, olivine becomes unstable while pyroxene becomes stable. The early-forming olivine crystals *react* with silica in the remaining liquid magma and are converted into pyroxene, something like this:

$Mg_2SiO_4 + SiO_2$ ——> $2MgSiO_3$

olivine pyroxene

In cases where cooling happens relatively quickly, individual plagioclase crystals can be zoned from calcium-rich in the centre to more sodium-rich around the outside. This occurs when calcium-rich early-forming plagioclase crystals become coated with progressively more sodium-rich plagioclase as the magma cools. Figure 9 shows a zoned plagioclase under a microscope.

Figure 9: A zoned plagioclase crystal. The central part is calcium-rich and the outside part is sodium-rich: [Sandra Johnstone, used with permission]

Finally, if the magma is quite silica-rich to begin with, there will still be some left at around 750° to 800°C, and from this last magma, potassium feldspar, quartz, and maybe muscovite mica will form.

The composition of the original magma is critical to magma crystallization because it determines how far the reaction process can continue before all of the silica is used up. The compositions of typical mafic, intermediate, and felsic magmas are shown in Figure 10. Note that, unlike Figure 4, these compositions are expressed in terms of "oxides" (e.g., Al_2O_3 rather than just Al). There are two reasons for this: one is that in the early analytical procedures, the results were always expressed that way, and the other is that all of these elements combine readily with oxygen to form oxides.

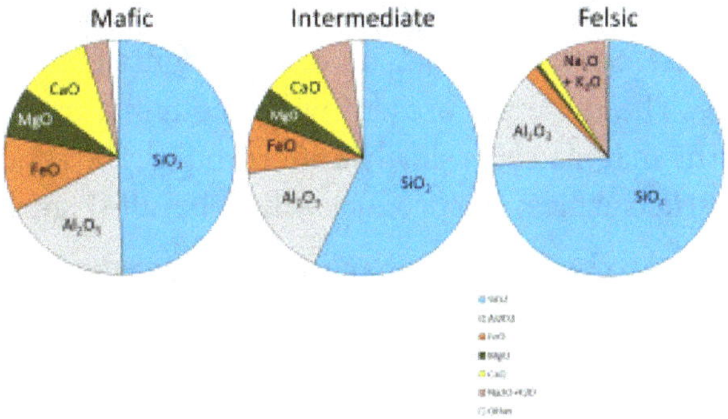

Figure 10: The chemical compositions of typical mafic, intermediate, and felsic magmas and the types of rocks that form from them.

Mafic magmas have 45% to 55% SiO_2, about 25% total of FeO and MgO plus CaO, and about 5% Na_2O + K_2O. Felsic magmas, on the other hand, have much more SiO_2 (65% to 75%) and Na_2O + K_2O (around 10%) and much less FeO and MgO plus CaO (about 5%).

Exercises

Exercise 3 Rock Types Based on Magma Composition

The proportions of the main chemical components of felsic, intermediate, and mafic magmas are listed in the table below.

Oxide	Felsic Magma	Intermediate Magma	Mafic Magma
SiO_2	65-75%	55-65%	45-55%
Al_2O_3	12-16%	14-18%	14-18%
FeO	2-4%	4-8%	8-12%
CaO	1-4%	4-7%	7-11%
MgO	0-3%	2-6%	5-9%
Na_2O	2-6%	3-7%	1-3%
K_2O	3-5%	2-4%	0.5-3%

Chemical data for four rock samples are shown in the following table. Compare these with those in the table above to determine whether each of these samples is felsic, intermediate, or mafic.

SiO_2	Al_2O_3	FeO	CaO	MgO	Na_2O	K_2O	Type?
55%	17%	5%	6%	3%	4%	3%	
74%	14%	3%	3%	0.5%	5%	4%	
47%	14%	8%	10%	8%	1%	2%	
65%	14%	4%	5%	4%	3%	3%	

As a mafic magma starts to cool, some of the silica combines with iron and magnesium to make olivine. As it cools further, much of the remaining silica goes into calcium-rich plagioclase, and any silica left may be used to convert some of the olivine to pyroxene. Soon after that, all of the magma is used up and no further changes takes place. The minerals present will be olivine, pyroxene, and calcium-rich plagioclase. If the magma cools slowly underground, the product will be gabbro; if it cools quickly at the surface, the product will be basalt (Figure 11).

Felsic magmas tend to be cooler than mafic magmas when crystallization begins (because they don't have to be as hot to remain liquid), and so they may start out crystallizing pyroxene (not olivine) and plagioclase. As cooling continues, the various reactions on the discontinuous branch will proceed because silica is abundant, the plagioclase will become increasingly sodium-rich, and eventually potassium feldspar and quartz will form. Commonly even very felsic rocks will not have biotite or muscovite because they may not have enough aluminum or enough hydrogen to make the OH complexes that are necessary for mica minerals. Typical felsic rocks are granite and rhyolite .

The cooling behaviour of intermediate magmas lie somewhere between those of mafic and felsic magmas. Typical intermediate rocks are diorite and andesite .

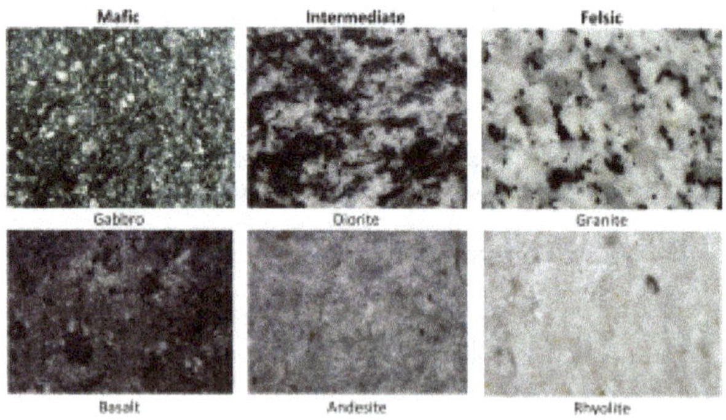

Figure 11: Examples of the igneous rocks that form from mafic, intermediate, and felsic magmas.

A number of processes that take place within a magma chamber can affect the types of rocks produced in the end. If the magma has a low viscosity (i.e., it's runny) —

which is likely if it is mafic — the crystals that form early, such as olivine (Figure 12a), may slowly settle toward the bottom of the magma chamber (Figure 12b). The means that the overall composition of the magma near the top of the magma chamber will become more felsic, as it is losing some iron- and magnesium-rich components. This process is known as fractional crystallization. The crystals that settle might either form an olivine-rich layer near the bottom of the magma chamber, or they might remelt because the lower part is likely to be hotter than the upper part (remember, from Chapter 1, that temperatures increase steadily with depth in Earth because of the geothermal gradient). If any melting takes place, crystal settling will make the magma at the bottom of the chamber more mafic than it was to begin with (Figure 12c).

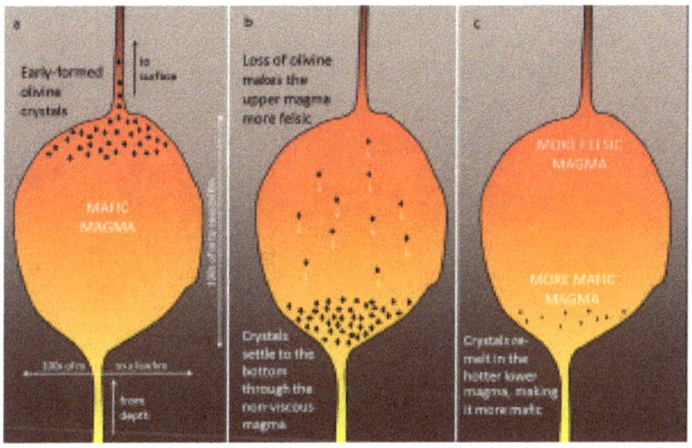

Figure 12: An example of crystal settling and the formation of a zoned magma chamber [SE]

If crystal settling does not take place, because the magma is too viscous, then the process of cooling will continue as predicted by the Bowen reaction series. In some cases, however, partially cooled but still liquid magma, with crystals in it, will either move farther up into a cooler part of the crust, or all the way to the surface during a volcanic eruption. In either of these situations, the magma that has moved toward the surface is likely to cool much faster than it did within the magma chamber, and the rest of the rock will have a finer crystalline texture. An igneous rock with large crystals embedded in a matrix of finer crystals is indicative of a two-stage cooling process, and the texture is porphyritic (Figure 13).

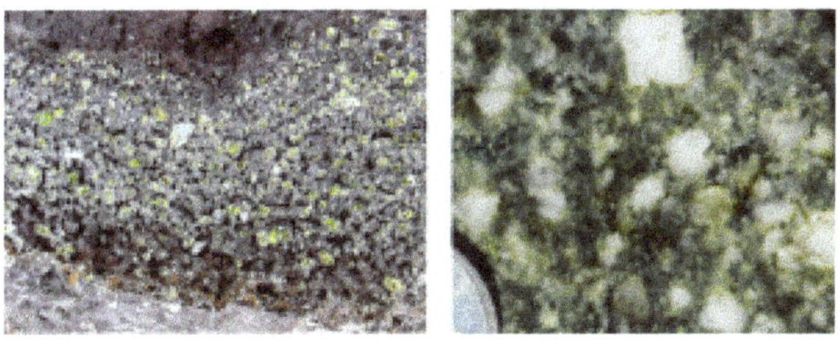

Figure 13: Porphyritic textures: volcanic porphyry (left – olivine crystals in Hawaiian basalt) and intrusive porphyry (right) [SE]

Exercises

Exercise 4 Porphyritic Minerals

As a magma cools below 1300°C, minerals start to crystallize within it. If that magma is then involved in a volcanic eruption, the rest of the liquid will cool quickly to form a porphyritic texture. The rock will have some relatively large crystals (phenocrysts) of the minerals that crystallized early, and the rest will be very fine grained or even glassy. Using the diagram shown here, predict what phenocrysts might be present where the magma cooled as far as line a in one case, and line b in another.

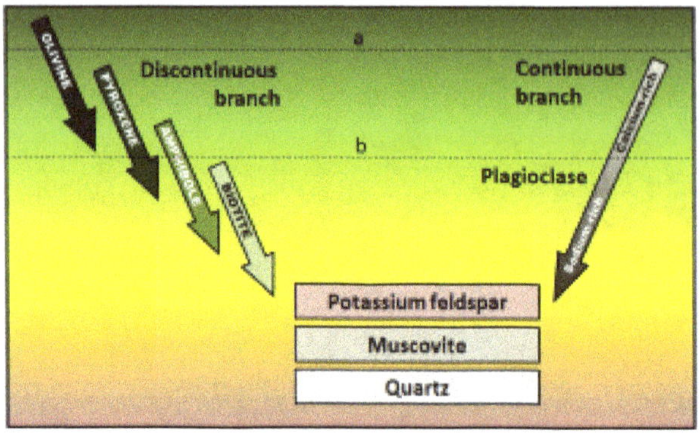

3.4 IGNEOUS ROCK CLASSIFICATION

As has already been described, igneous rocks are classified into four categories, based on either their chemistry or their mineral composition: felsic, intermediate, mafic, and ultramafic. The diagram in Figure 14 can be used to help classify igneous rocks by their mineral composition. An important feature to note on this diagram is the red line separating the non-ferromagnesian silicates in the lower left (K-feldspar, quartz, and plagioclase feldspar) from the ferromagnesian silicates in the upper right (biotite, amphibole, pyroxene, and olivine). In classifying intrusive igneous rocks, the first thing to consider is the percentage of ferromagnesian silicates. That's relatively easy in most igneous rocks because the ferromagnesian minerals are clearly darker than the others. At the same time, it's quite difficult to estimate the proportions of minerals in a rock.

Based on the position of the red line in Figure 14, it is evident that felsic rocks can have about 1% to 20% ferromagnesian silicates (the red line intersects the left side of the felsic zone 1% of the distance from the top of the diagram, and it intersects the right side of the felsic zone 20% of the distance from the top). Intermediate rocks have between 20% and 50% ferromagnesian silicates, and mafic rocks have 50% to 100% ferromagnesian silicates. To be more specific, felsic rocks typically have biotite and/or amphibole; intermediate rocks have amphibole and, in some cases, pyroxene; and mafic rocks have pyroxene and, in some cases, olivine.

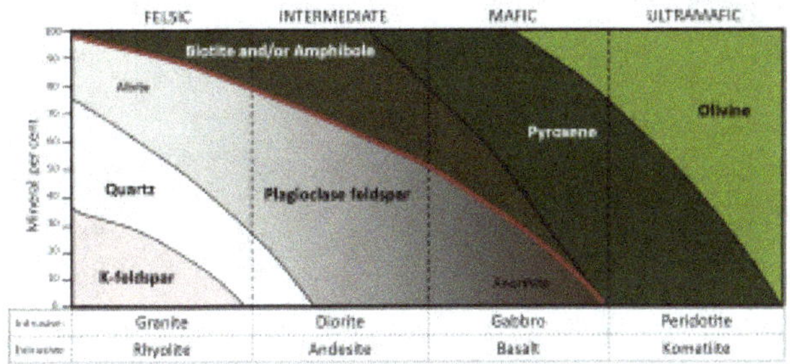

Figure 14: A simplified classification diagram for igneous rocks based on their mineral compositions

If we focus on the non-ferromagnesian silicates, it is evident that felsic rocks can have from 0% to 35% K-feldspar, from 25% to 35% quartz (the vertical thickness of the quartz field varies from 25% to 35%), and from 25% to 50% plagioclase (and that plagioclase will be sodium-rich, or albitic). Intermediate rocks can have up to 25% quartz and 50% to 75% plagioclase. Mafic rocks only have plagioclase (up to 50%), and that plagioclase will be calcium-rich, or anorthitic.

Figure 15 provides a diagrammatic representation of the proportions of dark minerals in light-coloured rocks. You can use that when trying to estimate the ferromagnesian mineral content of actual rocks, and you can get some practice doing that by completing Exercise 3.6.

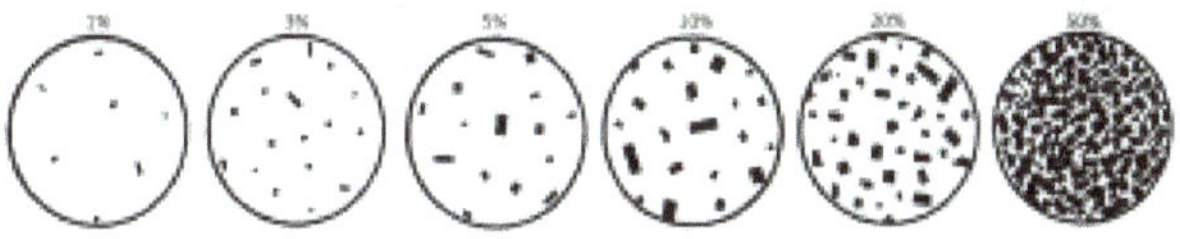

Figure 15: A guide to estimating the proportions of dark minerals in light-coloured rocks

3.5 INTRUSIVE IGNEOUS BODIES

In most cases, a body of hot magma is less dense than the rock surrounding it, so it has a tendency to move very slowly up toward the surface. It does so in a few different ways, including filling and widening existing cracks, melting the surrounding rock (called country rock"Country rock" is not necessarily music to a geologist's ears. The term refers to the original "rock of the country" or region, and hence the rock into which the magma intruded to form a pluton.), pushing the rock aside (where it is somewhat plastic), and breaking the rock. Where some of the country rock is broken off, it may fall into the magma, a process called stoping. The resulting fragments, illustrated in Figure 16, are known as xenoliths (Greek for "strange rocks").

Figure 16: Xenoliths of mafic rock in granite, Victoria, B.C. The fragments of dark rock have been broken off and incorporated into the light-coloured granite. [SE]

Some upward-moving magma reaches the surface, resulting in volcanic eruptions, but most cools within the crust. The resulting body of rock is known as a pluton. Plutons can have various different shapes and relationships to the surrounding country rock as shown in Figure 17.

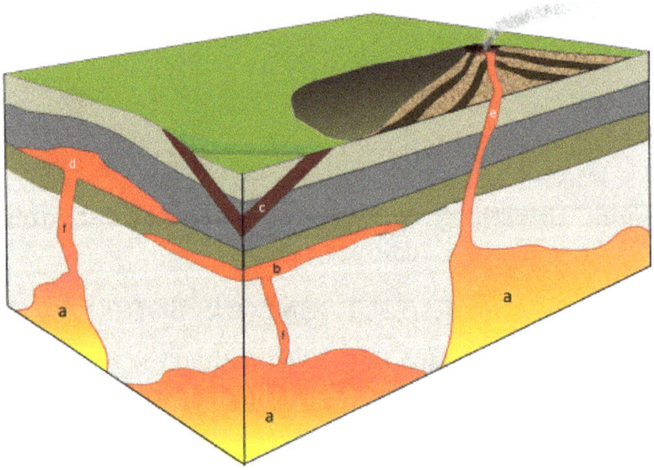

Figure 17: Depiction of some of the types of plutons. a: stocks (if they coalesce at depth then they might constitute a batholith), b: sill (a tabular body, in this case parallel to bedding), c: dyke (cross-cuts bedding), d: laccolith (a sill that has pushed up the overlying rock layers), e: pipe (a cylindrical conduit feeding a volcano). The two features labelled f could be pipes or dykes, but from this perspective it's not possible to determine if they are cylindrical or tabular. [SE drawing]

Large irregular-shaped plutons are called either stocks or batholiths. The distinction between the two is made on the basis of the area that is exposed at the surface: if the body has an exposed surface area greater than 100 km^2, then it's a batholith; smaller than 100 km^2 and it's a stock. Batholiths are typically formed only when a number of stocks coalesce beneath the surface to create one large body. One of the largest batholiths in the world is the Coast Range Plutonic Complex, which extends all the way from the Vancouver region to southeastern Alaska (Figure 18). More accurately, it's many batholiths.

Tabular (sheet-like) plutons are distinguished on the basis of whether or not they are concordant with (parallel to) existing layering (e.g., sedimentary bedding or metamorphic foliation) in the country rock. A sill is concordant with existing layering, and a dyke is discordant. If the country rock has no bedding or foliation, then any tabular body within it is a dyke. Note that the sill-versus-dyke designation is not determined simply by the orientation of the feature. A dyke can be horizontal and a sill can be vertical (if the bedding is vertical). A large dyke can be seen in Figure 18.

A laccolith is a sill-like body that has expanded upward by deforming the overlying rock.

Finally, a pipe is a cylindrical body (with a circular, ellipitical, or even irregular cross-section) that served as a conduit for the movement of magma from one location to another. Most known pipes fed volcanoes, although pipes can also connect plutons. It is also possible for a dyke to feed a volcano.

Figure 18: The Stawamus Chief, part of the Coast Range Plutonic Complex, near to Squamish, B.C. The cliff is about 600 m high. Most of the dark stripes are a result of algae and lichen growth where the surface is frequently wet, but there is a large (about 10 m across) vertical dyke that extends from bottom to top. [SE photo]

As discussed already, plutons can interact with the rocks into which they are intruded, sometimes leading to partial melting of the country rock or to stoping and formation of xenoliths. And, as we'll see in Chapter 7, the heat of a body of magma can lead to metamorphism of the country rock. The country rock can also have an effect on the magma within a pluton. The most obvious such effect is the formation of a chilled margin along the edges of the pluton, where it came in contact with country rock that was significantly colder than the magma. Within the chilled margin, the magma cooled more quickly than in the centre of the dyke, so the texture is finer and the colour may be different. An example is shown in Figure 19.

Figure 19: A mafic dyke with chilled margins within basalt at Nanoose, B.C. The coin is 24 mm in diameter. The dyke is about 25 cm across and the chilled margins are 2 cm wide.

AT A GLANCE

The topics covered in this chapter can be summarized as follows:

3.1	The Rock Cycle	The three types of rocks are igneous, formed from magma; sedimentary, formed from fragments of other rocks or precipitations from solution; and metamorphic, formed when existing rocks are altered by heat, pressure, and/or chemical action. The rock cycle summarizes the processes that contribute to cycling of rock material among these three types. The rock cycle is driven by Earth's internal heat, and by processes happening at the surface, which are driven by solar energy.
3.2	Magma and Magma Formation	Magma is molten rock, and in most cases, it forms from partial melting of existing rock. The two main processes of magma formation are decompression melting and flux melting. Magmas range in composition from ultramafic to felsic. Mafic rocks are rich in iron, magnesium, and calcium and have around 50% silica. Felsic rocks are rich in silica (~75%) and have lower levels of iron, magnesium, and calcium and higher levels of sodium and potassium than mafic rocks.

3.3	Crystallization of Magma	As a body of magma starts to cool, the first process to take place is the polymerization of silica tetrahedra into chains. This increases the magma's viscosity (makes it thicker) and because felsic magmas have more silica than mafic magmas, they tend to be more viscous. The Bowen reaction series allows us to predict the order of crystallization of magma as it cools. Magma can be modified by fractional crystallization (separation of early-forming crystals) and by incorporation of material from the surrounding rocks by partial melting.
3.4	Classification of Igneous Rocks	Igneous rocks are classified based on their mineral composition and texture. Felsic igneous rocks have less than 20% ferromagnesian silicates (amphibole and/or biotite) plus varying amounts of quartz and both potassium and plagioclase feldspars. Mafic igneous rocks have more than 50% ferromagnesian silicates (primarily pyroxene) plus plagioclase feldspar. Most intrusive igneous rocks are phaneritic (crystals are visible to the naked eye). If there were two stages of cooling (slow then fast), the texture may be porphyritic (large crystals in a matrix of smaller crystals). If water was present during cooling, the texture may be pegmatitic (very large crystals).
3.5	Intrusive Igneous Bodies	Magma intrudes into country rock by pushing it aside or melting through it. Intrusive igneous bodies tend to be either irregular (stocks and batholiths), tabular (dykes and sills), or pipe-like. Batholiths have exposed areas of greater than 100 km^2, while stocks are smaller. Sills are parallel to existing layering in the country rock, while dykes cut across layering. A pluton that intruded into cold rock it is likely to have a chilled margin.

CHAPTER

MAGMA AND VOLCANISM 4

A volcano is any location where magma comes to the surface, or has done so within the past several million years. This can include eruptions on the ocean floor (or even under the water of lake), where they are called subaqueous eruptions, or on land, where they are called subaerial eruptions. Not all volcanic eruptions produce the volcanic mountains with which we are familiar; in fact most of Earth's volcanism takes place along the spreading ridges on the sea floor and does not produce volcanic mountains at all — not even sea-floor mountains.

Canada has a great deal of volcanic rock, but most of it is old, some of it billions of years old. Only in B.C. and the Yukon are there volcanoes that have been active within the past 2.6 Ma (Pleistocene or younger), and the vast majority of these are in B.C. We'll look at those in some detail toward the end of this chapter, but a few of them are shown on Figures 1 and 2.

The study of volcanoes is critical to our understanding of the geological evolution of Earth, and to our understanding of significant changes in climate. But, most important of all, understanding volcanic eruptions allows us to save lives and property. Over the past few decades, volcanologists have made great strides in their ability to forecast volcanic eruptions and predict the consequences — this has already saved thousands of lives.

Figure 1: Mt. Garibaldi, near Squamish B.C., is one of Canada's tallest (2,678 m) and most recently active volcanoes. It last erupted approximately 10,000 years ago. [SE photo]

Figure 2: Mt. Garibaldi (background left, looking from the north) with Garibaldi Lake in the foreground. The volcanic peak in the centre is Mt. Price and the dark flat-topped peak is The Table. All three of these volcanoes were active during the last glaciation. [SE photo]

4.1 RELATIONSHIP BETWEEN PLATE TECTONICS AND VOLCANISM

The relationships between plate tectonics and volcanism are shown on Figure 3. As summarized in Chapter 3, magma is formed at three main plate-tectonic settings: divergent boundaries (decompression melting), convergent boundaries (flux melting), and mantle plumes (decompression melting).

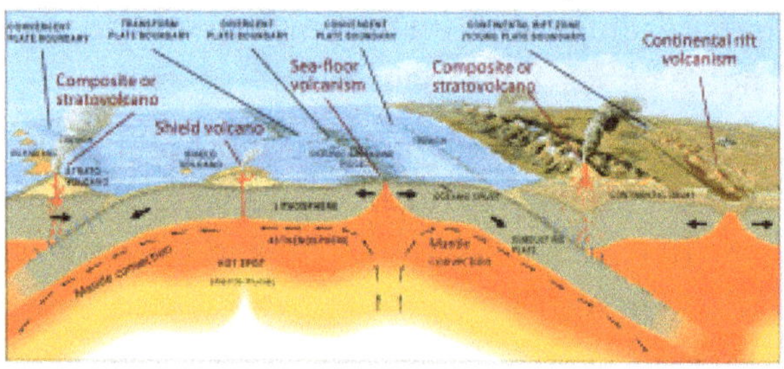

Figure 3: The plate-tectonic settings of common types of volcanism. Composite volcanoes form at subduction zones, either on ocean-ocean convergent boundaries (left) or ocean-continent convergent boundaries (right). Both shield volcanoes and cinder cones form in areas of continental rifting. Shield volcanoes form above mantle plumes, but can also form at other tectonic settings. Sea-floor volcanism can take place at divergent boundaries, mantle plumes and ocean-ocean-convergent boundaries. [SE, after USGS (http://pubs.usgs.gov/gip/dynamic/Vigil.html)]

The mantle and crustal processes that take place in areas of volcanism are illustrated in Figure 4. At a spreading ridge, hot mantle rock moves slowly upward by convection (cm/year), and within about 60 km of the surface, partial melting starts because of decompression. Over the triangular area shown in Figure 4a, about 10% of the ultramafic mantle rock melts, producing mafic magma that moves upward toward the axis of spreading (where the two plates are moving away from each other). The magma fills vertical fractures produced by the spreading and spills out onto the sea floor to form basaltic pillows (more on that later) and lava flows. There is spreading-ridge volcanism taking place about 200 km offshore from the west coast of Vancouver Island.

4.2 COMPOSITION OF THE MAGMA AND ERUPTION STYLE

The types of magma produced in the various volcanic settings can differ significantly. At divergent boundaries and oceanic mantle plumes, where there is little interaction with crustal materials and magma fractionation to create felsic melts does not take place, the magma tends to be consistently mafic. At subduction zones, where the magma ascends through significant thicknesses of crust, interaction between the magma and the crustal rock – some of which is quite felsic – leads to increases in the felsic character of the magma.

As shown in Figure 7, several processes can make magma that is stored in a chamber within the crust more felsic, and can also contribute to development of vertical zonation from more mafic at the bottom to more felsic at the top. Partial melting of country rock and country-rock xenoliths increases the overall felsic character of the magma; first, because the country rocks tends to be more felsic than the magma, and second, because the more felsic components of the country rock melt preferentially. Settling of ferromagnesian crystals from the upper part of the magma, and possible remelting of those crystals in the lower part can both contribute to the vertical zonation from relatively mafic at the bottom to more felsic at the top.

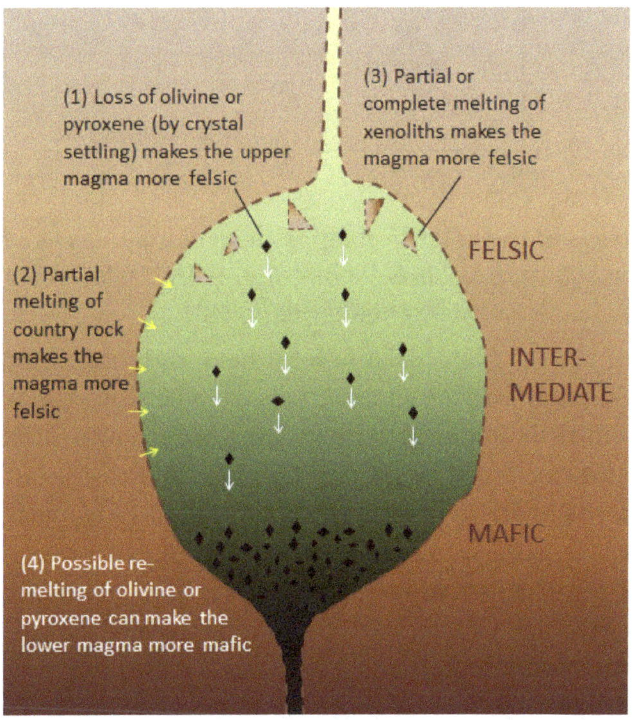

Figure 7: The important processes that lead to changes in the composition of magmas stored within magma chambers within relatively felsic rocks of the crust. [SE]

From the perspective of volcanism there are some important differences between felsic and mafic magmas. First, as we've already discussed, felsic magmas tend to be more viscous because they have more silica, and hence more polymerization. Second, felsic magmas tend to have higher levels of volatiles; that is, components that behave as gases during volcanic eruptions. The most abundant volatile in magma is water (H_2O), followed typically by carbon dioxide (CO_2), and then by sulphur dioxide (SO_2). The general relationship between the SiO_2 content of magma and the amount of volatiles is shown in Figure 8. Although there are many exceptions to this trend, mafic magmas typically have 1% to 3% volatiles, intermediate magmas have 3% to 4% volatiles, and felsic magmas have 4% to 7% volatiles.

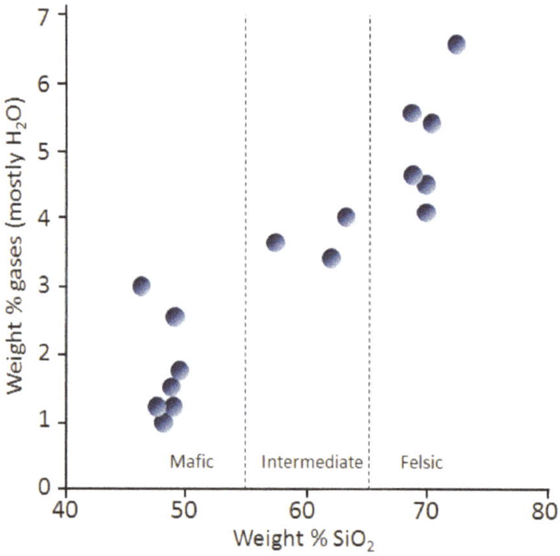

Figure 8: Variations in the volatile compositions of magmas as a function of silica content [SE after Schminke, 2004, (Schminke, H-U., 2004, Volcanism, Springer-Verlag, Heidelberg)]

Differences in viscosity and volatile level have significant implications for the nature of volcanic eruptions. When magma is deep beneath the surface and under high pressure from the surrounding rocks, the gases remain dissolved. As magma approaches the surface, the pressure exerted on it decreases. Gas bubbles start to form, and the more gas there is in the magma, the more bubbles form. If the gas content is low or the magma is runny enough for gases to rise up through it and escape to surface, the pressure will not become excessive. Assuming that it can break through to the surface, the magma will flow out relatively gently. An eruption that involves a steady non-violent flow of magma is called effusive.

4.3 NUMEROUS TYPES OF VOLCANOES

There are numerous types of volcanoes or volcanic sources; some of the more common ones are summarized in Table 1.

Table 1: A summary of the important types of volcanism

Type	Tectonic Setting	Size and Shape	Magma and Eruption Characteristics	Example
Cinder cone	Various; some form on the flanks of larger volcanoes	Small (10s to 100s of m) and steep (>20°)	Most are mafic and form from the gas-rich early stages of a shield- or rift-associated eruption	Eve Cone, northern B.C.
Composite volcano	Almost all are at subduction zones	Medium size (1000s of m) and moderate steepness (10° to 30°)	Magma composition varies from felsic to mafic, and from explosive to effusive	Mt. St. Helens
Shield volcano	Most are at mantle plumes; some are on spreading ridges	Large (up to several 1,000 m high and 200 km across), not steep (typically 2° to 10°)	Magma is almost always mafic, and eruptions are typically effusive, although cinder cones are common on the flanks of shield volcanoes	Kilauea, Hawaii
Large igneous provinces	Associated with "super" mantle plumes	Enormous (up to millions of km^2) and 100s of m thick	Magma is always mafic and individual flows can be 10s of m thick	Columbia River basalts
Sea-floor volcanism	Generally associated with spreading ridges but also with mantle plumes	Large areas of the sea floor associated with spreading ridges	At typical eruption rates, pillows form; at faster rates, lava flows develop	Juan de Fuca ridge
Kimberlite	Upper-mantle sourced	The remnants are typically 10s to 100s of m across	Most appear to have had explosive eruptions forming cinder cones; the youngest one is over 10 ka old, and all others are over 30 Ma old.	Lac de Gras Kimberlite Field, N.W.T.

The sizes and shapes of typical shield, composite, and cinder-cone volcanoes are compared in Figure 9, although, to be fair, Mauna Loa is the largest shield volcano on Earth; all others are smaller. Mauna Loa rises from the surrounding flat sea floor, and its diameter is in the order of 200 km. Its elevation is 4,169 m above sea level. Mt. St. Helens, a composite volcano, rises above the surrounding hills of the Cascade Range. Its diameter is about 6 km, and its height is 2,550 m above sea level. Cinder cones are much smaller. On this drawing, even a large cinder cone is just a dot.

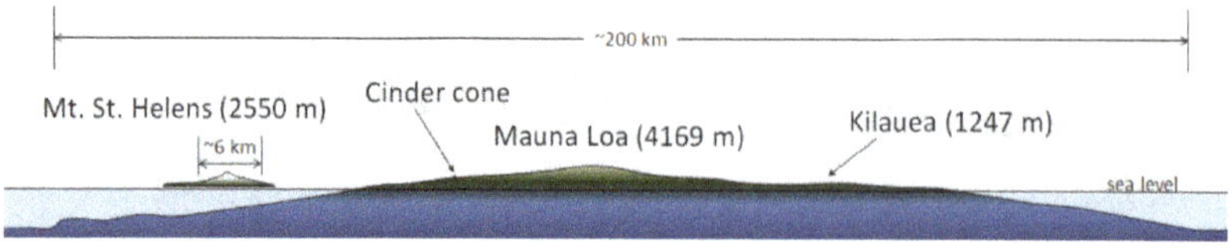

Figure 9: Profiles of Mauna Loa shield volcano, Mt. St. Helens composite volcano, and a large cinder cone [SE]

Cinder Cones

Cinder cones, like Eve Cone in northern B.C. (Figure 10), are typically only a few hundred metres in diameter, and few are more than 200 m high. Most are made up of fragments of vesicular mafic rock (scoria) that were expelled as the magma boiled when it approached the surface, creating fire fountains. In many cases, these later became effusive (lava flows) when the gases were depleted. Most cinder cones are monogenetic, meaning that they formed during a single eruptive phase that might have lasted weeks or months. Because cinder cones are made up almost exclusively of loose fragments, they have very little strength. They can be easily, and relatively quickly, eroded away.

Figure 10: Eve Cone, situated near to Mt. Edziza in northern B.C., formed approximately 700 years ago

Composite Volcanoes

Composite volcanoes, like Mt. St. Helens in Washington State (Figure 11), are almost all associated with subduction at convergent plate boundaries — either ocean-continent or ocean-ocean boundaries (Figure 4b). They can extend up to several thousand metres from the surrounding terrain, and, with slopes ranging up to 30°, are typically up to 10 km across. At many such volcanoes, magma is stored in a magma chamber in the upper part of the crust. For example, at Mt. St. Helens, there is evidence of a magma chamber that is approximately 1 km wide and extends from about 6 km to 14 km below the surface (Figure 12). Systematic variations in the composition of volcanism over the past several thousand years at Mt. St. Helens imply that the magma chamber is zoned, from more felsic at the top to more mafic at the bottom.

Figure 11: The north side of Mt. St. Helens in southwestern Washington State, 2003 [SE photo]. The large 1980 eruption reduced the height of the volcano by 400 m, and a sector collapse removed a large part of the northern flank. Between 1980 and 1986 the slow eruption of more mafic and less viscous lava led to construction of a dome inside the crater.

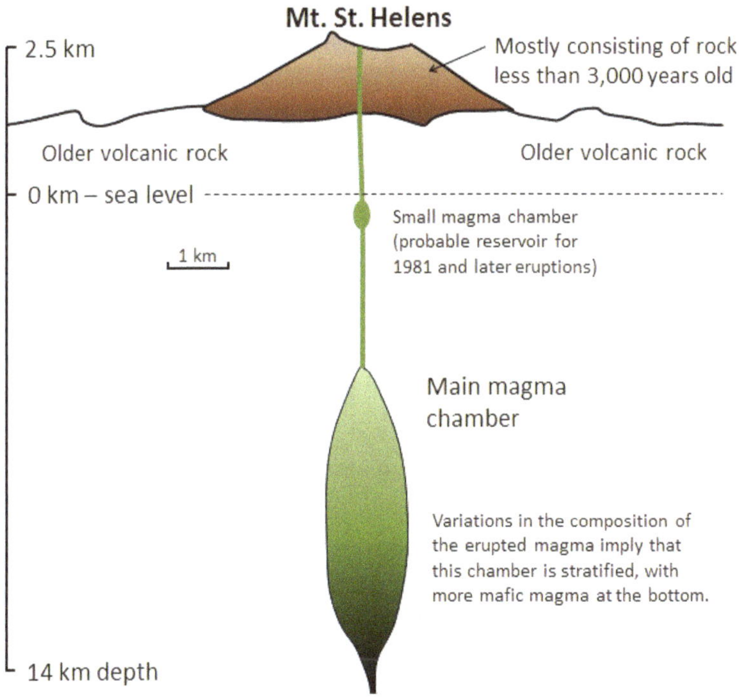

Figure 12: A cross-section through the upper part of the crust at Mt. St. Helens showing the zoned magma chamber. [SE, after Pringle, 1993]

Mafic eruptions (and some intermediate eruptions), on the other hand, produce lava flows; the one shown in Figure 13b is thick enough (about 10 m in total) to have cooled in a columnar jointing pattern (Figure 14). Lava flows both flatten the profile of the volcano (because the lava typically flows farther than pyroclastic debris falls) and protect the fragmental deposits from erosion. Even so, composite volcanoes tend to erode quickly. Patrick Pringle, a volcanologist with the Washington State Department of Natural Resources, describes Mt. St. Helens as a "pile of junk." The rock that makes up Mt. St. Helens ranges in composition from rhyolite (Figure 13a) to basalt (Figure 13b); this implies that the types of past eruptions have varied widely in character. As already noted, felsic magma doesn't flow easily and doesn't allow gases to escape easily. Under these circumstances, pressure builds up until a conduit opens, and then an explosive eruption results from the gas-rich upper part of the magma chamber, producing pyroclastic debris, as shown on Figure 13a. This type of eruption can also lead to rapid melting of ice and snow on a volcano, which typically triggers large mudflows known as lahars (Figure 13a). Hot, fast-moving pyroclastic flows and lahars are the two main causes of casualties in volcanic eruptions. Pyroclastic flows killed approximately 30,000 people during the 1902 eruption of Mt. Pelée on the Caribbean island of Martinique. Most were incinerated in their homes. In 1985 a massive lahar, triggered by the eruption of Nevado del Ruiz, killed 23,000 people in the Colombian town of Armero, about 50 km from the volcano.

In a geological context, composite volcanoes tend to form relatively quickly and do not last very long. Mt. St. Helens, for example, is made up of rock that is all younger than 40,000 years; most of it is younger than 3,000 years. If its volcanic activity ceases, it might erode away within a few tens of thousands of years. This is largely because of the presence of pyroclastic eruptive material, which is not strong.

Figure 13: Mt. St. Helens volcanic deposits: (a) lahar deposits (L) and felsic pyroclastic deposits (P) and (b) a columnar basalt lava flow. The two photos were taken at locations only about 500 m apart. [SE]

Shield Volcanoes

Most shield volcanoes are associated with mantle plumes, although some form at divergent boundaries, either on land or on the sea floor. Because of their non-viscous mafic magma they tend to have relatively gentle slopes (2 to 10°) and the larger ones can be over 100 km in diameter. The best-known shield volcanoes are those that make up the Hawaiian Islands, and of these, the only active ones are on the big island of Hawaii. Mauna Loa, the world's largest volcano and the world's largest mountain (by volume) last erupted in 1984. Kilauea, arguably the world's most active volcano, has been erupting, virtually without interruption, since 1983. Loihi is an underwater volcano on the southeastern side of Hawaii. It is last known to have erupted in 1996, but may have erupted since then without being detected.

All of the Hawaiian volcanoes are related to the mantle plume that currently lies beneath Mauna Loa, Kilauea, and Loihi (Figure 14). In this area, the Pacific Plate is moving northwest at a rate of about 7 cm/year. This means that the earlier formed — and now extinct — volcanoes have now moved well away from the mantle plume. As shown on Figure 14, there is evidence of crustal magma chambers beneath all three active Hawaiian volcanoes. At Kilauea, the magma chamber appears to be several kilometres in diameter, and is situated between 8 km and 11 km below surface.Lin, G, Amelung, F, Lavallee, Y, and Okubo, P, 2014, Seismic evidence for a crustal magma reservoir beneath the upper east rift zone of Kilauea volcano, Hawaii. Geology. V.

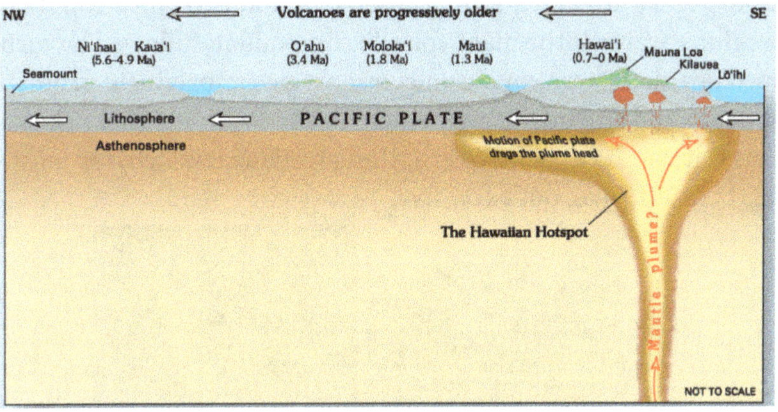

Figure 14: Mauna Kea from near to the summit of Mauna Loa, Hawaii [http://upload.wikimedia.org/wikipedia/commons/f/f1/Hawaii_hotspot_cross-sectional_diagram.jpg]

Although it is not a prominent mountain (Figure 9), Kilauea volcano has a large caldera in its summit area (Figure 15). A caldera is a volcanic crater that is more than 2 km in diameter; this one is 4 km long and 3 km wide. It contains a smaller feature called Halema'uma'u crater, which has a total depth of over 200 m below the surrounding area. Most volcanic craters and calderas are formed above magma chambers, and the level of the crater floor is influenced by the amount of pressure exerted by the magma body. During historical times, the floors of both Kilauea caldera and Halema'uma'u crater have moved up during expansion of the magma chamber and down during deflation of the chamber.

Figure 15: Aerial view of the Kilauea caldera. The caldera is about 4 km across, and up to 120 m deep. It encloses a smaller and deeper crater known as Halema'uma'u. [http://upload.wikimedia.org/wikipedia/commons/b/b4/Kilauea_ali_2012_01_28.jpg]

One of the conspicuous features of Kilauea caldera is rising water vapour (the white cloud in Figure 16) and a strong smell of sulphur (Figure 16). As is typical in magmatic regions, water is the main volatile component, followed by carbon dioxide and sulphur dioxide. These, and some minor gases, originate from the magma chamber at depth and rise up through cracks in the overlying rock. This degassing of the magma is critical to the style of eruption at Kilauea, which, for most of the past 30 years, has been effusive, not explosive.

Figure 16: A gas-composition monitoring station (left) within the Kilauea caldera and at the edge of Halema'uma'u crater. The rising clouds are mostly composed of water vapour, but also include carbon dioxide and sulphur dioxide. Sulphur crystals (right) have formed around a gas vent in the caldera. [SE photos]

The Kilauea eruption that began in 1983 started with the formation of a cinder cone at Pu'u 'O'o, approximately 15 km east of the caldera (Figure 17). The magma feeding this eruption flowed along a major conduit system known as the East Rift, which extends for about 20 km from the caldera, first southeast and then east. Lava fountaining and construction of the Pu'u 'O'o cinder cone (Figure 18a) continued until 1986 at which time the flow became effusive. From 1986 to 2014, lava flowed from a gap in the southern flank of Pu'u 'O'o down the slope of Kilauea through a lava tube (Figure 18d), emerging at or near the ocean. Since June 2014, the lava has flowed northeast (see Exercise 4).

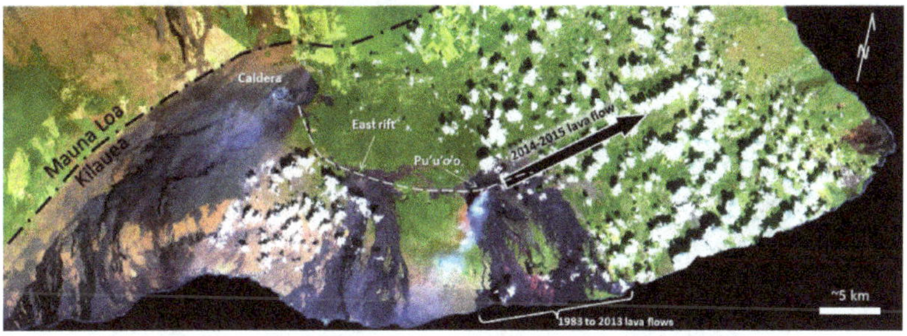

Figure 17: Satellite image of Kilauea volcano showing the East rift and Pu'u 'O'o, the site of the eruption that started in 1983. The puffy white blobs are clouds. [SE after, http://en.wikipedia.org/wiki/Hawaii_(island)#mediaviewer/File:Island_of_Hawai%27i_-_Landsat_mosaic.jpg]

The two main types of textures created during effusive subaerial eruptions are pahoehoe and aa. Pahoehoe, ropy lava that forms as non-viscous lava, flows gently, forming a skin that gels and then wrinkles because of ongoing flow of the lava below the surface (Figure 19b, and "lava flow video"). Aa, or blocky lava, forms when magma is forced to flow faster than it is able to (down a slope for example) (Figure 18c). Tephra (lava fragments) is produced during explosive eruptions, and accumulates in the vicinity of cinder cones.

Figure 19d is a view into an active lava tube on the southern edge of Kilauea. The red glow is from a stream of very hot lava (~1200°C) that has flowed underground for most of the 8 km from the Pu'u 'O'o vent. Lava tubes form naturally and readily on both shield and composite volcanoes because flowing mafic lava preferentially cools near its margins, forming solid lava levées that eventually close over the top of the flow. The magma within a lava tube is not exposed to the air, so it remains hot and fluid and can flow for tens of kilometres, thus contributing to the large size and low slopes of shield volcanoes. The Hawaiian volcanoes are riddled with thousands of old lava tubes, some as long as 50 km.

Kilauea is approximately 300 ka old, while neighbouring Mauna Loa is over 700 ka and Mauna Kea is over 1 Ma. If volcanism continues above the Hawaii mantle plume in the same manner that it has for the past 85 Ma, it is likely that Kilauea will continue to erupt for at least another 500,000 years. By that time, its neighbour, Loihi, will have emerged from the sea floor, and its other neighbours, Mauna Loa and Mauna Kea, will have become significantly eroded, like their cousins, the islands to the northwest (Figure 14).

Figure 18 Images of Kilauea volcano taken in 2002 (b & c) and 2007 (a & d) [SE photos] (a) Pu'u'O'o cinder cone in the background with tephra in the foreground and aa lava in the middle, (b) Formation of pahoehoe on the southern edge of Kilauea, (c) Formation of aa on a steep slope on Kilauea, (d) Skylight in an active lava tube, Kilauea.

Large Igneous Provinces

While the Hawaii mantle plume has produced a relatively low volume of magma for a very long time (~85 Ma), other mantle plumes are less consistent, and some generate massive volumes of magma over relatively short time periods. Although their origin is still controversial, it is thought that the volcanism leading to large igneous provinces (LIP) is related to very high volume but relatively short duration bursts of magma from mantle plumes. An example of an LIP is the Columbia River Basalt Group (CRGB), which extends across Washington, Oregon, and Idaho (Figure 19). This volcanism, which covered an area of about 160,000 km2 with basaltic rock up to several hundred metres thick, took place between 17 and 14 Ma.

Figure 19: A part of the Columbia River Basalt Group at Frenchman Coulee, eastern Washington. All of the flows visible here have formed large (up to two metres in diameter) columnar basalts, a result of relatively slow cooling of flows

that are tens of m thick. The inset map shows the approximate extent of the 17 to 14 Ma Columbia River Basalts, with the location of the photo shown as a star. [SE – photo and drawing]

Most other LIP eruptions are much bigger. The Siberian Traps (also basalt), which erupted at the end of the Permian period at 250 Ma, are estimated to have produced approximately 40 times as much lava as the CRBG.

The mantle plume that is assumed to be responsible for the CRBG is now situated beneath the Yellowstone area, where it leads to felsic volcanism. Over the past 2 Ma three very large explosive eruptions at Yellowstone have yielded approximately 900 km^3 of felsic magma, about 900 times the volume of the 1980 eruption of Mt. St. Helens, but only 5% of the volume of mafic magma in the CRBG.

Sea-Floor Volcanism

Some LIP eruptions occur on the sea floor, the largest being the one that created the Ontong Java plateau in the western Pacific Ocean at around 122 Ma. But most sea-floor volcanism originates at divergent boundaries and involves relatively low-volume eruptions. Under these conditions, hot lava that oozes out into the cold seawater quickly cools on the outside and then behaves a little like toothpaste. The resulting blobs of lava are known as pillows, and they tend to form piles around a sea-floor lava vent (Figure 20). In terms of area, there is very likely more pillow basalt on the sea floor than any other type of rock on Earth.

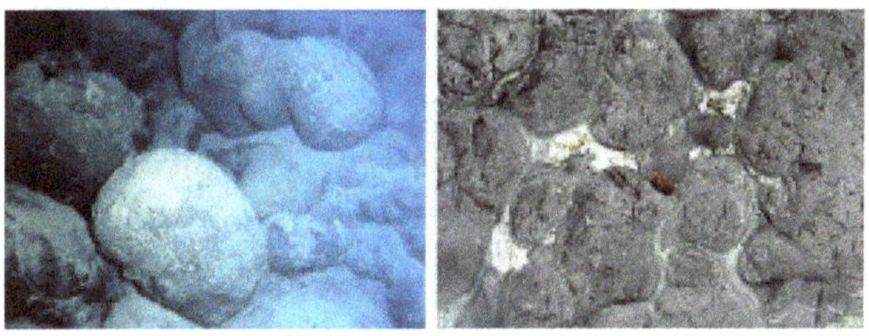

Figure 20: Modern and ancient sea-floor pillow basalts (left) Modern sea-floor pillows in the south Pacific [NOAA, from http://en.wikipedia.org/wiki/Basalt#mediaviewer/File:Pillow_basalt_crop_l.jpg] (right) Eroded 40 to 50 Ma pillows on the shore of Vancouver Island, near to Sooke. The pillows are 30 to 40 cm in diameter. [SE]

Kimberlites

While all of the volcanism discussed so far is thought to originate from partial melting in the upper mantle or within the crust, there is a special class of volcanoes called kimberlites that have their origins much deeper in the mantle, at depths of 150 km to 450 km. During a kimberlite eruption, material from this depth may make its way to surface quickly (hours to days) with little interaction with the surrounding rocks. As a result, kimberlite eruptive material is representative of mantle compositions: it is ultramafic.

Kimberlite eruptions that originate at depths greater than 200 km, within areas beneath old thick crust (shields), traverse the region of stability of diamond in the mantle, and in some cases, bring diamond-bearing material to the surface. All of the diamond deposits on Earth are assumed to have formed in this way; an example is the rich Ekati Mine in the Northwest Territories (Figure 21).

Figure 21: Ekati diamond mine, Northwest Territories, part of the Lac de Gras kimberlite field [http://upload.wikimedia.org/wikipedia/commons/8/88/Ekati_mine_640px.jpg]

The kimberlites at Ekati erupted between 45 and 60 Ma. Many kimberlites are older, some much older. There have been no kimberlite eruptions in historic times. The youngest known kimberlites are in the Igwisi Hills in Tanzania and are only about 10,000 years old. The next youngest known are around 30 Ma old.

4.4 CLASSES VOLCANIC HAZARDS

There are two classes of volcanic hazards, direct and indirect. Direct hazards are forces that directly kill or injure people, or destroy property or wildlife habitat. Indirect hazards are volcanism-induced environmental changes that lead to distress, famine, or habitat destruction. Indirect effects of volcanism have accounted for approximately 8 million deaths during historical times, while direct effects have accounted for fewer than 200,000, or 2.5% of the total. Some of the more important types of volcanic hazards are summarized in Table 2.

Table 2: A summary of the important volcanic hazards

Type	Description	Risk
Tephra emissions	Small particles of volcanic rock emitted into the atmosphere	Respiration problems for some individuals Significant climate cooling and famine Damage to aircraft
Gas emissions	The emission of gases before, during, and after an eruption	Climate cooling leading to crop failure and famine In some cases, widespread poisoning

Type	Description	Risk
Pyroclastic density current	A very hot (several 100°C) mixture of gases and volcanic tephra that flows rapidly (up to 100s of km/h) down the side of a volcano	Extreme hazard – destroys anything in the way
Pyroclastic fall	Vertical fall of tephra in the area surrounding an eruption	Thick tephra coverage of areas close to the eruption (km to 10s of km) Collapsed roofs
Lahar	A flow of mud and debris down a channel leading away from a volcano, triggered either by an eruption or a severe rain event	Severe risk of destruction for anything within the channel – lahar mud flows can move at 10s of km/h
Sector collapse/ debris avalanche	The failure of part of a volcano, either due to an eruption or for some other reason, leading to the failure of a large portion of the volcano	Severe risk of destruction for anything in the path of the debris avalanche
Lava flow	The flow of lava away from a volcanic vent	People and infrastructure at risk, but lava flows tend to be slow (km/h) and are relatively easy to avoid

The Emissions of Tephra and Volcanic Gas

Large volumes of tephra (rock fragments, mostly pumice) and gases are emitted during major plinian eruptions (large explosive eruptions with hot gas a tephra columns extending into the stratosphere) at composite volcanoes, and a large volume of gas is released during some very high-volume effusive eruptions. One of the major effects is cooling of the climate by 1° to 2°C for several months to a few years because the dust particles and tiny droplets and particles of sulphur compounds block the sun. The last significant event of this type was in 1991 and 1992 following the large eruption of Mt. Pinatubo in the Philippines. A drop of 1° to 2°C may not seem like very much, but that is the global average amount of cooling, and cooling was much more severe in some regions and at some times.

Over an eight-month period in 1783 and 1784, a massive effusive eruption took place at the Laki volcano in Iceland. Although there was relatively little volcanic ash involved, a massive amount of sulphur dioxide was released into the atmosphere, along with a significant volume of hydrofluoric acid (HF). The sulphate aerosols that formed in the atmosphere led to dramatic cooling in the northern hemisphere. There were serious crop failures in Europe and North America, and a total of 6 million people are estimated to have died from famine and respiratory complications. In Iceland, poisoning from the HF resulted in the death of 80% of sheep, 50% of cattle, and the ensuing famine, along with HF poisoning, resulted in more than 10,000 human deaths, about 25% of the population.

Volcanic ash can also have serious implications for aircraft because it can destroy jet engines. For example, over 5 million airline passengers had their travel disrupted by the 2010 Eyjafjallajökull volcanic eruption in Iceland.

Currents of Pyroclastic Density

In a typical explosive eruption at a composite volcano, the tephra and gases are ejected with explosive force and are hot enough to be forced high up into the atmosphere. As the eruption proceeds, and the amount of gas in the rising magma starts to decrease, parts will become heavier than air, and they can then flow downward along the flanks of the volcano (Figure 22). As they descend, they cool more and flow faster, reaching speeds up to several hundred km/h. A pyroclastic density current (PDC) consists of tephra ranging in size from boulders to microscopic shards of glass (made up of the edges and junctions of the bubbles of shattered pumice), plus gases (dominated by water vapour, but also including other gases). The temperature of this material can be as high as 1000°C. Among the most famous PDCs are the one that destroyed Pompeii in the year 79 CE, killing an estimated 18,000 people, and the one that destroyed the town of St. Pierre, Martinique, in 1902, killing an estimated 30,000.

The buoyant upper parts of pyroclastic density currents can flow over water, in some cases for several kilometres. The 1902 St. Pierre PDC flowed out into the harbour and destroyed several wooden ships anchored there.

Figure 22: The plinian eruption of Mt. Mayon, Philippines. in 1984. Although most of the eruption column is ascending into the atmosphere, there are pyroclastic density currents flowing down the sides of the volcano in several places. [USGS photo from: http://upload.wikimedia.org/wikipedia/commons/7/73/Pyroclastic_flows_at_Mayon_Volcano.jpg]

Pyroclastic Fall

Most of the tephra from an explosive eruption ascends high into the atmosphere, and some of it is distributed around Earth by high-altitude winds. The larger components (larger than 0.1 mm) tend to fall relatively close to the volcano, and the amount produced by large eruptions can cause serious damage and casualties. The large 1991 eruption of Mt. Pinatubo in the Philippines resulted in the accumulation of tens of centimetres of ash in fields and on rooftops in the surrounding populated region. Heavy typhoon rains that hit the island at the same time added to the weight of the tephra, leading to the collapse of thousands of roofs and to at least 300 of the 700 deaths attributed to the eruption.

Lahar

A lahar is any mudflow or debris flow that is related to a volcano. Most are caused by melting snow and ice during an eruption, as was the case with the lahar that destroyed the Colombian town of Armero in 1985 (described earlier). Lahars can also happen when there is no volcanic eruption, and one of the reasons is that, as we've seen, composite volcanoes tend to be weak and easily eroded.

In October 1998, category 5 hurricane Mitch slammed into the coast of central America. Damage was extensive and 19,000 people died, not so much because of high winds but because of intense rainfall – some regions received almost 2 m of rain over a few days! Mudflows and debris flows occurred in many areas, especially in Honduras and Nicaragua. An example is Casita Volcano in Nicaragua, where the heavy rains weakened rock and volcanic debris on the upper slopes, resulting in a debris flow that rapidly built in volume as it raced down the steep slope, and then ripped through the towns of El Porvenir and Rolando Rodriguez killing more than 2,000 people (Figure 23). El Porvenir and Rolando Rodriguez were new towns that had been built without planning approval in an area that was known to be at risk of lahars.

Figure 23: Part of the path of the lahar from Casita Volcano, October 30, 1998. [USGS photo from: http://volcanoes.usgs.gov/hazards/lahar/casita.php]

Sector Collapse and Debris Avalanche

In the context of volcanoes, sector collapse or flank collapse is the catastrophic failure of a significant part of an existing volcano, creating a large debris avalanche. This hazard was first recognized with the failure of the north side of Mt. St. Helens immediately prior to the large eruption on May 18, 1980. In the weeks before the eruption, a large bulge had formed on the side of the volcano, the result of magma transfer from depth into a satellite magma body within the mountain itself. Early on the morning of May 18, a moderate earthquake struck nearby; this is thought to have destabilized the bulge, leading to Earth's largest observed landslide in historical times. The failure of this part of the volcano exposed the underlying satellite magma chamber, causing it to explode sideways, which exposed the conduit leading to the magma chamber below. The resulting plinian eruption — with a 24 km high eruption column — lasted for nine hours.

In August 2010, a massive part of the flank of B.C.'s Mt. Meager gave way and about 48 million cubic metres of rock rushed down the valley, one of the largest slope failures in Canada in historical times (Figure 24). More than 25 slope failures have taken place at Mt. Meager in the past 8,000 years, some of them more than 10 times larger than the 2010 failure.

Figure 24: The August 2010 Mt. Meager rock avalanche, showing where the slide originated (arrow, 4 km upstream), its path down a steep narrow valley, and the debris field (and the stream that eventually cut through it) in the foreground.

(Mika McKinnon photo, http://www.geomika.com/blog/2011/ 01/05/the-trouble-with-landslides/ Used with permission (mika@geomika.com)

Lava Flows

As we saw in Exercise 4, lava flows at volcanoes like Kilauea do not advance very quickly, and in most cases, people can get out of the way. Of course, it is more difficult to move infrastructure, and so buildings and roads are typically the main casualties of lava flows.

4.5 VOLCANO MONITORING AND ERUPTION PREDICTION

In 2005 USGS geologist Chris Newhall made a list of the six most important signs of an imminent volcanic eruption. They are as follows:

- Gas leaks— the release of gases (mostly H_2O, CO_2, and SO_2) from the magma into the atmosphere through cracks in the overlying rock
- Bit of a bulge— the deformation of part of the volcano, indicating that a magma chamber at depth is swelling or becoming more pressurized
- Getting shaky— many (hundreds to thousands) of small earthquakes, indicating that magma is on the move. The quakes may be the result of the magma forcing the surrounding rocks to crack, or a harmonic vibration that is evidence of magmatic fluids moving underground.
- Dropping fast — a sudden decrease in the rate of seismicity, which may indicate that magma has stalled, which could mean that something is about to give way
- Big bump — a pronounced bulge on the side of the volcano (like the one at Mt. St. Helens in 1980), which may indicate that magma has moved close to surface
- Blowing off steam — steam eruptions (a.k.a. phreatic eruptions) that happen when magma near the surface heats groundwater to the boiling point. The water eventually explodes, sending fragments of the overlying rock far into the air.

With these signs in mind, we can make a list of the equipment we should have and the actions we can take to monitor a volcano and predict when it might erupt.

Assessing seismicity: The simplest and cheapest way to monitor a volcano is with seismometers. In an area with several volcanoes that have the potential to erupt (e.g., the Squamish-Pemberton area), a few well-placed seismometers can provide us with an early warning that something is changing beneath one of the volcanoes, and that we need to take a closer look. There are currently enough seismometers in the Lower Mainland and on Vancouver Island to provide this information.

If there is seismic evidence that a volcano is coming to life, more seismometers should be placed in locations within a few tens of kilometres of the source of the activity (Figure 25). This will allow geologists to determine the exact location and depth of the seismic activity so that they can see where the magma is moving.

Figure 25: A seismometer installed in 2007 in the vicinity of the Nazco Cone, B.C. [photo Cathie Hickson, used with permission]

Detecting gases: Water vapour quickly turns into clouds of liquid water droplets and is relatively easy to detect just by looking, but CO_2 and SO_2 are not as obvious. It's important to be able to monitor changes in the composition of volcanic gases, and we need instruments to do that. Some can be monitored from a distance (from the ground or even from the air) using infrared devices, but to obtain more accurate data, we need to sample the air and do chemical analysis. This can be achieved with instruments placed on the ground close to the source of the gases (see Figure 17), or by collecting samples of the air and analyzing them in a lab.

Measuring deformation: There are two main ways to measure ground deformation at a volcano. One is known as a tiltmeter, which is a sensitive three-directional level that can sense small changes in the tilt of the ground at a specific location. Another is through the use of GPS (global positioning system) technology (Figure 26). GPS is more effective than a tiltmeter because it provides information on how far the ground has actually moved — east-west, north-south, and up-down.

By combining information from these types of sources, along with careful observations made on the ground and from the air, and a thorough knowledge of how volcanoes work, geologists can get a good idea of the potential for a volcano to erupt in the near future (months to weeks, but not days). They can then make recommendations to authorities about the need for evacuations and restricting transportation corridors. Our ability to predict volcanic eruptions has increased dramatically in recent decades because of advances in our understanding of how volcanoes behave and in monitoring technology. Providing that careful work is done, there is no longer a large risk of surprise eruptions, and providing that public warnings are issued and heeded, it is less and less likely that thousands will die from sector collapse, pyroclastic flows, ash falls, or lahars. Indirect hazards are still very real, however, and we can expect the next eruption like the one at Laki in 1783 to take an even greater toll than it did then, especially since there are now roughly eight times as many people on Earth.

Figure 26: A GPS unit installed at Hualalai volcano, Hawaii. The dish-shaped antenna on the right is the GPS receiver. The antenna on the left is for communication with a base station. [from USGS at: http://hvo.wr.usgs.gov/volcanowatch/view.php?id=173]

Exercise 6 Volcano Alert!

You're the chief volcanologist for the Geological Survey of Canada (GSC), based in Vancouver. At 10:30 a.m. on a Tuesday, you receive a report from a seismologist at the GSC in Sidney saying that there has been a sudden increase in the number of small earthquakes in the vicinity of Mt. Garibaldi. You have two technicians available, access to some monitoring equipment, and a four-wheel-drive vehicle. At noon, you meet with your technicians and a couple of other geologists. By the end of the day, you need to have a plan to implement, starting tomorrow morning, and a statement to release to the press. What should your first day's fieldwork include? What should you say later today in your press release?

4.6 THE BRITISH COLUMBIAN VOLCANOES

As shown on the Figure 27, three types of volcanic environments are represented in British Columbia:

- The Cascade Arc (a.k.a. the Garibaldi Volcanic Belt in Canada) is related to subduction of the Juan de Fuca Plate beneath the North America plate.
- The Anahim Volcanic Belt is assumed to be related to a mantle plume.
- The Stikine Volcanic Belt and the Wells Gray-Clearwater Volcanic Field are assumed to be related to crustal rifting.

Figure 27: Major volcanic centres in British Columbia (base map from Wikipedia (http://commons.wikimedia.org/wiki/File:South-West_Canada.jpg). Volcanic locations from Wood, D., 1993, Waiting for another big blast – probing B.C.'s volcanoes, Canadian Geographic, based on the work of Cathie Hickson)

Subduction Volcanism

Southwestern British Columbia is at the northern end of the Juan de Fuca (Cascadia) subduction zone, and the volcanism there is related to magma generation by flux melting in the upper mantle above the subducting plate. In general, there has been a much lower rate and volume of volcanism in the B.C. part of this belt than in the U.S. part. One reason for this is that the northern part of the Juan de Fuca Plate (i.e., the Explorer Plate) is either not subducting, or is subducting at a slower rate than the rest of the plate. There are several volcanic centres in the Garibaldi Volcanic Belt: the Garibaldi centre (including Mt. Garibaldi and the Black Tusk-Mt. Price area adjacent to Garibaldi Lake (Figures 1 and 2), Mt. Cayley, and Mt. Meager (Figure 24). The most recent volcanic activity in this area was at Mt. Meager. Approximately 2,400 years ago, an explosive eruption of about the same magnitude as the 1980 Mt. St. Helens eruption took place at Mt. Meager. Ash spread as far east as Alberta. There was also significant eruptive activity at Mts. Price and Garibaldi approximately 12,000 and 10,000 years ago during the last glaciation; in both cases, lava and tephra built up against glacial ice in the adjacent valley (Figure 28). The Table in Figure 2 at the beginning of this chapter is a tuya, a volcano that formed beneath glacial ice and had its top eroded by the lake that formed around it in the ice.

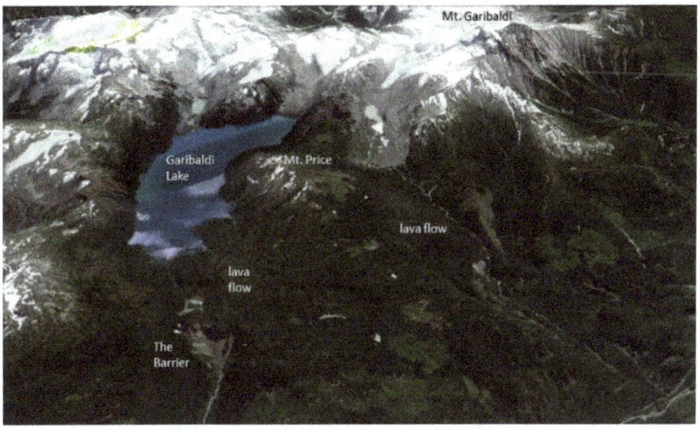

Figure 28: Perspective view of the Garibaldi region (looking east) showing the outlines of two lava flows from Mt. Price. Volcanism in this area last took place when the valley in the foreground was filled with glacial ice. The cliff known as the Barrier formed when part of the Mt. Price lava flow failed after deglaciation. The steep western face of Mt. Garibaldi formed by sector collapse, also because the rocks were no longer supported by glacial ice. [SE after Google Earth]

Mantle Plume Volcanism

The chain of volcanic complexes and cones extending from Milbanke Sound to Nazko Cone is interpreted as being related to a mantle plume currently situated close to the Nazko Cone, just west of Quesnel. The North America Plate is moving in a westerly direction at about 2 cm per year with respect to this plume, and the series of now partly eroded shield volcanoes between Nazco and the coast is interpreted to have been formed by the plume as the continent moved over it.

The Rainbow Range, which formed at approximately 8 Ma, is the largest of these older volcanoes. It has a diameter of about 30 km and an elevation of 2,495 m (Figure 29). The name "Rainbow" refers to the bright colours displayed by some of the volcanic rocks as they weather.

Figure 29: Rainbow Range, Chilcotin Plateau, B.C. (http://upload.wikimedia.org/wikipedia/commons/f/fd/Rainbow_Range_Colors.jpg).

Rift-Related Volcanism

While B.C. is not about to split into pieces, two areas of volcanism are related to rifting — or at least to stretching-related fractures that might extend through the crust. These are the Wells Gray-Clearwater volcanic field southeast of Quesnel, and the Northern Cordillera Volcanic Field, which ranges across the northwestern corner of the province (as already discussed in section 1). This area includes Canada's most recent volcanic eruption, a cinder cone and mafic lava flow that formed around 250 years ago at the Tseax River Cone in the Nass River area north of Terrace. According to Nisga'a oral history, as many as 2,000 people died during that eruption, in which lava overran their village on the Nass River. Most of the deaths are attributed to asphyxiation from volcanic gases, probably carbon dioxide.

Figure 30: Mount Edziza, in the Stikine area, B.C., with Eve Cone in the foreground. (http://upload.wikimedia.org/wikipedia/commons/5/54/Mount_Edziza%2C_British_Columbia.jpg).

The Mount Edziza Volcanic Field near the Stikine River is a large area of lava flows, sulphurous ridges, and cinder cones. The most recent eruption in this area was about 1,000 years ago. While most of the other volcanism in the Edziza region is mafic and involves lava flows and cinder cones, Mt. Edziza itself (Figure 30) is a composite volcano with rock compositions ranging from rhyolite to basalt. A possible explanation for the presence of composite volcanism in an area dominated by mafic flows and cinder cones is that there is a magma chamber beneath this area, within which magma differentiation is taking place.

AT A GLANCE

The topics covered in this chapter can be summarized as follows:

Plate Tectonics and Volcanism	Volcanism is closely related to plate tectonics. Most volcanoes are associated with convergent plate boundaries (at subduction zones), and there is also a great deal of volcanic activity at divergent boundaries and areas of continental rifting. At convergent boundaries magma is formed where water from a subducting plate acts as a flux to lower the melting temperature of the adjacent mantle rock. At divergent boundaries magma forms because of decompression melting. Decompression melting also takes place within a mantle plume.
Magma Composition and Eruption Style	The initial magmas in most volcanic regions are mafic in composition, but they can evolve into more felsic types through interaction with crustal rock, and as a result of crystal settling within a magma chamber. Felsic magmas tend to have higher gas contents than mafic magmas, and they are also more viscous. The higher viscosity prevents gases from escaping from the magma, and so felsic magmas are more pressurized and more likely to erupt explosively.
Types of Volcanoes	Cinder cones, which can form in various volcanic settings, are relatively small volcanoes that are composed mostly of mafic rock fragments that were formed during a single eruptive event. Composite volcanoes are normally associated with subduction, and while their magma tends to be intermediate on average, it can range all the way from felsic to mafic. The corresponding differences in magma viscosity lead to significant differences in eruptions style. Most shield volcanoes are associated with mantle plumes, and have consistently mafic magma which generally erupts as lava flows.
Volcanic Hazards	Most direct volcanic hazards are related to volcanoes that erupt explosively, especially composite volcanoes. Pyroclastic density currents, some as hot as 1000°C can move at hundreds of km/h and will kill anything in the way. Lahars, volcano-related mudflows, can be large enough to destroy entire towns. Lava flows will destroy anything in their paths, but tend to move slowly enough so that people can get to safety.
Monitoring Volcanoes and Predicting Eruptions	We have the understanding and technology to predict volcanic eruptions with some success, and to ensure that people are not harmed. The prediction techniques include monitoring seismicity in volcanic regions, detecting volcanic gases, and measuring deformation of the flanks of a volcano.

Volcanoes in British Columbia	There are examples of all of the important types of volcanoes in British Columbia, including subduction volcanism north of Vancouver, mantle-plume volcanism along the Nazco trend, and rift-related volcanism in the Wells Gray and Stikine regions.

CHAPTER

SOIL FORMATION AND WEATHERING 5

Weathering is what takes place when a body of rock is exposed to the "weather" – in other words, to the forces and conditions that exist at Earth's surface. With the exception of volcanic rocks and some sedimentary rocks, most rocks are formed at some depth within the crust. There they experience relatively constant temperature, high pressure, no contact with the atmosphere, and little or no moving water. Once a rock is exposed at the surface, which is what happens when the overlying rock is eroded away, conditions change dramatically. Temperatures vary widely, there is much less pressure, oxygen and other gases are plentiful, and in most climates, water is abundant (Figure 1).

Figure 1: The Hoodoos, near Drumheller, Alberta, have formed from the differential weathering of sedimentary rock that was buried beneath other rock for close to 100 Ma [SE photo]

Weathering includes two main processes that are quite different. One is the mechanical breakdown of rock into smaller fragments, and the other is the chemical change of the minerals within the rock to forms that are stable in the surface environment. Mechanical weathering provides fresh surfaces for attack by chemical processes, and chemical weathering weakens the rock so that it is more susceptible to mechanical weathering. Together, these processes create two very important products, one being the sedimentary clasts and ions in solution that can eventually become sedimentary rock, and the other being the soil that is necessary for our existence on Earth.

The various processes related to uplift and weathering are summarized in the rock cycle in Figure 2.

Figure 2: Weathering can take place once a rock is exposed at surface by uplift and the removal of the overlying rock. [SE]

5.1PROCESS OF MECHANICAL WEATHERING

Intrusive igneous rocks form at depths of several hundreds of metres to several tens of kilometres. Sediments are turned into sedimentary rocks only when they are buried by other sediments to depths in excess of several hundreds of metres. Most metamorphic rocks are formed at depths of kilometres to tens of kilometres. Weathering cannot even begin until these rocks are uplifted through various processes of mountain building — most of which are related to plate tectonics — and the overlying material has been eroded away and the rock is exposed as an outcrop. To a geologist, an outcrop is an exposure of bedrock, the solid rock of the crust.

The important agents of mechanical weathering are:

- The decrease in pressure that results from removal of overlying rock
- Freezing and thawing of water in cracks in the rock
- Formation of salt crystals within the rock
- Cracking from plant roots and exposure by burrowing animals

When a mass of rock is exposed by weathering and removal of the overlying rock, there is a decrease in the confining pressure on the rock, and the rock expands. This unloading promotes cracking of the rock, known as exfoliation, as shown in the granitic rock in Figure 5.3.

Figure 3: Exfoliation fractures in granitic rock exposed on the west side of the Coquihalla Highway north of Hope, B.C. [SE]

Figure 4: Exfoliation of slate at a road cut in the Columbia Mountains west of Golden, B.C. [SE photo]

Granitic rock tends to exfoliate parallel to the exposed surface because the rock is typically homogenous, and it doesn't have predetermined planes along which it must fracture. Sedimentary and metamorphic rocks, on the other hand, tend to exfoliate along predetermined planes (Figure 4).

Frost wedging is the process by which water seeps into cracks in a rock, expands on freezing, and thus enlarges the cracks (Figure 5). The effectiveness of frost wedging is related to the frequency of freezing and thawing. Frost wedging is most effective in a climate like Canada's. In warm areas where freezing is infrequent, in very cold areas where thawing is infrequent, or in very dry areas, where there is little water to seep into cracks, the role of frost wedging is limited.

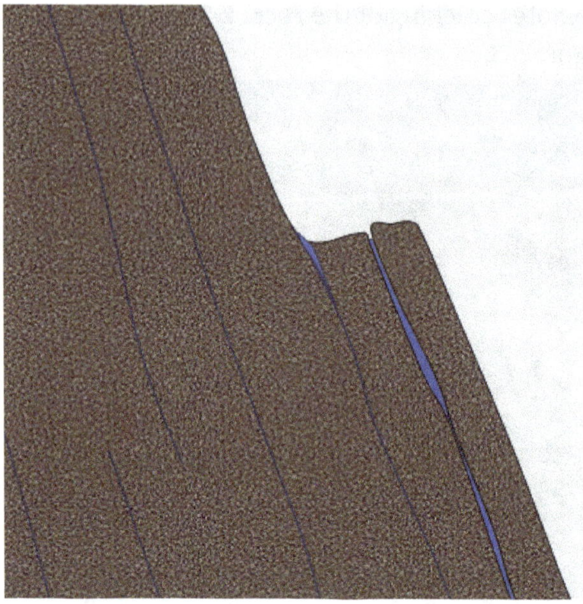

Figure 5: The process of frost wedging on a steep slope. Water gets into fractures and then freezes, expanding the fracture a little. When the water thaws it seeps a little farther into the expanded crack. The process is repeated many times, and eventually a piece of rock will be wedged away. [SE]

Figure 6: An area with very effective frost-wedging near Keremeos, B.C. The fragments that have been wedged away from the cliffs above have accumulated in a talus deposit at the base of the slope. The rocks in this area have quite varied colours, and those are reflected in the colours of the talus. [SE]

In many parts of Canada, the transition between freezing nighttime temperatures and thawing daytime temperatures is frequent — tens to hundreds of times a year. Even in warm coastal areas of southern B.C., freezing and thawing transitions are common at higher elevations. A common feature in areas of effective frost wedging is a talus slope — a fan-shaped deposit of fragments removed by frost wedging from the steep rocky slopes above (Figure 6).

A related process, frost heaving, takes place within unconsolidated materials on gentle slopes. In this case, water in the soil freezes and expands, pushing the overlying material up. Frost heaving is responsible for winter damage to roads all over North America.

When salt water seeps into rocks and then evaporates on a hot sunny day, salt crystals grow within cracks and pores in the rock. The growth of these crystals exerts pressure on the rock and can push grains apart, causing the rock to weaken and break. There are many examples of this on the rocky shorelines of Vancouver Island and the Gulf Islands, where sandstone outcrops are common and salty seawater is readily available (Figure 7). Salt weathering can also occur away from the coast, because most environments have some salt in them.

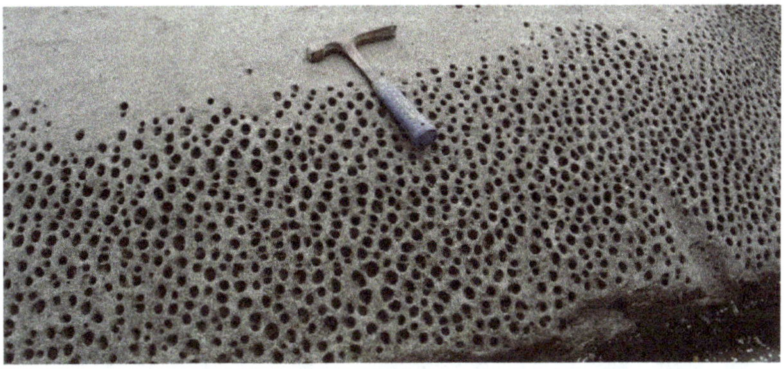

Figure 7: Honeycomb weathering of sandstone on Gabriola Island, B.C. The holes are caused by crystallization of salt within rock pores, and the seemingly regular pattern is related to the original roughness of the surface. It's a positive-feedback process because the holes collect salt water at high tide, and so the effect is accentuated around existing holes. This type of weathering is most pronounced on south-facing sunny exposures. [SE]

The effects of plants and animals are significant in mechanical weathering. Roots can force their way into even the tiniest cracks, and then they exert tremendous pressure on the rocks as they grow, widening the cracks and breaking the rock (Figure 8). Although animals do not normally burrow through solid rock, they can excavate and remove huge volumes of soil, and thus expose the rock to weathering by other mechanisms.

Figure 8: Conifers growing on granitic rocks at The Lions, near Vancouver, B.C. [SE]

Mechanical weathering is greatly facilitated by erosion, which is the removal of weathering products, allowing for the exposure of more rock for weathering.

Exercise 5.1 Process of Mechanical Weathering

This photo shows granitic rock at the top of Stawamus Chief near Squamish, B.C. Identify the mechanical weathering processes that you can see taking place, or you think probably take place at this location.

5.2 PROCESS OF CHEMICAL WEATHERING

Chemical weathering results from chemical changes to minerals that become unstable when they are exposed to surface conditions. The kinds of changes that take place are highly specific to the mineral and the environmental conditions. Some minerals, like quartz, are virtually unaffected by chemical weathering, while others, like feldspar, are easily altered. In general, the degree of chemical weathering is greatest in warm and wet climates, and least in cold and dry climates. The important characteristics of surface conditions that lead to chemical weathering are the presence of water (in the air and on the ground surface), the abundance of oxygen, and the presence of carbon dioxide, which produces weak carbonic acid when combined with water. That process, which is fundamental to most chemical weathering, can be shown as follows:

$H_2O + CO_2$ —>H_2CO_3 then H_2CO_3—> $H^+ + HCO_3^-$, water + carbon dioxide —> carbonic acid then carbonic acid —> hydrgen ion + carbonate ion

Here we have water (e.g., as rain) plus carbon dioxide in the atmosphere, combining to create carbonic acid. Then carbonic acid dissociates (comes apart) to form hydrogen and carbonate ions. The amount of CO_2 in the air is enough to make only very weak carbonic acid, but there is typically much more CO_2 in the soil, so water that percolates through the soil can become significantly more acidic.

There are two main types of chemical weathering. On the one hand, some minerals become altered to other minerals. For example, feldspar is altered — by hydrolysis — to clay minerals. On the other hand, some minerals dissolve completely, and their components go into solution. For example, calcite ($CaCO_3$) is soluble in acidic solutions.

The hydrolysis of feldspar can be written like this:

$CaAl_2Si_2O_8 + H_2CO_3 + ½O_2$ —> $Al_2Si_2O_5(OH)_4 + Ca^{2+} + CO_3^{2-}$

plagioclase + carbonic acid —> kaolinite + dissolved calcium + carbonate ions

Figure 9: Unweathered (left) and weathered (right) surfaces of the same piece of granitic rock. On the unweathered surfaces the feldspars are still fresh and glassy-looking. On the weathered surface the feldspar has been altered to the chalky-looking clay mineral kaolinite. [SE]

This reaction shows calcium plagioclase feldspar, but similar reactions could also be written for sodium or potassium feldspars. In this case, we end up with the mineral kaolinite, along with calcium and carbonate ions in solution. Those ions can eventually combine (probably in the ocean) to form the mineral calcite. The hydrolysis of feldspar to clay is illustrated in Figure 9, which shows two images of the same granitic rock, a recently broken fresh surface on the left and a clay-altered weathered surface on the right. Other silicate minerals can also go through hydrolysis, although the end results will be a little different. For example, pyroxene can be converted to the clay minerals chlorite or smectite, and olivine can be converted to the clay mineral serpentine.

Oxidation is another very important chemical weathering process. The oxidation of the iron in a ferromagnesian silicate starts with the dissolution of the iron. For olivine, the process looks like this, where olivine in the presence of carbonic acid is converted to dissolved iron, carbonate, and silicic acid:

$Fe_2SiO_4 + 4H_2CO_3 \longrightarrow 2Fe_2^{+} + 4HCO_3^{-} + H_4SiO_4$

olivine + (carbonic acid) —> dissolved iron + dissolved carbonate + dissolved silicic acid

In the presence of oxygen, the dissolved iron is then quickly converted to hematite:

$2Fe_2^{+} + 4HCO_3 - + \frac{1}{2} O_2 + 2H_2O \longrightarrow Fe_2O_3 + 4H_2CO_3$

dissolved iron + bicarbonate + oxygen + water—->hematite + carbonic acid

Figure 10: A granitic rock containing biotite and amphibole which have been altered near to the rock's surface to limonite, which is a mixture of iron oxide minerals. [SE]

The equation shown here is for olivine, but it could apply to almost any other ferromagnesian silicate, including pyroxene, amphibole, or biotite. Iron in the sulphide minerals (e.g., pyrite) can also be oxidized in this way. And the mineral hematite is not the only possible end result, as there is a wide range of iron oxide minerals that can form in this way. The results of this process are illustrated in Figure 10, which shows a granitic rock in which some of the biotite and amphibole have been altered to form the iron oxide mineral limonite.

A special type of oxidation takes place in areas where the rocks have elevated levels of sulphide minerals, especially pyrite (FeS_2). Pyrite reacts with water and oxygen to form sulphuric acid, as follows:

$$2FeS_2 + 7O_2 + 2H_2O \longrightarrow 2Fe^{2+} \; H_2SO_4 + 2H^+$$

pyrite + oxygen + water —-> iron ions + sulphuric acid + hydrogen ions

The runoff from areas where this process is taking place is known as acid rock drainage (ARD), and even a rock with 1% or 2% pyrite can produce significant ARD. Some of the worst examples of ARD are at metal mine sites, especially where pyrite-bearing rock and waste material have been mined from deep underground and then piled up and left exposed to water and oxygen. One example of that is the Mt. Washington Mine near Courtenay on Vancouver Island (Figure 11), but there are many similar sites across Canada and around the world.

Figure 11: Exposed oxidizing and acid generating rocks and mine waste at the abandoned Mt. Washington Mine, B.C. (left), and an example of acid drainage downstream from the mine site (right). [SE]

At many ARD sites, the pH of the runoff water is less than 4 (very acidic). Under these conditions, metals such as copper, zinc, and lead are quite soluble, which can lead to toxicity for aquatic and other organisms. For many years, the river downstream from the Mt. Washington Mine had so much dissolved copper in it that it was toxic to salmon. Remediation work has since been carried out at the mine and the situation has improved.

The hydrolysis of feldspar and other silicate minerals and the oxidation of iron in ferromagnesian silicates all serve to create rocks that are softer and weaker than they were to begin with, and thus more susceptible to mechanical weathering.

The weathering reactions that we've discussed so far involved the transformation of one mineral to another mineral (e.g., feldspar to clay), and the release of some ions in solution (e.g., Ca^{2+}). Some weathering processes involve the complete dissolution of a mineral. Calcite, for example, will dissolve in weak acid, to produce calcium and bicarbonate ions. The equation is as follows:

$$CaCO_3 + H^+ + HCO_3^- \longrightarrow Ca^{2+} + 2HCO_3^-$$

calcite + hydrogen ions + bicarbonate —-> calcium ions + bicarbonate

Calcite is the major component of limestone (typically more than 95%), and under surface conditions, limestone will dissolve to varying degrees (depending on which minerals it contains, other than calcite), as shown in Figure 12. Limestone also dissolves at relatively shallow depths underground, forming limestone caves. This is discussed in more detail in Chapter 14, where we look at groundwater.

Figure 12: A limestone outcrop on Quadra Island, B.C. The limestone, which is primarily made up of the mineral calcite, has been dissolved to different degrees in different areas because of compositional differences. The buff-coloured bands are volcanic rock, which is not soluble. [SE]

Exercise 2 Process of Chemical Weathering

The main processes of chemical weathering are *hydrolysis*, *oxidation*, and *dissolution*. Complete the following table by indicating which process is primarily responsible for each of the described chemical weathering changes:

Chemical Change	Process?
Pyrite to hematite	
Calcite to calcium and bicarbonate ions	
Feldspar to clay	
Olivine to serpentine	
Pyroxene to iron oxide	

5.3 The Products of Weathering and Erosion

The products of weathering and erosion are the unconsolidated materials that we find around us on slopes, beneath glaciers, in stream valleys, on beaches, and in deserts. The nature of these materials — their composition, size, degree of sorting, and degree of rounding — is determined by the type of rock that is being weathered, the nature of the weathering, the erosion and transportation processes, and the climate.

A summary of the weathering products of some of the common minerals present in rocks is provided in Table 1.

Table 1: A list of the typical weathering products of some of the minerals in common rocks [SE]

Common Mineral	Typical Weathering Products
Quartz	Quartz as sand grains
Feldspar	Clay minerals plus potassium, sodium, and calcium in solution
Biotite and amphibole	Chlorite plus iron and magnesium in solution
Pyroxene and olivine	Serpentine plus iron and magnesium in solution
Calcite	Calcium and carbonate in solution
Pyrite	Iron oxide minerals plus iron in solution and sulphuric acid

Some examples of the products of weathering are shown in Figure 13. They range widely in size and shape depending on the processes involved. If and when deposits like these are turned into sedimentary rocks, the textures of those rocks will vary significantly. Importantly, when we describe sedimentary rocks that formed millions of years in the past, we can use those properties to make inferences about the conditions that existed during their formation.

We'll talk more about the nature and interpretation of sediments and sedimentary rocks in Chapter 6, but it's worth considering here why the sandy sediments shown in Figure 13 are so strongly dominated by the mineral quartz, even though quartz makes up less than 20% of Earth's crust. The explanation is that quartz is highly resistant to the types of weathering that occur at Earth's surface. It is not affected by weak acids or the presence of oxygen. This makes it unique among the minerals that are common in igneous rocks. Quartz is also very hard, and doesn't have cleavage, so it is resistant to mechanical erosion.

As weathering proceeds, the ferromagnesian silicates and feldspar are very likely to be broken into small pieces and converted into clay minerals and dissolved ions (e.g., Ca^{2+}, Na^{+}, K^{+}, Fe^{2+}, Mg^{2+}, and H_4SiO_4). In other words, quartz, clay minerals, and dissolved ions are the most common products of weathering. Quartz and some of the clay minerals tend to form sedimentary deposits on and at the edges of continents, while the rest of the clay minerals and the dissolved ions tend to be washed out into the oceans to form sediments on the sea floor.

Figure 13: Products of weathering and erosion formed under different conditions. [SE]

Exercises

5.4 WEATHERING AND THE FORMATION OF SOIL

Weathering is a key part of the process of soil formation, and soil is critical to our existence on Earth. In other words, we owe our existence to weathering, and we need to take care of our soil!

Many people refer to any loose material on Earth's surface as soil, but to geologists (and geology students) soil is the material that includes organic matter, lies within the top few tens of centimetres of the surface, and is important in sustaining plant growth.

Soil is a complex mixture of minerals (approximately 45%), organic matter (approximately 5%), and empty space (approximately 50%, filled to varying degrees with air and water). The mineral content of soils is variable, but is dominated by clay minerals and quartz, along with minor amounts of feldspar and small fragments of rock. The types of weathering that take place within a region have a major influence on soil composition and texture. For example, in a warm climate, where chemical weathering dominates, soils tend to be richer in clay. Soil scientists describe soil texture in terms of the relative proportions of sand, silt, and clay, as shown in Figure 14. The sand and silt components in this diagram are dominated by quartz, with lesser amounts of feldspar and rock fragments, while the clay component is dominated by the clay minerals.

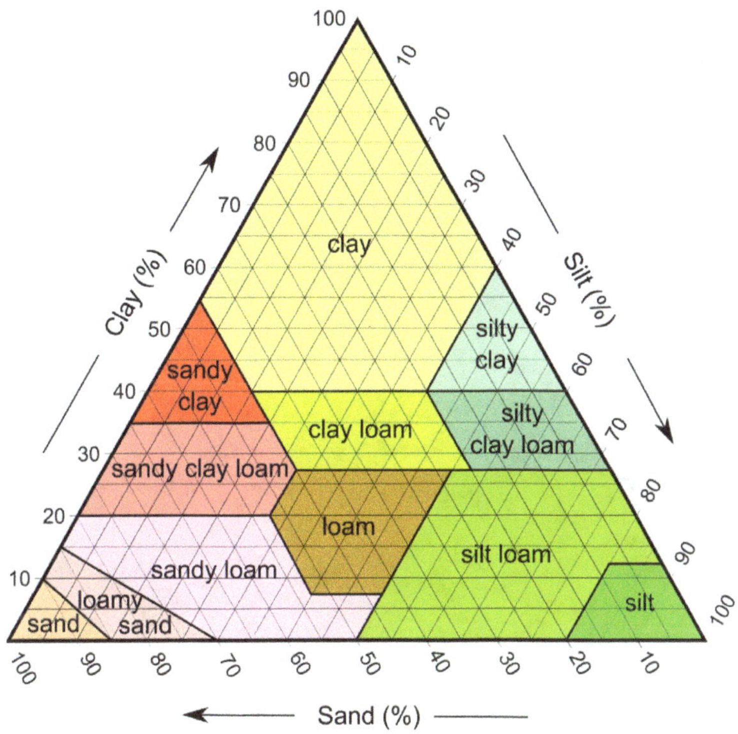

Figure 14: The U.S. Department of Agriculture soil texture diagram. This diagram applies only to the mineral component of soils, and the names are textural descriptions, not soil classes. [http://en.wikipedia.org/wiki/ Soil#media viewer/ File:SoilTexture_USDA.png]

Soil forms through accumulation and decay of organic matter and through the mechanical and chemical weathering processes described above. The factors that affect the nature of soil and the rate of its formation include climate (especially average temperature and precipitation amounts, and the consequent types of vegetation), the type of parent material, the slope of the surface, and the amount of time available.

Climate

Soils develop because of the weathering of materials on Earth's surface, including the mechanical breakup of rocks, and the chemical weathering of minerals. Soil development is facilitated by the downward percolation of water. Soil forms most readily under temperate to tropical conditions (not cold) and where precipitation amounts are moderate (not dry, but not too wet). Chemical weathering reactions (especially the formation of clay minerals) and biochemical reactions proceed fastest under warm conditions, and plant growth is enhanced in warm climates. Too much water (e.g., in rainforests) can lead to the leaching of important chemical nutrients and hence to acidic soils. In humid and poorly drained regions, swampy conditions may prevail, producing soil that is dominated by organic matter. Too little water (e.g., in deserts and semi-deserts), results in very limited downward chemical transportation and the accumulation of salts and carbonate minerals (e.g., calcite) from upward-moving water. Soils in dry regions also suffer from a lack of organic material (Figure 15).

Figure 15: Poorly developed soil on wind-blown silt (loess) in an arid part of northeastern Washington State [SE]

Parent Material

Soil parent materials can include all different types of bedrock and any type of unconsolidated sediments, such as glacial deposits and stream deposits. Soils are described as residual soils if they develop on bedrock, and transported soils if they develop on transported material such as glacial sediments. But the term "transported soil" is misleading because it implies that the soil itself has been transported, which is not the case. When referring to such soil, it is better to be specific and say "soil developed on unconsolidated material," because that distinguishes it from soil developed on bedrock.

Quartz-rich parent material, such as granite, sandstone, or loose sand, leads to the development of sandy soils. Quartz-poor material, such as shale or basalt, generates soils with little sand.

Parent materials provide important nutrients to residual soils. For example, a minor constituent of granitic rocks is the calcium-phosphate mineral apatite, which is a source of the important soil nutrient phosphorus. Basaltic parent material tends to generate very fertile soils because it also provides phosphorus, along with significant amounts of iron, magnesium, and calcium.

Some unconsolidated materials, such as river-flood deposits, make for especially good soils because they tend to be rich in clay minerals. Clay minerals have large surface areas with negative charges that are attractive to positively charged elements like calcium, magnesium, iron, and potassium — important nutrients for plant growth.

Slope

Soil can only develop where surface materials remain in place and are not frequently moved away by mass wasting. Soils cannot develop where the rate of soil formation is less than the rate of erosion, so steep slopes tend to have little or no soil.

Time

Even under ideal conditions, soil takes thousands of years to develop. Virtually all of southern Canada was still glaciated up until 14 ka, and most of the central and northern parts of B.C., the prairies, Ontario, and Quebec were still glaciated at 12 ka. Glaciers still dominated the central and northern parts of Canada until around 10 ka, and so, at that time, conditions were still not ideal for soil development even in the southern regions. Therefore, soils in Canada, and especially in central and northern Canada, are relatively young and not well developed.

The same applies to soils that are forming on newly created surfaces, such as recent deltas or sand bars, or in areas of mass wasting.

Soil Horizons

The process of soil formation generally involves the downward movement of clay, water, and dissolved ions, and a common result of that is the development of chemically and texturally different layers known as soil horizons. The typically developed soil horizons, as illustrated in Figure 16, are:

O — the layer of organic matter

A — the layer of partially decayed organic matter mixed with mineral material

E— the eluviated (leached) layer from which some of the clay and iron have been removed to create a pale layer that may be sandier than the other layers

B — the layer of accumulation of clay, iron, and other elements from the overlying soil

C — the layer of incomplete weathering

Although rare in Canada, another type of layer that develops in hot arid regions is known as caliche (pronounced ca-lee-chee). It forms from the downward (or in some cases upward) movement of calcium ions, and the precipitation of calcite within the soil. When well developed, caliche cements the surrounding material together to form a layer that has the consistency of concrete.

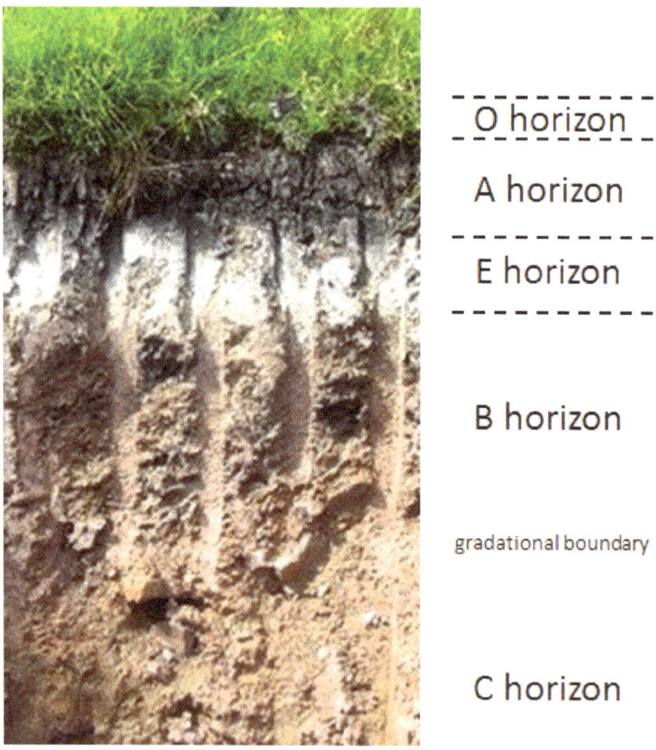

Figure 16: Soil horizons in a podsol from a site in northeastern Scotland. O: organic matter A: organic matter and mineral material E: leached layer B:

accumulation of clay, iron etc. C: incomplete weathering of parent material [SE after http://commons.wikimedia.org/wiki/File:Podzol_-_geograph.org.uk_-_218892.jpg]

Like all geological materials, soil is subject to erosion, although under natural conditions on gentle slopes, the rate of soil formation either balances or exceeds the rate of erosion. Human practices related to forestry and agriculture have significantly upset this balance.

Soils are held in place by vegetation. When vegetation is removed, either through cutting trees or routinely harvesting crops and tilling the soil, that protection is either temporarily or permanently lost. The primary agents of the erosion of unprotected soil are water and wind.

Water erosion is accentuated on sloped surfaces because fast-flowing water obviously has greater eroding power than still water (Figure 17). Raindrops can disaggregate exposed soil particles, putting the finer material (e.g., clays) into suspension in the water. Sheetwash, unchannelled flow across a surface carries suspended material away, and channels erode right through the soil layer, removing both fine and coarse material.

Figure 17: Soil erosion by rain and channelled runoff on a field in Alberta. [from Alberta Agriculture and Rural Development, http://www1.agric.gov.ab.ca/$department/deptdocs.nsf/all/agdex9313, used with permission]

Wind erosion is exacerbated by the removal of trees that act as wind breaks and by agricultural practices that leave bare soil exposed (Figure 18).

Tillage is also a factor in soil erosion, especially on slopes, because each time the soil is lifted by a cultivator, it is moved a few centimetres down the slope.

Figure 18: Soil erosion by wind in Alberta. [from Alberta Agriculture and Rural Development, http://www1.agric.gov.ab.ca/$department/deptdocs.nsf/all/agdex9313, used with permission]

5.5 THE SOILS OF CANADA

Up until the 1950s, the classification of soils in Canada was based on the system used in the United States. However, it was long recognized that the U.S .system did not apply well to many parts of Canada because of climate and environmental differences. The Canadian System of Soil Classification was first outlined in 1955 and has been refined and modified numerous times since then.

There are 10 orders of soil recognized in Canada. Each one is divided into groups, and then families, and then series, but we will only look at the orders, some of which are summarized in Table 5.2. The distribution of these types of soils (and a few others) in Canada is shown in Figure 19.

Table 2: The nature, origins and distributions of the more important soil orders in Canada

Order	Brief Description	Environment
Forest soils		
Podsol	Well-developed A and B horizons	Coniferous forests throughout Canada
Luvisol	Clay rich B horizon	Northern prairies and central B.C., mostly on sedimentary rocks
Brunisol	Poorly developed or immature soil, that does not have the well-defined horizons of podsol or luvisol	Boreal-forest soils in the discontinuous permafrost areas of central and western Canada, and also in southern B.C.
Grassland soils		

Order	Brief Description	Environment
Chernozem	High levels of organic matter and an A horizon at least 10 cm thick	Southern prairies (and parts of B.C.'s southern interior), in areas that experience water deficits during the summer
Solonetzic	A clay-rich B horizon, commonly with a salt-bearing C horizon	Southern prairies, in areas that experience water deficits during the summer
Other important soils		
Organic	Dominated by organic matter; mineral horizons are typically absent	Wetland areas, especially along the western edge of Hudson Bay, and in the area between the prairies and the boreal forest
Cryosol	Poorly developed soil, mostly C horizon	Permafrost areas of northern Canada

As we've discussed, the processes of soil formation are dominated by the downward transportation of clays and certain elements, and the nature of those processes depends in large part on the climate. In Canada's predominantly cool and humid climate (which applies to most places other than the far north), podsolization is the norm. This involves downward transportation of hydrogen, iron, and aluminum (and other elements) from the upper part of the soil profile, and accumulation of clay, iron, and aluminum in the B horizon. Most of the podsols, luvisols, and brunisols of Canada form through various types of podsolization.

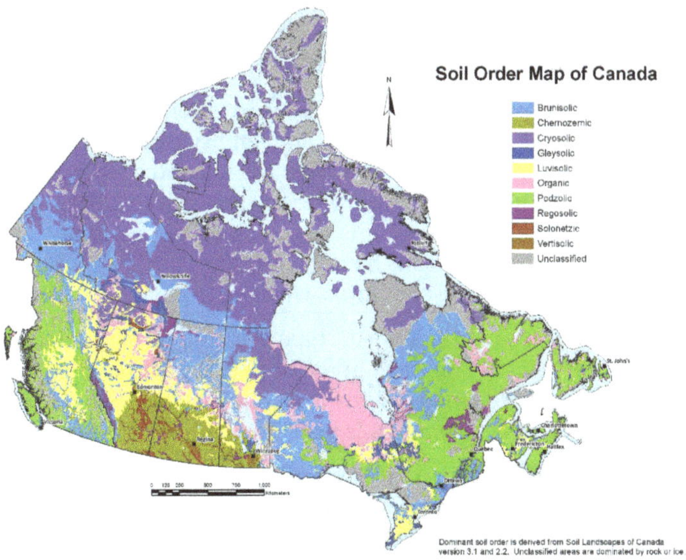

Figure 19: The soil order map of Canada. [from The Department of Soil Science, University of Saskatchewan, http://www.soilsofcanada.ca/ used with permission]

In the grasslands of the dry southern parts of the prairie provinces and in some of the drier parts of southern B.C., dark brown organic-rich chernozem soils are dominant. In some parts of these areas, weak calcification takes place with leaching of calcium from the upper layers and accumulation of calcium in the B layer. Development of caliche layers is rare in Canada.

Organic soils form in areas with poor drainage (i.e., swamps) and a rich supply of organic matter. These soils have very little mineral matter.

In the permafrost regions of the north, where glacial retreat was most recent, the time available for soil formation has been short and the rate of soil formation is very slow. The soils are called cryosols (*cryo* means "ice cold"). Permafrost areas are also characterized by the churning of the soil by freeze-thaw processes, and as a result, development of soil horizons is very limited.

AT A GLANCE

The topics covered in this chapter can be summarized as follows:

Mechanical Weathering	Rocks weather when they are exposed to surface conditions, which in most case are quite different from those at which they formed. The main processes of mechanical weathering include exfoliation, freeze-thaw, salt crystallization, and the effects of plant growth.
Chemical Weathering	Chemical weathering takes place when minerals within rocks are not stable in their existing environment. Some of the important chemical weathering processes are hydrolysis of silicate minerals to form clay minerals, oxidation of iron in silicate and other minerals to form iron oxide minerals, and dissolution of calcite.
The Products of Weathering and Erosion	The main products of weathering and erosion are grains of quartz (because quartz is resistant to chemical weathering), clay minerals, iron oxide minerals, rock fragments, and a wide range of ions in solution.
Weathering and the Formation of Soil	Soil is a mixture of fine mineral fragments (including quartz and clay minerals), organic matter, and empty spaces that may be partially filled with water. Soil formation is controlled by climate (especially temperature and humidity), the nature of the parent material, the slope (because soil can't accumulate on steep slopes), and the amount of time available. Typical soils have layers called horizons which form because of differences in the conditions with depth.
The Soils of Canada	Canada has a range of soil types related to our unique conditions. The main types of soil form in forested and grassland regions, but there are extensive wetlands in Canada that produce organic soils, and large areas where soil development is poor because of cold conditions.
Weathering and Climate Change	The geological carbon cycle plays a critical role in balancing Earth's climate. Carbon is released to the atmosphere during volcanic eruptions. Carbon is extracted from the atmosphere during weathering of silicate minerals and this is eventually stored in the ocean and in sediments. Atmospheric carbon is also transferred to organic matter and some of that is later stored in soil, permafrost, and rocks. Our use of geologically stored carbon (fossil fuels) upsets this climate balance.

CHAPTER

SEDIMENTARY ROCKS AND SEDIMENTS 6

Transportation is the movement of sediments or dissolved ions from the site of erosion to a site of deposition; this can be by wind, flowing water, glacial ice, or mass movement down a slope. Deposition takes place where the conditions change enough so that sediments being transported can no longer be transported (e.g., a current slows). Burial occurs when more sediments are piled onto existing sediments, and layers formed earlier are covered and compacted. Lithification is what happens — at depths of hundreds to thousands of metres — when those compacted sediments become cemented together to form solid sedimentary rock.

Figure 1: The Cretaceous Dinosaur Park Formation at Dinosaur Provincial Park, Alberta, one the world's most important sites for dinosaur fossils. The rocks in the foreground show cross-bedding, indicative of deposition in a fluvial (river) environment

In this textbook, we divide sedimentary rocks into two main types: clastic and chemical. Clastic sedimentary rocks are mainly composed of material that has been transported as solid fragments (clasts). Chemical sedimentary rocks are mainly composed of material that has been transported as ions in solution. It's important not to assume that mechanical weathering leads only to clastic sedimentary rocks, while chemical weathering leads only to chemical sedimentary rocks. In most cases, millions of years separate the weathering and depositional processes, and both types of sedimentary rocks tend to include at least some material derived from both types of weathering.

Figure 2: The rock cycle, showing the processes related to sedimentary rocks on the right-hand side.

6.1 CLASTIC SEDIMENTARY ROCKS

A clast is a fragment of rock or mineral, ranging in size from less than a micronA micron is a millionth of a metre. There are 1,000 microns in a millimetre. (too small to see) to as big as an apartment block. Various types of clasts are shown in Figure 12 and in Exercise 3. The smaller ones tend to be composed of a single mineral crystal, and the larger ones are typically composed of pieces of rock. As we've seen in Chapter 5, most sand-sized clasts are made of quartz because quartz is more resistant to weathering than any other common mineral. Most clasts that are smaller than sand size (<1/16 mm) are made of clay minerals. Most clasts larger than sand size (>2 mm) are actual fragments of rock, and commonly these might be fine-grained rock like basalt or andesite, or if they are bigger, coarse-grained rock like granite or gneiss.

Grain-Size Classification

Geologists that study sediments and sedimentary rocks use the Udden-Wentworth grain-size scale for describing the sizes of the grains in these materials (Table 1).

Table 1: The Udden-Wentworth grain-size scale for classifying sediments and the grains that make up sedimentary rocks

Description		Size Range in mm			
from		to	no limit		
Boulder	large	1,024	1024		
	medium	512	512		
	small	256	256		
Cobble	large	128	128		
	small	64	64		
Pebble (Granule)	very coarse	32	32		
	coarse	16	16		
	medium	8			
	fine	4	8	**Size in microns**	
	very fine	2	4	from	to
Sand	very coarse	1	2	1,000	2,000
	coarse	0.5	1	500	1,000
	medium	0.25	0.5	250	500
	fine	0.125	0.25	125	250
	very fine	0.063	0.125	63	125
Silt	very coarse			32	63
	coarse			16	32
	medium			8	16
	fine			4	8
	v. fine			2	4
Clay	clay			0	2

There are six main grain-size categories; five are broken down into subcategories, with clay being the exception. The diameter limits for each successive subcategory are twice as large as the one beneath it. In general, a boulder is bigger than a toaster and difficult to lift. There is no upper limit to the size of boulder. The largest known free-standing rock (i.e., not part of bedrock) is Giant Rock in the Mojave Desert, California. It's about as big as an apartment building — seven storeys high! A small cobble will fit in one hand, a large one in two hands. A pebble is something that you could throw quite easily. The smaller ones — known as granules — are gravel size, but still you could throw one. But you can't really throw a single grain of sand. Sand ranges from 2 mm down to 0.063 mm, and its key characteristic is that it feels "sandy" or gritty between your fingers — even the finest sand grains feel that way. Silt is essentially too small for individual grains to be visible, and while sand feels sandy to your fingers, silt feels smooth to your fingers but gritty in your mouth. Clay is so fine that it feels smooth even in your mouth.

Transportation

One of the key principles of sedimentary geology is that the ability of a moving medium (air or water) to move sedimentary particles, and keep them moving, is dependent on the velocity of flow. The faster the medium flows, the larger the particles it can move. This is illustrated in Figure 6.4. Parts of the river are moving faster than other parts, especially where the slope is greatest and the channel is narrow. Not only does the velocity of a river change from place to place, but it changes from season to season.

During peak discharge: Discharge of a stream is the volume of flow passing a point per unit time. It's normally measured in cubic metres per second (m^3/s). at this location, the water is high enough to flow over the embankment on the right, and it flows fast enough to move the boulders that cannot be moved during low flows.

Figure 4: Variations in flow velocity on the Englishman River near Parksville, B.C. When the photo was taken the river was not flowing fast enough anywhere to move the boulders and cobbles visible here, but it is fast enough when the discharge is higher.

Clasts within streams are moved in several different ways, as illustrated in Figure 5. Large bedload clasts are pushed (by traction) or bounced along the bottom (saltation), while smaller clasts are suspended in the water and kept there by the turbulence of the flow. As the flow velocity changes, different-sized clasts may be either incorporated into the flow or deposited on the bottom. At various places along a river, there are always some clasts being deposited, some staying where they are, and some being eroded and transported. This changes over time as the discharge of the river changes in response to changing weather conditions.

Other sediment transportation media, such as waves, ocean currents, and wind, operate under similar principles, with flow velocity as the key underlying factor that controls transportation and deposition.

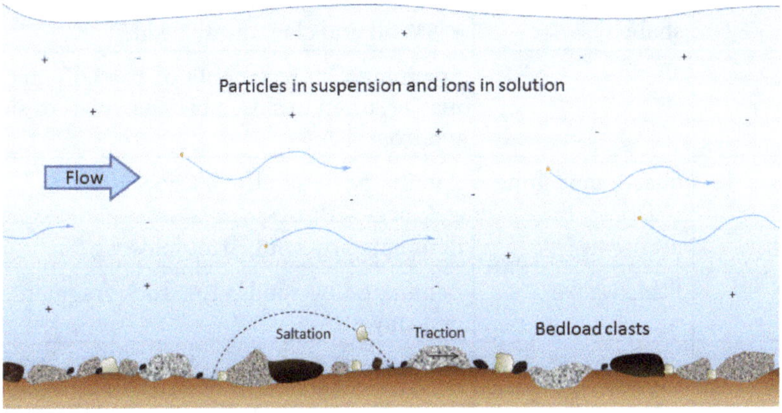

Figure 5: Transportation of sediment clasts by stream flow. The larger clasts, resting on the bottom (bedload), are moved by traction (sliding) or by saltation (bouncing). Smaller clasts are kept in suspension by turbulence in the flow. Ions (depicted as + and - in the image, but invisible in real life) are dissolved in the water.

Clastic sediments are deposited in a wide range of environments, including glaciers, slope failures, rivers — both fast and slow, lakes, deltas, and ocean environments — both shallow and deep. Depending on the grain size in particular, they may eventually form into rocks ranging from fine mudstone to coarse breccia and conglomerate.

Lithification is the term used to describe a number of different processes that take place within a deposit of sediment to turn it into solid rock. One of these processes is burial by other sediments, which leads to compaction of the material and removal of some of the intervening water and air. After this stage, the individual clasts are all touching one another. Cementation is the process of crystallization of minerals within the pores between the small clasts, and also at the points of contact between the larger clasts (sand size and larger). Depending on the pressure, temperature, and chemical conditions, these crystals might include calcite, hematite, quartz, clay minerals, or a range of other minerals.

The characteristics and distinguishing features of clastic sedimentary rocks are summarized in Table 2. Mudrock is composed of at least 75% silt- and clay-sized fragments. If it is dominated by clay, it is called claystone. If it shows evidence of bedding or fine laminations, it is shale; otherwise it is mudstone. Mudrocks form in very low energy environments, such as lakes, river backwaters, and the deep ocean.

Table 2: The main types of clastic sedimentary rocks and their characteristics.

Group	Examples	Characteristics
Mudrock	mudstone	>75% silt and clay, not bedded
	shale	>75% silt and clay, thinly bedded
Coal		dominated by fragments of partially decayed plant matter, often enclosed between beds of sandstone or mudrock
Sandstone	quartz sandstone	dominated by sand, >90% quartz
	arkose	dominated by sand, >10% feldspar
	lithic wacke	dominated by sand, >10% rock fragments, >15% silt and clay
Conglomerate		dominated by rounded clasts, pebble size and larger
Breccia		dominated by angular clasts, pebble size and larger

Most coal forms in fluvial or delta environments where vegetation growth is vigorous and where decaying plant matter accumulates in long-lasting swamps with low oxygen levels. To avoid oxidation and breakdown, the organic matter must remain submerged for centuries or millennia, until it is covered with another layer of either muddy or sandy sediments.

It is important to note that in some textbooks coal is described as an "organic sedimentary rock." In this book, coal is classified with the clastic rocks for two reasons: first, because it is made up of fragments of organic matter; and second, because coal seams (sedimentary layers) are almost always interbedded with layers of clastic rocks, such as mudrock or sandstone. In other words, coal accumulates in environments where other clastic rocks accumulate.

It's worth taking a closer look at the different types of sandstone because sandstone is a common and important sedimentary rock. Typical sandstone compositions are shown in Figure 6. The term arenite applies to a so-called clean sandstone, meaning one with less than 15% silt and clay. Considering the sand-sized grains only, arenites with 90% or more quartz are called quartz arenites. If they have more than 10% feldspar and more feldspar than rock fragments, they are called feldspathic arenites or arkosic arenites (or just arkose). If they have more than 10% rock fragments, and more rock fragments than feldspar, they are lithic"Lithic" means "rock." Lithic clasts are rock fragments, as opposed to mineral fragments. arenites. A sandstone with more than 15% silt or clay is called a wacke (pronounced wackie). The terms quartz wacke, lithic wacke, and feldspathic wacke are used. Another name for a lithic wacke is greywacke.

Some examples of sandstones, magnified in thin section are shown in Figure 7. (A thin section is rock sliced thin enough so that light can shine through.)

Clastic sedimentary rocks in which a significant proportion of the clasts are larger than 2 mm are known as conglomerate if the clasts are well rounded, and breccia if they are angular. Conglomerates form in high-energy environments where the particles can become rounded, such as fast-flowing rivers. Breccias typically form where the particles are not transported a significant distance in water, such as alluvial fans and talus slopes. Some examples of clastic sedimentary rocks are shown on Figure 8.

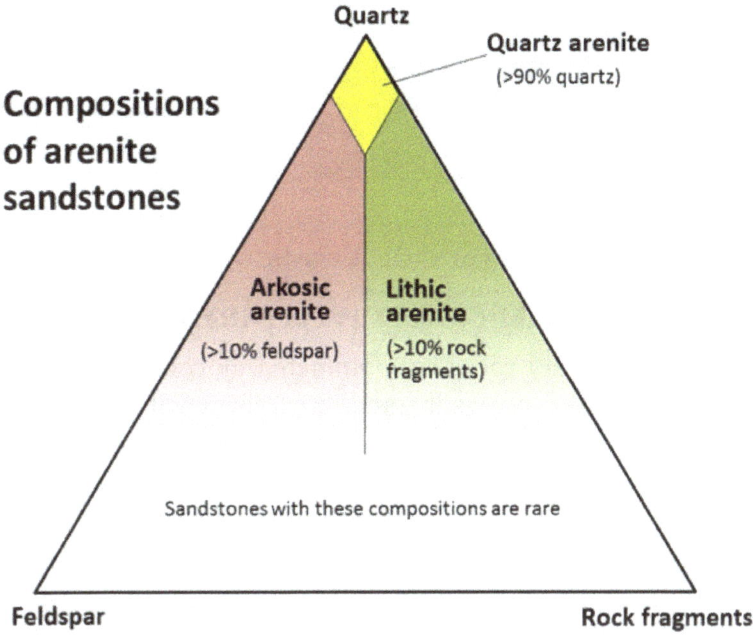

Figure 6: A compositional triangle for arenite sandstones, with the three most common components of sand-sized grains: quartz, feldspar, and rock fragments. Arenites have less than 15% silt or clay. Sandstones with more than 15% silt and clay are called wackes (e.g., quartz wacke, lithic wacke).

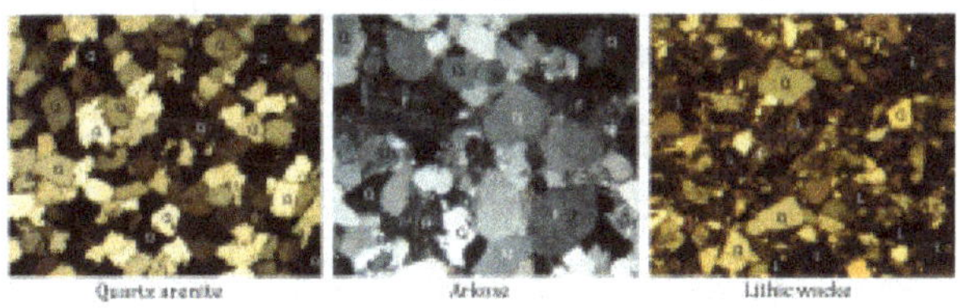

Figure 7: Photos of thin sections of three types of sandstone. Some of the minerals are labelled: Q=quartz, F=feldspar and L= lithic (rock fragments). The quartz arenite and arkose have relatively little silt-clay matrix, while the lithic wacke has abundant matrix.

(a) Mudrock with bivalve impressions, Cretaceous Nanaimo Group, Browns River, Vancouver Island

(b) Coarse sandstone with cross-bedding, Cambrian Tapeats Formation Chino Valley, Arizona

(c) Conglomerate with imbricate (aligned, tilted down to the left) cobbles, Cretaceous Geoffrey Formation, Hornby Island, BC

(d) Sedimentary breccia, the Pre-Cambrian Toby Formation, east of Castlegar, BC (image is approx. 1 m across)

Figure 8: Examples of various clastic sedimentary rocks.

6.2 CHEMICAL SEDIMENTARY ROCKS

Whereas clastic sedimentary rocks are dominated by components that have been transported as solid clasts (clay, silt, sand, etc.), chemical sedimentary rocks are dominated by components that have been transported as ions in solution (Na^+, Ca^{2+}, HCO_3^-, etc.). There is some overlap between the two because almost all clastic sedimentary rocks contain cement formed from dissolved ions, and many chemical sedimentary rocks include some clasts. Since ions can stay in solution for tens of thousands of years (some much longer), and can travel for tens of thousands of kilometres, it is virtually impossible to relate chemical sediments back to their source rocks.

The most common chemical sedimentary rock, by far, is limestone. Others include chert, banded iron formation, and a variety of rocks that form when bodies of water evaporate. Biological processes are important in the formation of some chemical sedimentary rocks, especially limestone and chert. For example, limestone is made up almost entirely of fragments of marineWe use the word marine when referring to salt water (i.e., oceanic) environments, and the word aquatic when referring to freshwater environments. organisms that manufacture calcite for their shells and other hard parts, and most chert includes at least some of the silica tests (shells) of tiny marine organisms (such as diatoms and radiolaria).

Limestone

Almost all limestone forms in the oceans, and most of that forms on the shallow continental shelves, especially in tropical regions with coral reefs. Reefs are highly productive ecosystems populated by a wide range of organisms, many of which use calcium and bicarbonate ions in seawater to make carbonate minerals (especially calcite) for their shells and other structures. These include corals, of course, but also green and red algae, urchins, sponges, molluscs, and crustaceans. Especially after they die, but even while they are still alive, these organisms are eroded by waves and currents to produce carbonate fragments that accumulate in the surrounding region, as illustrated in Figure 9.

Figure 9: Various corals and green algae on a reef at Ambergris, Belize. The light-coloured sand consists of carbonate fragments eroded from the reef organisms.

Figure 10 shows a cross-section through a typical reef in a tropical environment (normally between 40° N and 40° S). Reefs tend to form near the edges of steep drop-offs because the reef organisms thrive on nutrient-rich upwelling currents. As the reef builds up, it is eroded by waves and currents to produce carbonate sediments that are transported into the steep offshore fore-reef area and the shallower inshore back-reef area. These sediments are dominated by reef-type carbonate fragments of all sizes, including mud. In many such areas, carbonate-rich sediments also accumulate in quiet lagoons, where mud and mollusc-shell fragments predominate (Figure 11a) or in offshore areas with strong currents, where either foraminifera tests accumulate (Figure 11b) or calcite crystallizes inorganically to form ooids – spheres of calcite that form in shallow tropical ocean water with strong currents (Figure 11c).

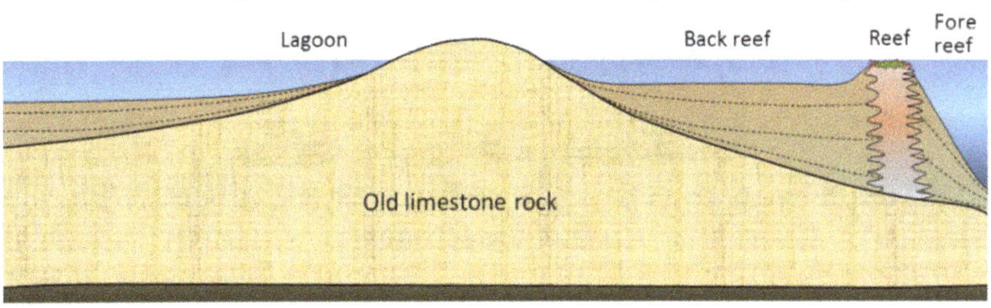

Figure 10: Schematic cross-section through a typical tropical reef.

Figure 11: Carbonate rocks and sediments: (a) mollusc-rich limestone formed in a lagoon area at Ambergris, Belize, (b) foraminifera-rich sediment from a submerged carbonate sandbar near to Ambergris, Belize (c) ooids from a beach at Joulters Cay, Bahamas.

Limestone also accumulates in deeper water, from the steady rain of the carbonate shells of tiny organisms that lived near the ocean surface. The lower limit for limestone accumulation is around 4,000 m. Beneath that depth, calcite is soluble so limestone does not accumulate.

Calcite can also form on land in a number of environments. Tufa forms at springs (Figure 12) and travertine (which is less porous) forms at hot springs. Similar material precipitates within limestone caves to form stalactites, stalagmites, and a wide range of other speleothems.

Figure 12: Tufa formed at a spring at Johnston Creek, Alberta. The rock to the left is limestone.

Dolomite ($CaMg(CO_3)_2$) is another carbonate mineral, but dolomite is also the name for a rock composed of the mineral dolomite (although some geologists use the term dolostone to avoid confusion). Dolomite rock is quite common (there's a whole Italian mountain range named after it), which is surprising since marine organisms don't make dolomite. All of the dolomite found in ancient rocks has been formed through magnesium replacing some of the calcium in the calcite in carbonate muds and sands. This process is known as dolomitization, and it is thought to take place where magnesium-rich water percolates through the sediments in carbonate tidal flat environments.

Chert (Brown Layer) and Banded Iron Formation

As we've seen, not all marine organisms make their hard parts out of calcite; some, like radiolaria and diatoms, use silica, and when they die their tiny shells (or tests) settle slowly to the bottom where they accumulate as chert. In some cases, chert is deposited along with limestone in the moderately deep ocean, but the two tend to remain separate, so chert beds within limestone are quite common (Figure 13), as are nodules, link the flint nodules of the Cretaceous chalk of southeastern England. In other situations, and especially in very deep water, chert accumulates on its own, commonly in thin beds.

Figure 13: Chert (brown layers) interbedded with Triassic Quatsino Fm. limestone on Quadra Island, B.C. All of the layers have been folded, and the chert, being insoluble and harder than limestone, stands out.

Some ancient chert beds — most dating to between 1800 and 2400 Ma — are also combined with a rock known as banded iron formation (BIF), a deep sea-floor deposit of iron oxide that is a common ore of iron (Figure 14). BIF forms when iron dissolved in seawater is oxidized, becomes insoluble, and sinks to the bottom in the same way that silica tests do to form chert. The prevalence of BIF in rocks dating from 2400 to 1800 Ma is due to the changes in the atmosphere and oceans that took place over that time period. Photosynthetic bacteria (i.e., cyanobacteria, a.k.a. blue-green algae) consume carbon dioxide from the atmosphere and use solar energy to convert it to oxygen. These bacteria first evolved around 3500 Ma, and for the next billion years, almost all of that free oxygen was used up by chemical and biological processes, but by 2400 Ma free oxygen levels started to increase in the atmosphere and the oceans. Over a period of 600 million years, that oxygen gradually converted soluble ferrous iron (Fe^{2+}) to insoluble ferric iron (Fe^{3+}), which combined with oxygen to form the mineral hematite (Fe_2O_3), leading to the accumulation of BIFs. After 1800 Ma, little dissolved iron was left in the oceans and the formation of BIF essentially stopped.

Figure 14: Banded iron formation (red) interbedded with chert (white), Dales Gorge, Australia

Evaporites

In arid regions, lakes and inland seas typically have no stream outlet and the water that flows into them is removed only by evaporation. Under these conditions, the water becomes increasingly concentrated with dissolved salts, and eventually some of these salts reach saturation levels and start to crystallize (Figure 15). Although all evaporite deposits are unique because of differences in the chemistry of the water, in most cases minor amounts of carbonates start to precipitate when the solution is reduced to about 50% of its original volume. Gypsum ($CaSO_4{\cdot}H_2O$) precipitates at about 20% of the original volume and halite (NaCl) precipitates at 10%. Other important evaporite minerals include sylvite (KCl) and borax ($Na_2B_4O_7{\cdot}10H_2O$). Sylvite is mined at numerous locations across Saskatchewan (Figure 16) from evaporites that were deposited during the Devonian (~385 Ma) when an inland sea occupied much of the region.

Figure 15: Spotted Lake, near Osoyoos, B.C. This photo was taken in May when the water was relatively fresh because of winter rains. By the end of the summer the surface of this lake is typically fully encrusted with salt deposits.

Figure 16: A mining machine at the face of potash ore (sylvite) in the Lanigan Mine near Saskatoon, Saskatchewan. The mineable potash layer is about 3 m thick.

6.3 IMPORTANT DEPOSITIONAL ENVIRONMENTS AND SEDIMENTARY BASINS

Sediments accumulate in a wide variety of environments, both on the continents and in the oceans. Some of the more important of these environments are illustrated in Figure 17.

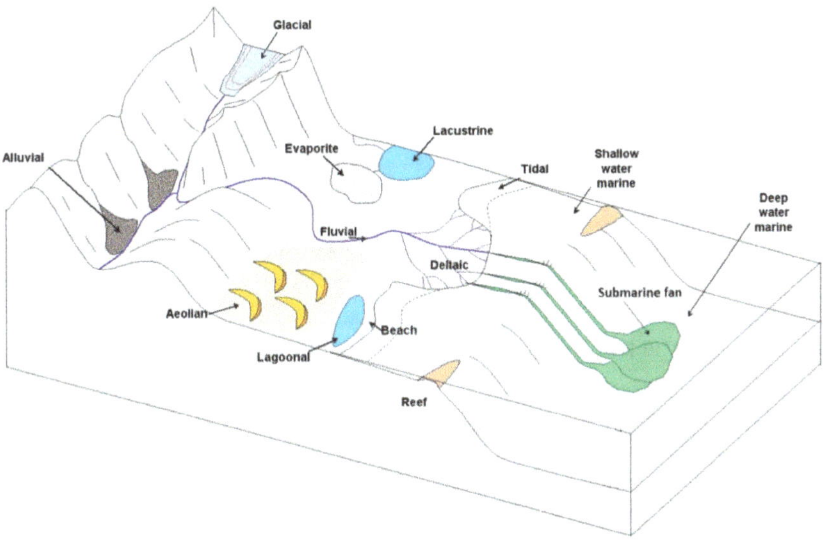

Figure 17: Some of the important depositional environments for sediments and sedimentary rocks

Table 3 provides a summary of the processes and sediment types that pertain to the various depositional environments illustrated in Figure 17. We'll look more closely at the types of sediments that accumulate in these environments in the last section of this chapter. The characteristics of these various environments, and the processes that take place within them, are also discussed in later chapters on glaciation, mass wasting, streams, coasts, and the sea floor.

Table 3: The important terrestrial and marine depositional environments and their characteristics

Environment	Important Transport Processes	Depositional Environments	Typical Sediment Types
Terrestrial Environments			
Glacial	gravity, moving ice, moving water	valleys, plains, streams, lakes	glacial till, gravel, sand, silt, and clay
Colluvial	gravity	steep-sided valleys	coarse angular fragments
Fluvial	moving water	streams	gravel, sand, silt, and OM*
Aeolian	wind	deserts and coastal regions	sand, silt
Lacustrine	moving water	lakes	sand, silt, clay, and OM*
Evaporite	moving water	lakes in arid regions	salts, clay
Marine Environments			
Deltaic	moving water	deltas	sand, silt, clay, and OM*
Beach	waves, longshore currents	beaches, spits, sand bars	gravel, sand
Tidal	tidal currents	tidal flats	silt, clay
Reefs	waves and tidal currents	reefs and adjacent basins	carbonates
Shallow water marine	waves and tidal currents	shelves and slopes, lagoons	carbonates (in tropical climates); sand/silt/clay (elsewhere)
Lagoonal	little transportation	lagoon bottom	carbonates (in tropical climates)
Submarine fan	underwater gravity flows	continental slopes and abyssal plains	gravel, sand, mud
Deep water marine	ocean currents	deep-ocean abyssal plains	clay, carbonate mud, silica mud

* OM (organic matter) only accumulates in swampy parts of these environments.

Most of the sediments that you might see around you, including talus on steep slopes, sand bars in streams, or gravel in road cuts, will never become sedimentary rocks because they have only been deposited relatively recently — perhaps a few centuries or millennia ago — and will be re-eroded before they are buried deep enough beneath other sediments to be lithified. In order for sediments to be preserved long enough to be turned into rock, a process that takes millions or tens of millions of years, they need to have been deposited in a basin that will last that long. Most such basins are formed by plate tectonic processes, and some of the more important examples are shown in Figure 18.

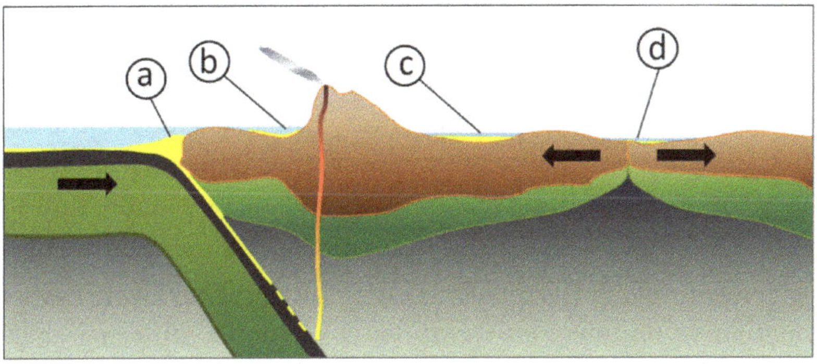

Figure 18: Some of the more important types of tectonically produced basins: (a) trench basin, (b) forearc basin, (c) foreland basin, and (d) rift basin

Trench basins form where a subducting oceanic plate dips beneath the overriding continental or oceanic crust. They can be several kilometres deep, and in many cases, host thick sequences of sediments from eroding coastal mountains. There is a well-developed trench basin off the west coast of Vancouver Island. A forearc basin lies between the subduction zone and the volcanic arc, and may be formed in part by friction between the subducting plate and the overriding plate, which pulls part of the overriding plate down. The Strait of Georgia is a forearc basin. A foreland basin is caused by the mass of the volcanic range depressing the crust on either side. Foreland basins are not only related to volcanic ranges, but can form adjacent to fold belt mountains like the Canadian Rockies. A rift basin forms where continental crust is being pulled apart, and the crust on both sides the rift subsides. As rifting continues this eventually becomes a narrow sea, and then an ocean basin. The East African rift basin represents an early stage in this process.

6.4 FOSSILS AND SEDIMENTARY STRUCTURES

Through careful observation over the past few centuries, geologists have discovered that the accumulation of sediments and sedimentary rocks takes place according to some important geological principles, as follows:

- The principle of original horizontality states that sediments accumulate in essentially horizontal layers. The implication is that tilted sedimentary layers observed to day must have been subjected to tectonic forces.
- The principle of superposition states that sedimentary layers are deposited in sequence, and that unless the entire sequence has been turned over by tectonic processes, the layers at the bottom are older than those at the top.

- The principle of inclusions states that any rock fragments in a sedimentary layer must be older than the layer. For example, the cobbles in a conglomerate must have been formed before the conglomerate.
- The principle of faunal succession states that there is a well-defined order in which organisms have evolved through geological time, and therefore the identification of specific fossils in a rock can be used to determine its age.

In addition to these principles that apply to all sedimentary rocks, a number of other important characteristics of sedimentary processes lead to the development of distinctive sedimentary features in specific sedimentary environments. By understanding the origins of these features, we can make some very useful inferences about the processes that led to deposition the rocks that we are studying.

Bedding, for example, is the separation of sediments into layers that either differ from one another in textures, composition, colour, or weathering characteristics, or are separated by partings — narrow gaps between adjacent beds (Figure 19). Bedding is an indication of changes in depositional processes that may be related to seasonal differences, changes in climate, changes in locations of rivers or deltas, or tectonic changes. Partings may represent periods of non-deposition that could range from a few decades to a few centuries. Bedding can form in almost any depositional environment.

Figure 19: The Triassic Sulphur Mt. Formation near Exshaw, Alberta. Bedding is defined by differences in colour and texture, and also by partings (gaps) between beds that may otherwise appear to be similar.

Cross-bedding is bedding that contains angled layers and forms when sediments are deposited by flowing water or wind. Some examples are shown in Figures 1,

8b and 20. Cross-beds in streams tend to be on the scale of centimetres to tens of centimetres, while those in aeolian (wind deposited) sediments can be on the scale of metres to several metres.

Figure 20: Cross-bedded Jurassic Navajo Formation aeolian sandstone at Zion National Park, Utah. In most of the layers the cross-beds dip down toward the right, implying wind direction from right to left during deposition. One bed dips in the opposite direction, implying an abnormal wind.

Cross-beds form as sediments are deposited on the leading edge of an advancing ripple or dune. Each layer is related to a different ripple that advances in the flow direction, and is partially eroded by the following ripple (Figure 21). Cross-bedding is a very important sedimentary structure to recognize because it can provide information on the direction of current flows and, when analyzed in detail, on other features like the rate of flow and the amount of sediment available.

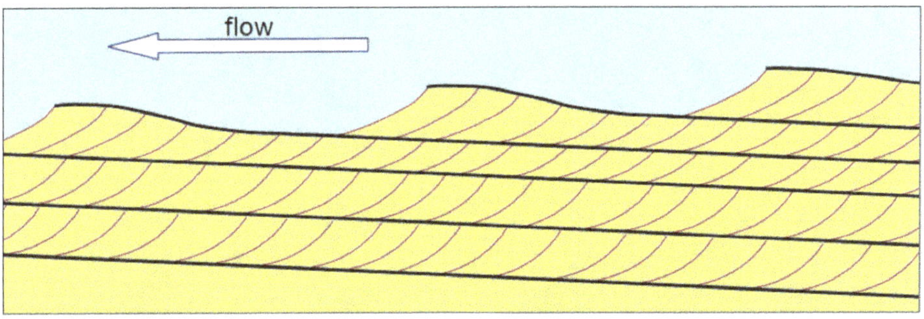

Figure 21: Formation of cross-beds as a series of ripples or dunes migrates with the flow. Each ripple advances forward (right to left in this view) as more sediment is deposited on its leading face.

Graded bedding is characterized by a gradation in grain size from bottom to top within a single bed. "Normal" graded beds are coarse at the bottom and become finer toward the top, a product of deposition from a slowing current (Figure 6.22). Some graded beds are reversed (coarser at the top), and this normally results from deposition by a fast-moving debris flow (see Chapter 15). Most graded beds form in a submarine-fan environment (see Figure 6.17), where sediment-rich flows descend periodically from a shallow marine shelf down a slope and onto the deeper sea floor.

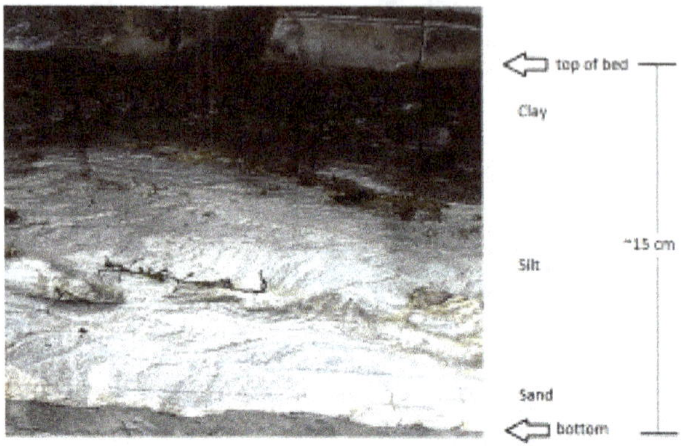

Figure 22: A graded turbidite bed in Cretaceous Spray Formation rocks on Gabriola Island, B.C. The lower several centimetres of sand and silt probably formed over the duration of an hour. The upper few centimetres of fine clay may have accumulated over a few hundred years.

Ripples, which are associated with the formation of cross-bedding, may be preserved on the surfaces of sedimentary beds. Ripples can also help to determine flow direction as they tend to have their steepest surface facing down flow.

In a stream environment, boulders, cobbles, and pebbles can become imbricated, meaning that they are generally tilted in the same direction. Clasts in streams tend to tilt with their upper ends pointing downstream because this is the most stable position with respect to the stream flow (Figure 23 and Figure 8c).

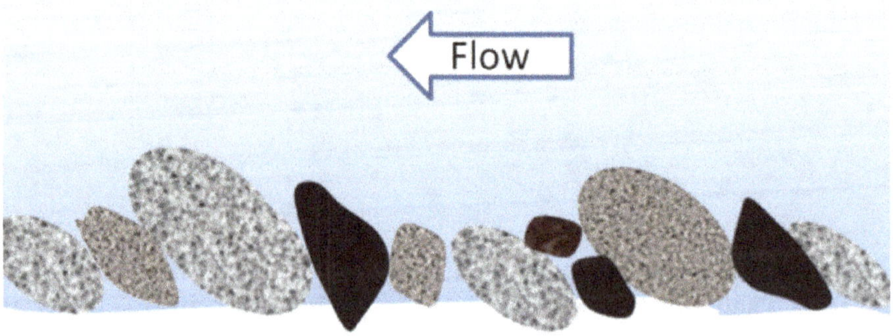

Figure 23: An illustration of imbrication of clasts in a fluvial environment.

Mud cracks form when a shallow body of water (e.g., a tidal flat or pond), into which muddy sediments have been deposited, dries up and cracks (Figure 6.24). This happens because the clay in the upper mud layer tends to shrink on drying, and so it cracks because it occupies less space when it is dry.

The various structures described above are critical to understanding and interpreting the formation of sedimentary rocks. In addition to these, geologists also look very closely at sedimentary grains to determine their mineralogy or lithology (in order to make inferences about the type of source rock and the weathering processes), their degree of rounding, their sizes, and the extent to which they have been sorted by transportation and depositional processes.

Figure 24: Mudcracks in volcanic mud at a hot-spring area near Myvatn, Iceland [SE]

We won't be covering fossils in any detail in this book, but they are extremely important for understanding sedimentary rocks. Of course, fossils can be used to date sedimentary rocks, but equally importantly, they tell us a great deal about the depositional environment of the sediments and the climate at the time. For example, they can help to differentiate marine, aquatic, and terrestrial environments; estimate the depth of the water; detect the existence of currents; and estimate average temperature and precipitation.

The tests of tiny marine organisms (mostly foraminifera) have been recovered from deep-ocean sediment cores from all over the world, and their isotopic signatures have been measured. As we'll see in Chapter 19, this provides us with information about the changes in average global temperatures over the past 65 million years.

6.5 GROUPS, FORMATIONS, AND MEMBERS

Geologists who study sedimentary rocks need ways to divide them into manageable units, and they also need to give those units names so that they can easily be referred to and compared with other rocks deposited in other places. The International Commission on Stratigraphy (ICS) (http://www.stratigraphy.org/) has established a set of conventions for grouping, describing, and naming sedimentary rock units.

The main stratigraphic unit is a formation, which according to the ICS, should be established with the following principles in mind:

The contrast in lithology between formations required to justify their establishment varies with the complexity of the geology of a region and the detail needed for geologic mapping and to work out its geologic history. No formation is considered justifiable and useful that cannot be delineated at the scale of geologic mapping practiced in the region. The thickness of formations may range from less than a meter to several thousand meters.

In other words, a formation is a series of beds that is distinct from other beds above and below, and is thick enough to be shown on the geological maps that are widely used within the area in question. In most parts of the world, geological mapping is done at a relatively coarse scale, and so most formations are in the order of a few hundred metres thick. At that thickness, a typical formation would appear on a typical geological map as an area that is at least a few millimetres thick.

A series of formations can be classified together to define a group, which could be as much as a few thousand metres thick, and represents a series of rocks that were deposited within a single basin (or a series of related and adjacent basins) over a few million to a few tens of millions of years.

In areas where detailed geological information is needed (for example, within a mining or petroleum district) a formation might be divided into members, where each member has a specific and distinctive lithology. For example, a formation that includes both shale and sandstone might be divided into members, each of which is either shale or sandstone. In some areas, where particular detail is needed, members may be divided into beds, but this is only applicable to beds that have a special geological significance. Groups, formations, and members are typically named for the area where they are found.

The sedimentary rocks of the Nanaimo Group provide a useful example for understanding groups, formations, and members. During the latter part of the Cretaceous Period, from about 90 Ma to 65 Ma, a thick sequence of clastic rocks was deposited in a foreland basin between what is now Vancouver Island and the B.C. mainland (Figure 25). The Nanaimo Group strata comprise a 5000 m thick sequence of conglomerate, sandstone, and mudstone layers. Coal was mined from Nanaimo Group rocks from around 1850 to 1950 in the Nanaimo region, and is still being mined in the Campbell River area.

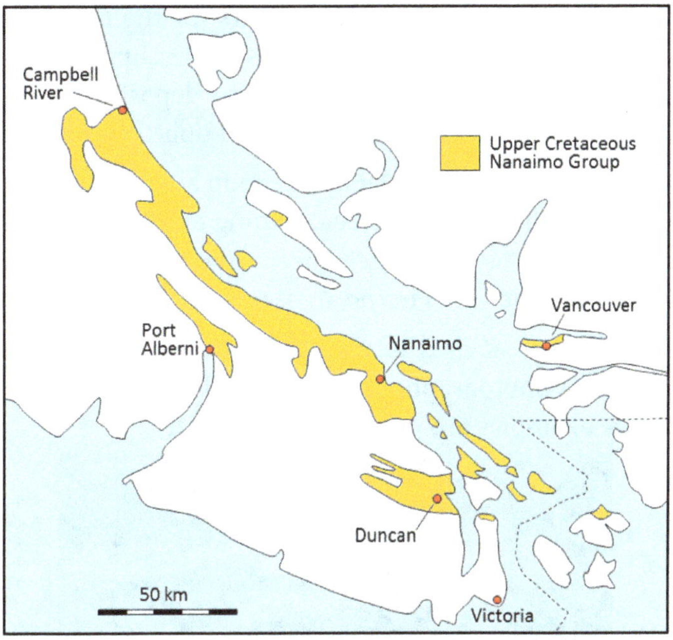

Figure 25: The distribution of the Upper Cretaceous Nanaimo Group rocks on Vancouver Island, the Gulf Islands, and in the Vancouver area.

Table 4: The formations of the Nanaimo Group. Formations that are predominantly fine-grained are shaded. In tables like this one the layers are always listed with the oldest at the bottom and the youngest at the top. [Based on data in Mustard, P., 1994, The Upper Cretaceous Nanaimo Group, Georgia Basin, in J. Monger (ed) Geology and Geological Hazards of the Vancouver Region, Geol. Survey of Canada, Bull. 481, p. 27-95.]

	Age (Ma)	Formation	Lithologies	Depositional Environment
NANAIMO GROUP	~ 65 - 66	Gabriola	sandstone with minor mudstone	submarine fan, high energy
	~66 - 67	Spray	mudstone/sandstone turbidites	submarine fan, low energy
	~ 67 - 68	Geoffrey	sandstone and conglomerate	submarine fan, high energy
	~68 - 70	Northumberland	mudstone turbidites	submarine fan, low energy
	~70	De Courcy	sandstone	submarine fan, high energy
	~70 - 72	Cedar District	mudstone turbidites	submarine fan, low energy
	~72 - 75	Protection	sandstone and minor coal	nearshore marine and onshore deltaic and fluvial
	~75 - 80	Pender	sandstone and minor coal	nearshore marine and onshore deltaic and fluvial
	~ 80	Extension	conglomerate, with minor sandstone and some coal	nearshore marine and onshore deltaic and fluvial
	~80 - 85	Haslam	mudstone and siltstone	shallow marine
	~85 - 90	Comox	conglomerate, sandstone, mudstone (coal in the Campbell River area)	nearshore fluvial and marine

The Nanaimo Group is divided into 11 formations as described in Table 4. In general, the boundaries between formations are based on major lithological differences. As can be seen in the far-right column of Table 4, a wide range of depositional environments existed during the accumulation of the Nanaimo Group rocks, from nearshore marine for the Comox and Haslam Formation, to fluvial and deltaic with backwater swampy environments for the coal-bearing Extension, Pender, and Protection Formations, to a deep-water submarine fan environment for the upper six formations. The differences in the depositional environments are probably a product of variations in tectonic-related uplift over time.

The five lower formations of the Nanaimo Group are all exposed in the Nanaimo area, and were well studied during the coal mining era between 1850 and 1950. All of these formations (except Haslam) have been divided into members, as that was useful for understanding the rocks in the areas where coal mining was taking place.

Although there is a great deal of variety in the Nanaimo Group rocks, and it would take hundreds of photographs to illustrate all of the different types of rocks, a few representative examples are provided in Figure 26.

Figure 26: Representative photos of Nanaimo Group rocks. (a) Turbidite layers in the Spray Formation on Gabriola Island. Each turbidite set consists of a lower sandstone layer (light colour) that grades upward into siltstone, and then into mudstone. (See Figure 6.21 for detail.)

(b) Two separate layers of fluvial sandstone with a thin (approx. 75 cm) coal seam in between. Pender Formation in Nanaimo.

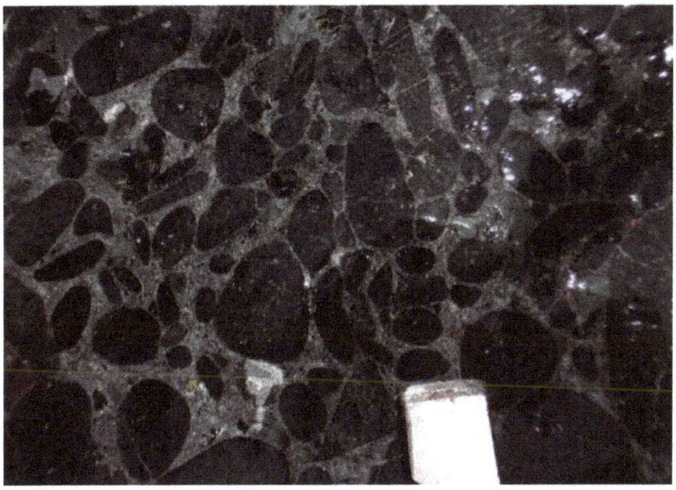

(c) Comox Formation conglomerate at the very base of the Nanaimo Group in Nanaimo. The metal object is the end of a rock hammer that is 3 cm wide. Almost all of the clasts in this view are well-rounded basalt pebbles cobbles eroded from the Triassic Karmutsen Formation which makes up a major part of Vancouver Island.

AT A GLANCE

The topics covered in this chapter can be summarized as follows:

6.1	Clastic Sedimentary Rocks	Sedimentary clasts are classified based on their size, and variations in clast size have important implications for transportation and deposition. Clastic sedimentary rocks range from conglomerate to mudstone. Clast size, sorting, composition, and shape are important features that allow us to differentiate clastic rocks and understand the processes that took place during their deposition.
6.2	Chemical Sedimentary Rocks	Chemical sedimentary rocks form from ions that were transported in solution, and then converted into minerals by biological and/or chemical processes. The most common chemical rock, limestone, typically forms in shallow tropical environments, where biological activity is a very important factor. Chert and banded iron formation are deep-ocean sedimentary rocks. Evaporites form where the water of lakes and inland seas becomes supersaturated due to evaporation.
6.3	Depositional Environments and Sedimentary Basins	There is a wide range of depositional environments, both on land (glaciers, lakes, rivers, etc.) and in the ocean (deltas, reefs, shelves, and the deep-ocean floor). In order to be preserved, sediments must accumulate in long-lasting sedimentary basins, most of which form through plate-tectonic processes.

6.4	Sedimentary Structures and Fossils	The deposition of sedimentary rocks takes place according to a series of important principles, including original horizontality, superposition, and faunal succession. Sedimentary rocks can also have distinctive structures that are important in determining their depositional environments. Fossils are useful for determining the age of a rock, the depositional environment, and the climate at the time of deposition.
6.5	Groups, Formations, and Members	Sedimentary sequences are classified into groups, formations, and members so that they can be referred to easily and without confusion.

CHAPTER

METAMORPHIC ROCKS AND METAMORPHISM 7

Metamorphism is the change that takes place within a body of rock as a result of it being subjected to conditions that are different from those in which it formed. In most cases, but not all, this involves the rock being deeply buried beneath other rocks, where it is subjected to higher temperatures and pressures than those under which it formed. Metamorphic rocks typically have different mineral assemblages and different textures from their parent rocks (Figure 1) but they may have the same overall composition.

Figure 1: Metamorphic rock (gneiss) of the Okanagan Metamorphic and Igneous Complex at Skaha Lake, B.C. The dark bands are amphibole-rich, the light bands are feldspar-rich. [SE photo]

Most metamorphism results from the burial of igneous, sedimentary, or pre-existing metamorphic to the point where they experience different pressures and temperatures than those at which they formed (Figure 2). Metamorphism can also take place if cold rock near the surface is intruded and heated by a hot igneous body. Although most metamorphism involves temperatures above 150°C, some metamorphism takes place at temperatures lower than those at which the parent rock formed.

Figure 2: The rock cycle. The processes related to metamorphic rocks are at the bottom of the cycle. [SE]

7.1 REGULATION OF METAMORPHIC PROCESSES

The main factors that control metamorphic processes are:

- The mineral composition of the parent rock
- The temperature at which metamorphism takes place
- The amount and type of pressure during metamorphism
- The types of fluids (mostly water) that are present during metamorphism
- The amount of time available for metamorphism

Parent Rock

The parent rock is the rock that exists before metamorphism starts. In most cases, this is sedimentary or igneous rock, but metamorphic rock that reaches the surface and is then reburied can also be considered a parent rock. On the other hand, if, for example, a mudstone is metamorphosed to slate and then buried deeper where it is metamorphosed to schist, the parent rock of the schist is mudstone, not slate. The critical feature of the parent rock is its mineral composition because it is the stability of minerals that counts when metamorphism takes place. In other words, when a rock is subjected to increased temperatures, certain minerals may become unstable and start to recrystallize into new minerals.

Temperature

The temperature that the rock is subjected to is a key variable in controlling the type of metamorphism that takes place. As we learned in the context of igneous rocks, mineral stability is a function of temperature, pressure, and the presence of fluids (especially water). All minerals are stable over a specific range of temperatures. For example, quartz is stable from environmental temperatures (whatever the weather can throw at it) all the way up to about 1800°C. If the pressure is higher, that upper limit will be higher. If there is water present, it will be lower. On the other hand, most clay minerals are only stable up to about 150° or 200°C; above that, they transform into micas. Most other common minerals have upper limits between 150°C and 1000°C.

Some minerals will crystallize into different **polymorphs** (same composition, but different crystalline structure) depending on the temperature and pressure. Quartz is a good example as slightly different forms are stable between 0°C and 1800°C. The minerals kyanite, andalusite, and sillimanite are polymorphs with the composition Al2SiO5. They are stable at different pressures and temperatures, and, as we will see later, they are important indicators of pressures and temperatures in metamorphic rocks (Figure 3).

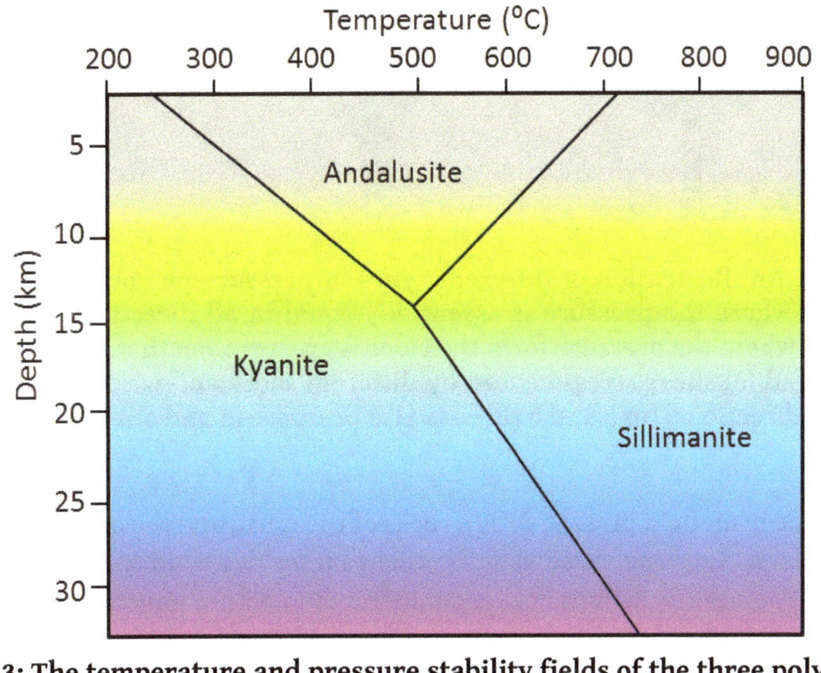

Figure 3: The temperature and pressure stability fields of the three polymorphs of Al2SiO5 (Pressure is equivalent to depth. Kyanite is stable at low to moderate temperatures and low to high pressures, andalusite at moderate temperatures and low pressures, and sillimanite at higher temperatures.) [SE]

Pressure

Pressure is important in metamorphic processes for two main reasons. First, it has implications for mineral stability (Figure 3). Second, it has implications for the texture of metamorphic rocks. Rocks that are subjected to very high confining pressures are typically denser than others because the mineral grains are squeezed together (Figure 4a), and because they may contain mineral polymorphs in which the atoms are more closely packed. Because of plate tectonics, pressures within the crust are typically not applied equally in all directions. In areas of plate convergence, the pressure in one direction (perpendicular to the direction of convergence) is typically greater than in the other directions (Figure 4b). In situations where different blocks of the crust are being pushed in different directions, the rocks will be subjected to sheer stress (Figure 4c).

One of the results of directed pressure and sheer stress is that rocks become foliated – meaning that they'll have a directional fabric. Foliation is described in more detail later in this chapter.

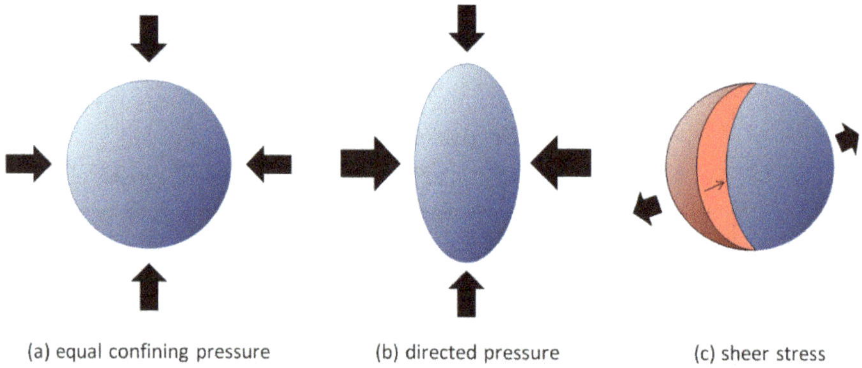

Figure 4: An illustration of different types of pressure on rocks. (a) confining pressure, where the pressure is essentially equal in all directions, (b) directed pressure, where the pressure form the sides is greater than that from the top and bottom, and (c) sheer stress caused by different blocks of rock being pushed in different directions. (In a and b there is also pressure in and out of the page.) [SE]

Fluids

Water is the main fluid present within rocks of the crust, and the only one that we'll consider here. The presence of water is important for two main reasons. First, water facilitates the transfer of ions between minerals and within minerals, and therefore increases the rates at which metamorphic reactions take place. So, while the water doesn't necessarily change the outcome of a metamorphic process, it speeds the process up so metamorphism might take place over a shorter time period, or metamorphic processes that might not otherwise have had time to be completed are completed.

Secondly, water, especially hot water, can have elevated concentrations of dissolved substances, and therefore it is an important medium for moving certain elements around within the crust. So not only does water facilitate metamorphic reactions on a grain-to-grain basis, it also allows for the transportation of ions from one place to another. This is very important in hydrothermal processes, which are discussed toward the end of this chapter, and in the formation of mineral deposits.

Time

Most metamorphic reactions take place at very slow rates. For example, the growth of new minerals within a rock during metamorphism has been estimated to be about 1 mm per million years. For this reason, it is very difficult to study metamorphic processes in a lab.

While the rate of metamorphism is slow, the tectonic processes that lead to metamorphism are also very slow, so in most cases, the chance for metamorphic reactions to be completed is high. For example, one important metamorphic setting is many kilometres deep within the roots of mountain ranges. A mountain range takes tens of millions of years to form, and tens of millions of years more to be eroded to the extent that we can see the rocks that were metamorphosed deep beneath it.

7.2 TYPES OF METAMORPHIC ROCKS

There are two main types of metamorphic rocks: those that are foliated because they have formed in an environment with either directed pressure or shear stress, and those that are not foliated because they have formed in an environment without directed pressure or relatively near the surface with very little pressure at all. Some types of metamorphic rocks, such as quartzite and marble, which also form in directed-pressure situations, do not necessarily exhibit foliation because their minerals (quartz and calcite respectively) do not tend to show alignment (see Figure 12).

When a rock is squeezed under directed pressure during metamorphism it is likely to be deformed, and this can result in a textural change such that the minerals are elongated in the direction perpendicular to the main stress (Figure 5). This contributes to the formation of foliation.

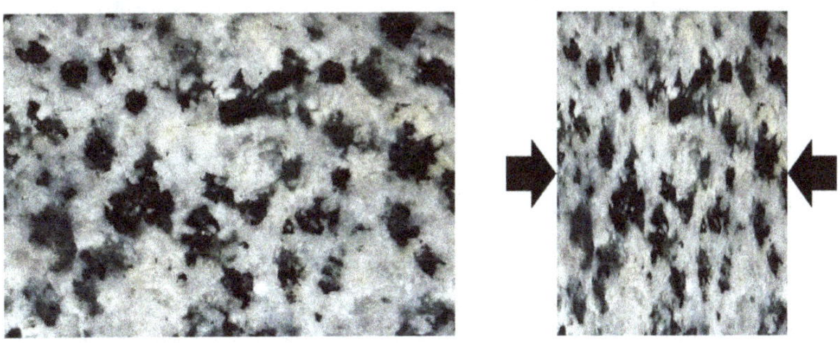

Figure 5: The textural effects of squeezing during metamorphism. [SE]

When a rock is both heated and squeezed during metamorphism, and the temperature change is enough for new minerals to form from existing ones, there is a likelihood that the new minerals will be forced to grow with their long axes perpendicular to the direction of squeezing. This is illustrated in Figure 6, where the parent rock is shale, with bedding as shown. After both heating and squeezing, new minerals have formed within the rock, generally parallel to each other, and the original bedding has been largely obliterated.

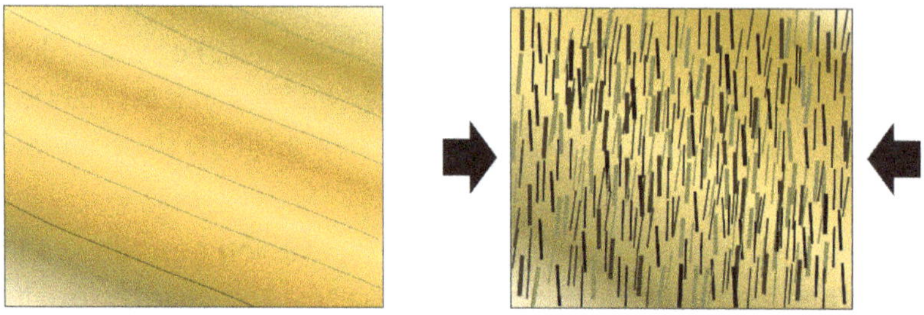

Figure 6: The textural effects of squeezing and aligned mineral growth during metamorphism. The left-hand diagram represents shale with bedding in the direction shown. The right-hand diagram represents schist (derived from that shale), with the mica crystals orientated perpendicular to the main stress direction and the original bedding no longer easily visible. [SE]

Figure 7 shows an example of this effect. This large boulder has bedding still visible as dark and light bands sloping steeply down to the right. The rock also has a strong slaty foliation, which is horizontal in this view, and has developed because the rock was being squeezed during metamorphism. The rock has split from bedrock along this foliation plane, and you can see that other weaknesses are present in the same orientation.

Figure 7: A slate boulder on the side of Mt. Wapta in the Rockies near Field, BC. Bedding is visible as light and dark bands sloping steeply to the right. Slaty cleavage is evident from the way the rock has broken and also from lines of weakness that same trend. [SE]

Squeezing and heating alone (as shown in Figure 5) and squeezing, heating, and formation of new minerals (as shown in Figure 6) can contribute to foliation, but most foliation develops when new minerals are forced to grow perpendicular to the direction of greatest stress (Figure 6). This effect is especially strong if the new minerals are platy like mica or elongated like amphibole. The mineral crystals don't have to be large to produce foliation. Slate, for example, is characterized by aligned flakes of mica that are too small to see.

The various types of foliated metamorphic rocks, listed in order of the grade or intensity of metamorphism and the type of foliation are slate, phyllite, schist, and gneiss (Figure 8). As already noted, slate is formed from the low-grade metamorphism of shale, and has microscopic clay and mica crystals that have grown perpendicular to the stress. Slate tends to break into flat sheets. Phyllite is similar to slate, but has typically been heated to a higher temperature; the micas have grown larger and are visible as a sheen on the surface. Where slate is typically planar, phyllite can form in wavy layers. In the formation of schist, the temperature has been hot enough so that individual mica crystals are visible, and other mineral crystals, such as quartz, feldspar, or garnet may also be visible. In gneiss, the minerals may have separated into bands of different colours. In the example shown in Figure 7.8d, the dark bands are largely amphibole while the light-coloured bands are feldspar and quartz. Most gneiss has little or no mica because it forms at temperatures higher than those under which micas are stable. Unlike slate and phyllite, which typically only form from mudrock, schist, and especially gneiss, can form from a variety of parent rocks, including mudrock, sandstone, conglomerate, and a range of both volcanic and intrusive igneous rocks.

Figure 8: Examples of foliated metamorphic rocks [a, b, and d: SE, c: Michael C. Rygel, http://en.wikipedia.org/wiki/Schist#mediaviewer/File:Schist_detail.jpg]

Schist and gneiss can be named on the basis of important minerals that are present. For example a schist derived from basalt is typically rich in the mineral chlorite, so we call it chlorite schist. One derived from shale may be a muscovite-biotite schist, or just a mica schist, or if there are garnets present it might be mica-garnet schist. Similarly, a gneiss that originated as basalt and is dominated by amphibole, is an amphibole gneiss or, more accurately, an amphibolite.

If a rock is buried to a great depth and encounters temperatures that are close to its melting point, it will partially melt. The resulting rock, which includes both metamorphosed and igneous material, is known as a migmatite (Figure 9).

Figure 9: Migmatite from Prague, Czech Republic

As already noted, the nature of the parent rock controls the types of metamorphic rocks that can form from it under differing metamorphic conditions. The kinds of rocks that can be expected to form at different metamorphic grades from various parent rocks are listed in Table 1. Some rocks, such as granite, do not change much at the lower metamorphic grades because their minerals are still stable up to several hundred degrees.

Table 1: A rough guide to the types of metamorphic rocks that form from different parent rocks at different grades of regional metamorphism

	Very Low Grade	Low Grade	Medium Grade	High Grade
Approximate Temperature Ranges				
Parent Rock	150-300°C	300-450°C	450-550°C	Above 550°C
Mudrock	slate	phyllite	schist	gneiss
Granite	no change	no change	no change	granite gneiss

	Very Low Grade	Low Grade	Medium Grade	High Grade
Basalt	chlorite schist	chlorite schist	amphibolite	amphibolite
Sandstone	no change	little change	quartzite	quartzite
Limestone	little change	marble	marble	marble

Metamorphic rocks that form under either low-pressure conditions or just confining pressure do not become foliated. In most cases, this is because they are not buried deeply, and the heat for the metamorphism comes from a body of magma that has moved into the upper part of the crust. This is contact metamorphism. Some examples of non-foliated metamorphic rocks are marble, quartzite, and hornfels.

Marble is metamorphosed limestone. When it forms, the calcite crystals tend to grow larger, and any sedimentary textures and fossils that might have been present are destroyed. If the original limestone was pure calcite, then the marble will likely be white (as in Figure 10), but if it had various impurities, such as clay, silica, or magnesium, the marble could be "marbled" in appearance.

Figure 10: Marble with visible calcite crystals (left) and an outcrop of banded marble (right) [SE (left) and http://gallery.usgs.gov/images/08_11_2010/a1Uh83Jww6_08_11_2010/large/DSCN2868.JPG (right)]

Figure 11: Quartzite from the Rocky Mountains, found in the Bow River at Cochrane, Alberta [SE]

Quartzite is metamorphosed sandstone (Figure 11). It is dominated by quartz, and in many cases, the original quartz grains of the sandstone are welded together with additional silica. Most sandstone contains some clay minerals and may also include other minerals such as feldspar or fragments of rock, so most quartzite has some impurities with the quartz.

Even if formed during regional metamorphism, quartzite does not tend to be foliated because quartz crystals don't align with the directional pressure. On the other hand, any clay present in the original sandstone is likely to be converted to mica during metamorphism, and any such mica is likely to align with the directional pressure. An example of this is shown in Figure 12. The quartz crystals show no alignment, but the micas are all aligned, indicating that there was directional pressure during regional metamorphism of this rock.

Figure 12: Magnified thin section of quartzite in polarized light. The irregular-shaped white, grey, and black crystals are all quartz. The small, thin, brightly coloured crystals are mica. This rock is foliated, even though it might not appear to be if examined without a microscope, and so it must have formed under directed-pressure conditions. [Photo by Sandra Johnstone, used with permission]

Hornfels is another non-foliated metamorphic rock that normally forms during contact metamorphism of fine-grained rocks like mudstone or volcanic rock (Figure 13). In some cases, hornfels has visible crystals of minerals like biotite or andalusite. If the hornfels formed in a situation without directed pressure, then these minerals would be randomly orientated, not foliated as they would be if formed with directed pressure.

Figure 13: Hornfels from the Novosibirsk region of Russia. The dark and light bands are bedding. The rock has been recrystallized during contact metamorphism and does not display foliation. (scale in cm) [http://en.wikipedia.org/wiki/Hornfels#mediaviewer/ File:Hornfels.jpg]

Exercise 1 Naming Metamorphic Rocks

Provide reasonable names for the following metamorphic rocks:

Rock Description	Name
A rock with visible minerals of mica and with small crystals of andalusite. The mica crystals are consistently parallel to one another.	
A very hard rock with a granular appearance and a glassy lustre. There is no evidence of foliation.	
A fine-grained rock that splits into wavy sheets. The surfaces of the sheets have a sheen to them.	
A rock that is dominated by aligned crystals of amphibole.	

7.3 PROCESS OF METAMORPHISM AND PLATE TECTONICS

All of the important processes of metamorphism that we are familiar with can be directly related to geological processes caused by plate tectonics. The relationships between plate tectonics and metamorphism are summarized in Figure 14, and in more detail in Figures 15, 16,17, and 19.

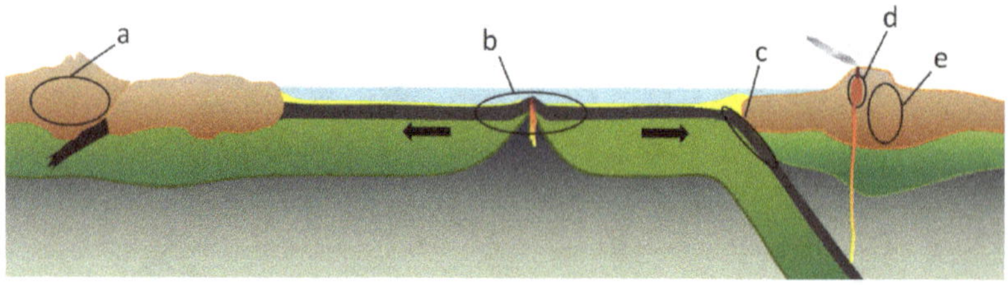

Figure 14: Environments of metamorphism in the context of plate tectonics: (a) regional metamorphism related to mountain building at a continent-continent convergent boundary, (b) regional metamorphism of oceanic crust in the area on either side of a spreading ridge, (c) regional metamorphism of oceanic crustal rocks within a subduction zone, (d) contact metamorphism adjacent to a magma body at a high level in the crust, and (e) regional metamorphism related to mountain building at a convergent boundary. [SE]

Most regional metamorphism takes place within continental crust. While rocks can be metamorphosed at depth in most areas, the potential for metamorphism is greatest in the roots of mountain ranges where there is a strong likelihood for burial of relatively young sedimentary rock to great depths, as depicted in Figure 15. An example would be the Himalayan Range. At this continent-continent convergent boundary, sedimentary rocks have been both thrust up to great heights (nearly 9,000 m above sea level) and also buried to great depths. Considering that the normal geothermal gradient (the rate of increase in temperature with depth) is around 30°C per kilometre, rock buried to 9 km below sea level in this situation could be close to 18 km below the surface of the ground, and it is reasonable to expect temperatures up to 500°C. Metamorphic rocks formed there are likely to be foliated because of the strong directional pressure of converging plates.

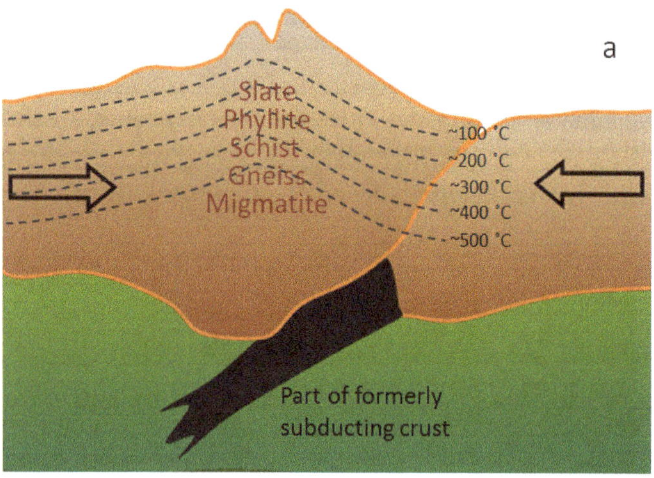

Figure 15: a: Regional metamorphism beneath a mountain range related to continent-continent collision (typical geothermal gradient). (Example: Himalayan Range) [SE]

At an oceanic spreading ridge, recently formed oceanic crust of gabbro and basalt is slowly moving away from the plate boundary (Figure 7.16). Water within the crust is forced to rise in the area close to the source of volcanic heat, and this draws more water in from farther out, which eventually creates a convective system where cold seawater is drawn into the crust and then out again onto the sea floor near the ridge. The passage of this water through the oceanic crust at 200° to 300°C promotes metamorphic reactions that change the original pyroxene in the rock to chlorite and serpentine. Because this metamorphism takes place at temperatures well below the temperature at which the rock originally formed (~1200°C), it is known as retrograde metamorphism. The rock that forms in this way is known as greenstone if it isn't foliated, or greenschist if it is. Chlorite ($(Mg_5Al)(AlSi_3)O_{10}(OH)_8$) and serpentine ($(Mg, Fe)_3Si_2O_5(OH)_4$) are both "hydrated minerals" meaning that they have water (as OH) in their chemical formulas. When metamorphosed ocean crust is later subducted, the chlorite and serpentine are converted into new non-hydrous minerals (e.g., garnet and pyroxene) and the water that is released migrates into the overlying mantle, where it contributes to flux melting (Chapter 3, section 3.2).

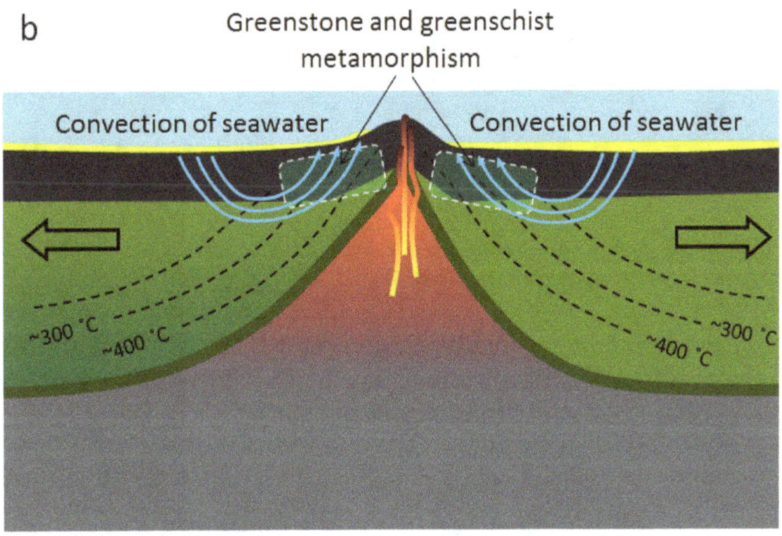

Figure 16 b: Regional metamorphism of oceanic crustal rock on either side of a spreading ridge. (Example: Juan de Fuca spreading ridge) [SE]

At a subduction zone, oceanic crust is forced down into the hot mantle. But because the oceanic crust is now relatively cool, especially along its sea-floor upper surface, it does not heat up quickly, and the subducting rock remains several hundreds of degrees cooler than the surrounding mantle (Figure 17). A special type of metamorphism takes place under these very high-pressure but relatively low-temperature conditions, producing an amphibole mineral known as glaucophane ($Na_2(Mg_3Al_2)Si_8O_{22}(OH)_2$), which is blue in colour, and is a major component of a rock known as blueschist.

If you've never seen or even heard of blueschist, it's not surprising. What is surprising is that anyone has seen it! Most blueschist forms in subduction zones, continues to be subducted, turns into eclogite at about 35 km depth, and then eventually sinks deep into the mantle — never to be seen again. In only a few places in the world, where the subduction process has been interrupted by some tectonic process, has partially subducted blueschist rock returned to the surface. One such place is the area around San Francisco; the rock is known as the Franciscan Complex (Figure 18).

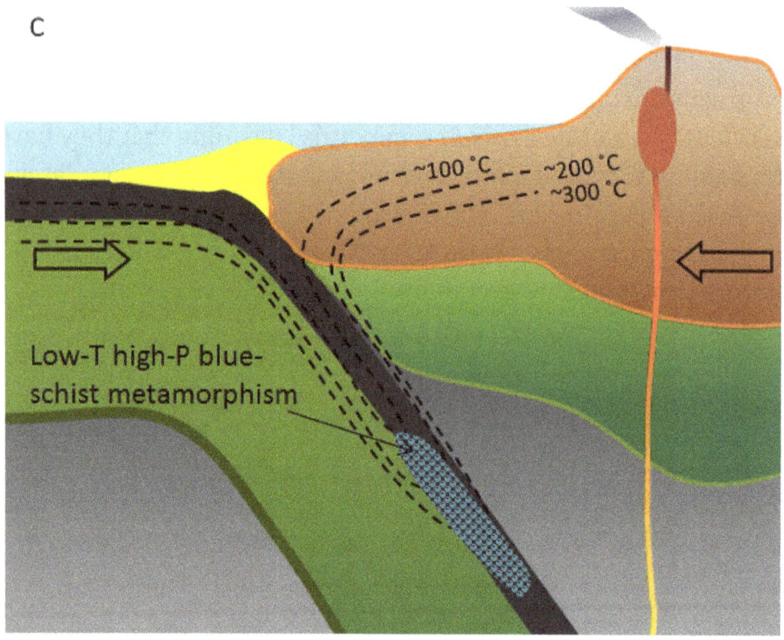

Figure 17 c: Regional metamorphism of oceanic crust at a subduction zone. (Example: Cascadia subduction zone. Rock of this type is exposed in the San Francisco area.) [SE]

Figure 18: Franciscan Complex blueschist rock exposed north of San Francisco. The blue colour of rock is due to the presence of the amphibole mineral glaucophane. [SE]

Magma is produced at convergent boundaries and rises toward the surface, where it can form magma bodies in the upper part of the crust. Such magma bodies, at temperatures of around 1000°C, heat up the surrounding rock, leading to contact metamorphism (Figure 19). Because this happens at relatively shallow depths, in the absence of directed pressure, the resulting rock does not normally develop foliation. The zone of contact metamorphism around an intrusion is very small (typically metres to tens of metres) compared with the extent of regional metamorphism in other settings (tens of thousands of square kilometres).

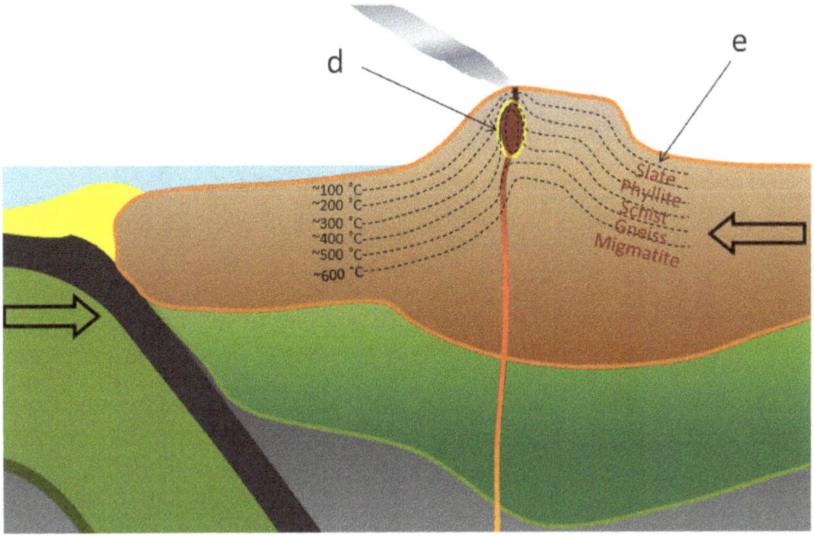

Figure 19: d: Contact metamorphism around a high-level crustal magma chamber (Example: the magma chamber beneath Mt. St. Helens.) e: Regional metamorphism in a volcanic-arc related mountain range (volcanic-region temperature gradient) (Example: The southern part of the Coast Range, B.C.) [SE]

Regional metamorphism also takes place within volcanic-arc mountain ranges, and because of the extra heat associated with the volcanism, the geothermal gradient is typically a little steeper in these settings (somewhere between 40° and 50°C/km). As a result higher grades of metamorphism can take place closer to surface than is the case in other areas (Figure 19).

Another way to understand metamorphism is by using a diagram that shows temperature on one axis and depth (which is equivalent to pressure) on the other (Figure 20). The three heavy dotted lines on this diagram represent Earth's geothermal gradients under different conditions. In most areas, the rate of increase in temperature with depth is 30°C/km. In other words, if you go 1,000 m down into a mine, the temperature will be roughly 30°C warmer than the average temperature at the surface. In most parts of southern Canada, the average surface temperature is about 10°C, so at 1,000 m depth, it will be about 40°C. That's uncomfortably hot, so deep mines must have effective ventilation systems. This typical geothermal gradient is shown by the green dotted line in Figure 20. At 10 km depth, the temperature is about 300°C and at 20 km it's about 600°C.

In volcanic areas, the geothermal gradient is more like 40° to 50°C/km, so the temperature at 10 km depth is in the 400° to 500°C range. Along subduction zones, as described above, the cold oceanic crust keeps temperatures low, so the gradient is typically less than 10°C/km. The various types of metamorphism described above are represented in Figure 20 with the same letters (a through e) used in Figures 14 to 17 and 19.

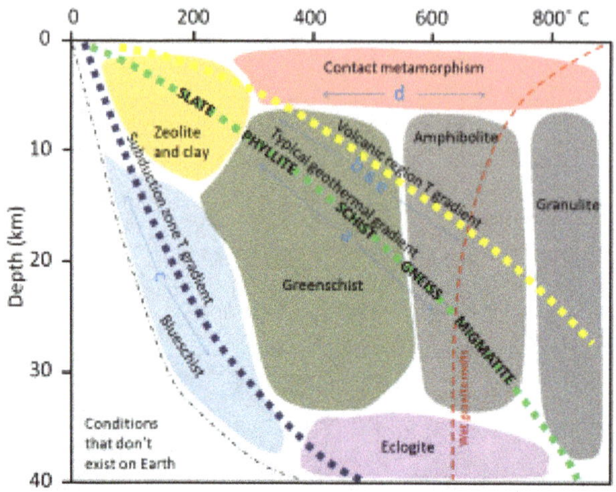

Figure 20: Types of metamorphism shown in the context of depth and temperature under different conditions. The metamorphic rocks formed from mudrock under regional metamorphosis with a typical geothermal gradient are listed. The letters a through e correspond with those shown in Figures 14 to 17 and 19. [SE]

By way of example, if we look at regional metamorphism in areas with typical geothermal gradients, we can see that burial in the 5 km to 10 km range puts us in the zeoliteZeolites are silicate minerals that typically form during low-grade metamorphism of volcanic rocks. and clay mineral zone (see Figure 20), which is equivalent to the formation of slate. At 10 km to 15 km, we are in the greenschist zone (where chlorite would form in mafic volcanic rock) and very fine micas form in mudrock, to produce phyllite. At 15 km to 20 km, larger micas form to produce schist, and at 20 km to 25 km amphibole, feldspar, and quartz form to produce gneiss. Beyond 25 km depth in this setting, we cross the partial melting line for granite (or gneiss) with water present, and so we can expect migmatite to form.

Exercise 7.2 Metamorphic Rocks in Areas with Higher Geothermal Gradients

Metamorphic Rock Type	Depth (km)
Slate	
Phyllite	
Schist	

Metamorphic Rock Type	Depth (km)
Gneiss	
Migmatite	

Figure 20 shows the types of rock that might form from mudrock at various points along the curve of the "typical" geothermal gradient (dotted green line). Looking at the geothermal gradient for volcanic regions (dotted yellow line in Figure 20), estimate the depths at which you would expect to find the same types of rock forming from a mudrock parent.

7.4 REGIONAL METAMORPHISM

Regional metamorphism occurs when rocks are buried deep in the crust. This is commonly associated with convergent plate boundaries and the formation of mountain ranges. Because burial to 10 km to 20 km is required, the areas affected tend to be large.

Rather than focusing on metamorphic rock textures (slate, schist, gneiss, etc.), geologists tend to look at specific minerals within the rocks that are indicative of different grades of metamorphism. Some common minerals in metamorphic rocks are shown in Figure 21, arranged in order of the temperature ranges within which they tend to be stable. The upper and lower limits of the ranges are intentionally vague because these limits depend on a number of different factors, such as the pressure, the amount of water present, and the overall composition of the rock.

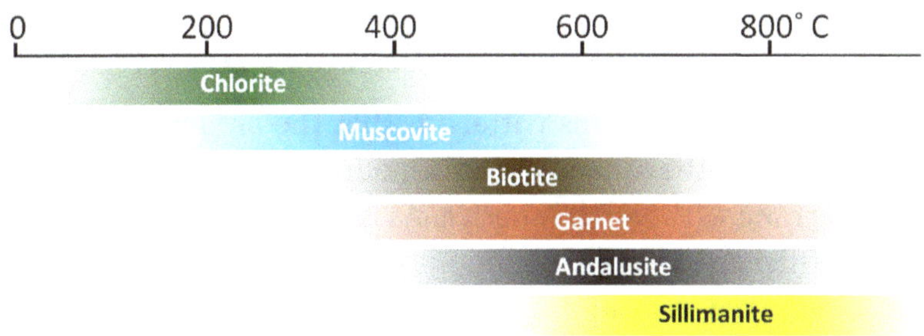

Figure 21: Metamorphic index minerals and their approximate temperature ranges [SE]

The southern and southwestern parts of Nova Scotia were regionally metamorphosed during the Devonian Acadian Orogeny (around 400 Ma), when a relatively small continental block (the Meguma TerraneNo, it's not a spelling mistake! A terrane is a distinctive block of crust that is now part of a continent, but is thought to have come from elsewhere, and was added on by plate-tectonic processes.) was pushed up against the existing eastern margin of North America. As shown in Figure 22, clastic sedimentary rocks within this terrane were variably metamorphosed, with the strongest metamorphism in the southwest (the sillimanite zone), and progressively weaker metamorphism toward the east and north. The rocks of the sillimanite zone were likely heated to over 700°C, and therefore must have buried to depths between 20 km and 25 km. The surrounding lower-grade rocks were not buried as deep, and the rocks within the peripheral chlorite zone were likely not buried to more than about 5 km.

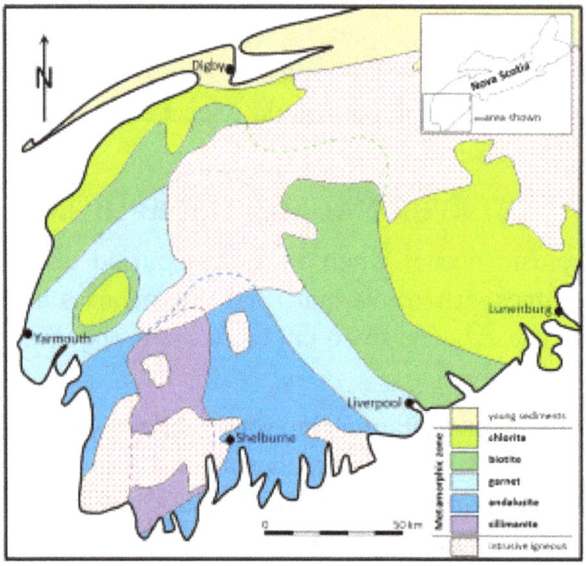

Figure 22: Regional metamorphic zones in the Meguma Terrane of southwestern Nova Scotia [SE, after Keppie, D, and Muecke, G, 1979, Metamorphic map of Nova Scotia, N.S. Dept. of Mines and Energy, Map 1979-006., and from White, C and Barr, S., 2012, Meguma Terrane revisted, Stratigraphy, metamorphism, paleontology and provenance, Geoscience Canada, V. 39, No.1]

A probable explanation for this pattern is that the area with the highest-grade rocks was buried beneath the central part of a mountain range formed by the collision of the Meguma Terrane with North America. As is the case with all mountain ranges, the crust became thickened as the mountains grew, and it was pushed farther down into the mantle than the surrounding crust. This happens because Earth's crust is floating on the underlying mantle. As the formation of mountains adds weight, the crust in that area sinks farther down into the mantle to compensate for the added weight. The likely pattern of metamorphism in this situation is shown in cross-section in Figure 23a. The mountains were eventually eroded (over tens of millions of years), allowing the crust to rebound upward and exposing the metamorphic rock (Figure 23b).

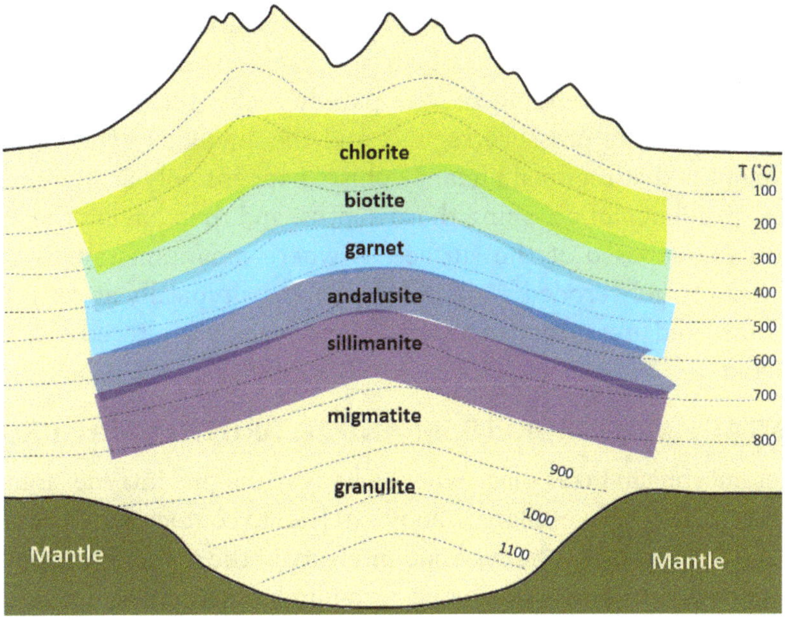

Figure 23: (a) Schematic cross-section through the Meguma Terrane during the Devonian. The crust is thickened underneath the mountain range to compensate for the added weight of the mountains above. Temperature contours are shown, and the metamorphic zones are depicted using colours similar to those in Figure 22.

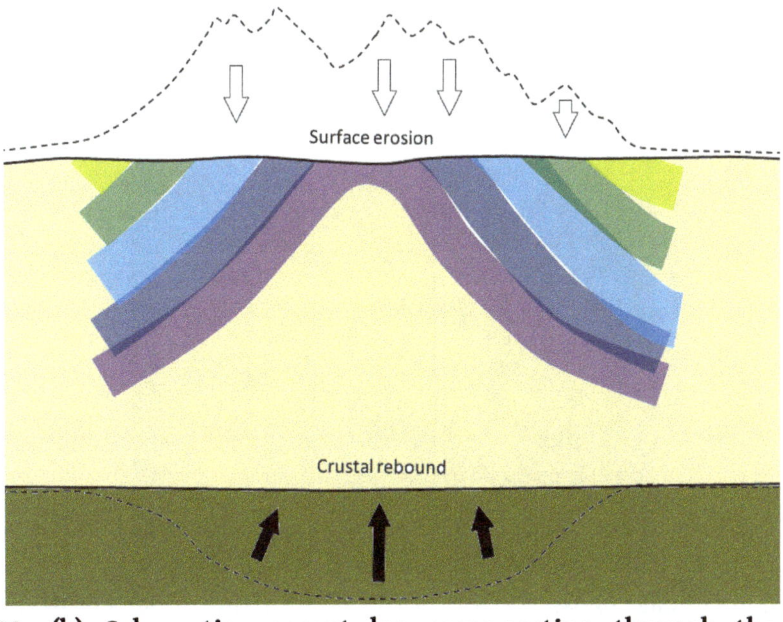

Figure 23: (b) Schematic present-day cross-section through the Meguma Terrane. The mountains have been eroded. As they lost mass the base of the crust gradually rebounded, pushing up the core of the metamorphosed region so that the once deeply buried metamorphic zones are now exposed at surface.

The metamorphism in Nova Scotia's Meguma Terrane is just one example of the nature of regional metamorphism. Obviously many different patterns of regional metamorphism exist, depending on the parent rocks, the geothermal gradient, the depth of burial, the pressure regime, and the amount of time available. The important point is that regional metamorphism happens only at significant depths. The greatest likelihood of attaining those depths, and then having the once-buried rocks eventually exposed at the surface, is where mountain ranges existed and have since been largely eroded away. As this happens typically at convergent plate boundaries, directed pressures can be strong, and regionally altered rocks are almost always foliated.

7.5 CONTACT METAMORPHISM AND HYDROTHERMAL PROCESSES

Contact metamorphism takes place where a body of magma intrudes into the upper part of the crust. Any type of magma body can lead to contact metamorphism, from a thin dyke to a large stock. The type and intensity of the metamorphism, and width of the metamorphic aureole will depend on a number of factors, including the type of country rock, the temperature of the intruding body and the size of the body (Figure 24). A large intrusion will contain more thermal energy and will cool much more slowly than a small one, and therefore will provide a longer time and more heat for metamorphism. That will allow the heat to extend farther into the country rock, creating a larger aureole.

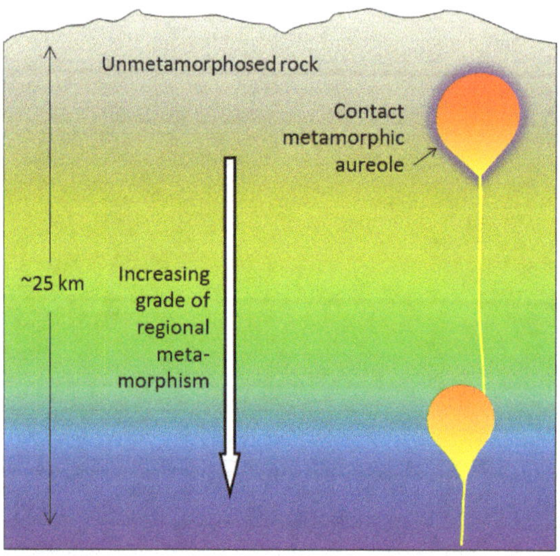

Figure 24: Schematic cross-section of the middle and upper crust showing two magma bodies. The upper body, which has intruded into cool unmetamorphosed rock, has created a zone of contact metamorphism. The lower body is surrounded by rock that is already hot (and probably already metamorphosed), and so it does not have a significant metamorphic aureole. [SE]

Contact metamorphic aureoles are typically quite small, from just a few centimetres around small dykes and sills, to as much as 100 m around a large stock. As was shown in Figure 20, contact metamorphism can take place over a wide range of temperatures — from around 300° to over 800°C — and of course the type of metamorphism, and new minerals formed, will vary accordingly. The nature of the country rock is also important. Mudrock or volcanic rock will be converted to hornfels. Limestone will be metamorphosed to marble, and sandstone to quartzite.

A hot body of magma in the upper crust can create a very dynamic situation that may have geologically interesting and economically important implications. In the simplest cases, water does not play a big role, and the main process is transfer of heat from the pluton to the surrounding rock, creating a zone of contact metamorphism (Figure 25a). In many cases, however, water is released from the magma body as crystallization takes place, and this water is dispersed along fractures in the country rock (Figure 25b). The water released from a magma chamber is typically rich in dissolved minerals. As this water cools, is chemically changed by the surrounding rocks, or boils because of a drop in pressure, minerals are deposited, forming veins within the fractures in the country rock. Quartz veins are common in this situation, and they might also include pyrite, hematite, calcite, and even silver and gold.

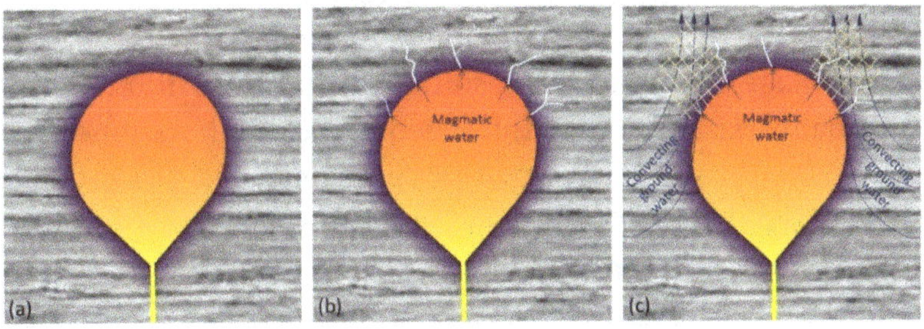

Figure 25: Depiction of metamorphism and alteration around a pluton in the upper crust. (a) Thermal metamorphism only (within the purple zone) (b) Thermal metamorphism plus veining (white) related to dispersal of magmatic fluids into the overlying rock (c) Thermal metamorphism plus veining from magmatic fluids plus alteration and possible formation of metallic minerals (hatched yellow areas) from convection of groundwater

Heat from the magma body will cause surrounding groundwater to expand and then rise toward the surface. In some cases, this may initiate a convection system where groundwater circulates past the pluton. Such a system could operate for thousands of years, resulting in the circulation of millions of tonnes of groundwater from the surrounding region past the pluton. Hot water circulating through the rocks can lead to significant changes in the mineralogy of the rock, including alteration of feldspars to clays, and deposition of quartz, calcite, and other minerals in fractures and other open spaces (Figure 26). As with the magmatic fluids, the nature of this circulating groundwater can also change adjacent to, or above, the pluton, resulting in deposition of other minerals, including ore minerals. Metamorphism in which much of the change is derived from fluids passing through the rock is known as metasomatism. When hot water contributes to changes in rocks, including mineral alteration and formation of veins, it is known as hydrothermal alteration.

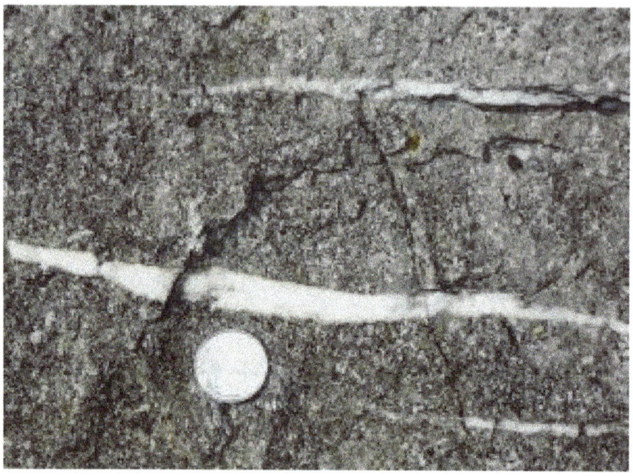

Figure 26: Calcite veins in limestone of the Comox Formation, Nanaimo, B.C. [SE]

A special type of metasomatism takes place where a hot pluton intrudes into carbonate rock such as limestone. When magmatic fluids rich in silica, calcium, magnesium, iron, and other elements flow through the carbonate rock, their chemistry can change dramatically, resulting in the deposition of minerals that would not normally exist in either the igneous rock or limestone. These include garnet, epidote (another silicate), magnetite, pyroxene, and a variety of copper and other minerals (Figure 27). This type of metamorphism is known as skarn.

Figure 27: A skarn rock from Mount Monzoni, Northern Italy, with recrystallized calcite (blue), garnet (brown), and pyroxene (green). The rock is 6 cm across.

[by Siim Sepp, from http://commons.wikimedia.org/wiki/File:00031_6_cm_grossular_calcite_augite_skarn.jpg]

AT A GLANCE

Controls over Metamorphic Processes	Metamorphism is controlled by five main factors: the composition of the parent rock, the temperature to which the rock is heated, the amount and type of pressure, the volumes and compositions of aqueous fluids that are present, and the amount of time available for metamorphic reactions to take place.
Classification of Metamorphic Rocks	Metamorphic rocks are classified on the basis of texture and mineral composition. Foliation is a key feature of metamorphic rocks formed under directed pressure; foliated metamorphic rocks include slate, phyllite, schist, and gneiss. Metamorphic rocks formed in environments without strong directed pressure include hornfels, marble, and quartzite.
Plate Tectonics and Metamorphism	Almost all metamorphism can be explained by plate-tectonic processes. Oceanic crustal rock can be metamorphosed near the spreading ridge where it was formed, but most other regional metamorphism takes place in areas where mountain ranges have formed, which are most common at convergent boundaries. Contact metamorphism takes place around magma bodies in the upper part of the crust, which are also most common above convergent boundaries.
Regional Metamorphism	Geologists classify metamorphic rocks based on some key minerals — such as chlorite, garnet, andalusite, and sillimanite — that only form at specific temperatures and pressures. Most regional metamorphism takes place beneath mountain ranges because the crust becomes thickened and rocks are pushed down to great depths. When mountains erode, those metamorphic rocks are uplifted by crustal rebound.
Contact Metamorphism and Hydrothermal Processes	Contact metamorphism takes place around magma bodies that have intruded into cool rocks at high levels in the crust. Heat from the magma is transferred to the surrounding country rock, resulting in mineralogical and textural changes. Water from a cooling body of magma, or from convection of groundwater produced by the heat of the pluton, can also lead to metasomatism, hydrothermal alteration, and accumulation of valuable minerals in the surrounding rocks.

Two things that a geologist first considers when looking at a metamorphic rock are what the parent rock might have been, and what type of metamorphism has taken place. This can be difficult to do, even if you have the actual rock in your hand, but give it a try for the following:

Metamorphic Rock	Likely Parent Rock	Grade and/or Type of Metamorphism
Chlorite schist		
Slate		
Mica-garnet schist		
Amphibolite		
Marble		

CHAPTER

GEOLOGICAL TIME SCALE 8

Time is the dimension that sets geology apart from most other sciences. Geological time is vast, and Earth has changed enough over that time that some of the rock types that formed in the past could not form today. Furthermore, as we've discussed, even though most geological processes are very, very slow, the vast amount of time that has passed has allowed for the formation of extraordinary geological features, as shown in Figure 1.

Figure 1: Arizona's Grand Canyon is an icon for geological time; 1,450 million years are represented by this photo. The light-coloured layered rocks at the top formed at around 250 Ma, and the dark ones at the bottom (within the steep canyon) at around 1,700 Ma. [SE]

We have numerous ways of measuring geological time. We can tell the relative ages of rocks (for example, whether one rock is older than another) based on their spatial relationships; we can use fossils to date sedimentary rocks because we have a detailed record of the evolution of life on Earth; and we can use a range of isotopic techniques to determine the actual ages (in millions of years) of igneous and metamorphic rocks.

But just because we can measure geological time doesn't mean that we understand it. One of the biggest hurdles faced by geology students, and geologists as well, in understanding geology, is to really come to grips with the slow rates at which geological processes happen and the vast amount of time involved.

8.1 THE GEOLOGICAL TIME SCALE

William "Strata" Smith worked as a surveyor in the coal-mining and canal-building industries in southwestern England in the late 1700s and early 1800s. While doing his work, he had many opportunities to look at the Paleozoic and Mesozoic sedimentary rocks of the region, and he did so in a way that few had done before. Smith noticed the textural similarities and differences between rocks in different locations, and more importantly, he discovered that fossils could be used to correlate rocks of the same age. Smith is credited with formulating the principle of faunal succession (the concept that specific types of organisms lived during different time intervals), and he used it to great effect in his monumental project to create a geological map of England and Wales, published in 1815.

Inset into Smith's great geological map is a small diagram showing a schematic geological cross-section extending from the Thames estuary of eastern England all the way to the west coast of Wales. Smith shows the sequence of rocks, from the Paleozoic rocks of Wales and western England, through the Mesozoic rocks of central England, to the Cenozoic rocks of the area around London (Figure 2). Although Smith did not put any dates on these — because he didn't know them — he was aware of the principle of superposition (the idea, developed much earlier by the Danish theologian and scientist Nicholas Steno, that young sedimentary rocks form on top of older ones), and so he knew that this diagram represented a stratigraphic column. And because almost every period of the Phanerozoic is represented along that section through Wales and England, it is a primitive geological time scale.

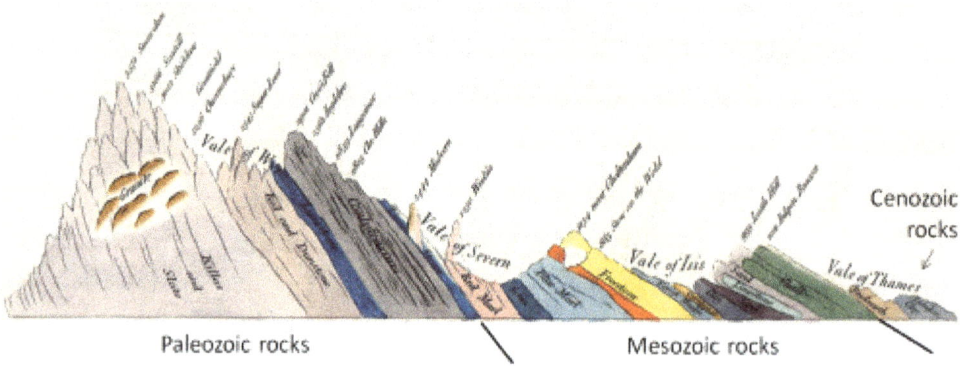

Figure 2: William Smith's "Sketch of the succession of strata and their relative altitudes," an inset on his geological map of England and Wales (with era names added). [SE after: http://earthobservatory.nasa.gov/Features/WilliamSmith/images/sketch_of_the_succession_of_strata.jpg]

Smith's work set the stage for the naming and ordering of the geological periods, which was initiated around 1820, first by British geologists, and later by other European geologists. Many of the periods are named for places where rocks of that age are found in Europe, such as Cambrian for Cambria (Wales), Devonian for Devon in England, Jurassic for the Jura Mountains in France and Switzerland, and Permian for the Perm region of Russia. Some are named for the type of rock that is common during that age, such as Carboniferous for the coal- and carbonate-bearing rocks of England, and Cretaceous for the chalks of England and France.

The early time scales were only relative because 19th century geologists did not know the ages of the rocks. That information was not available until the development of isotopic dating techniques early in the 20th century.

The geological time scale is currently maintained by the International Commission on Stratigraphy (ICS), which is part of the International Union of Geological Sciences. The time scale is continuously being updated as we learn more about the timing and nature of past geological events. You can view the ICS time scale at http://www.stratigraphy.org/index.php/ics-chart-timescale. It would be a good idea to print a copy (in colour) to put on your wall while you are studying geology.

Geological time has been divided into four eons: Hadean, Archean, Proterozoic, and Phanerozoic, and as shown in Figure 3, the first three of these represent almost 90% of Earth's history. The last one, the Phanerozoic (meaning "visible life"), is the time that we are most familiar with because Phanerozoic rocks are the most common on Earth, and they contain evidence of the life forms that we are all somewhat familiar with.

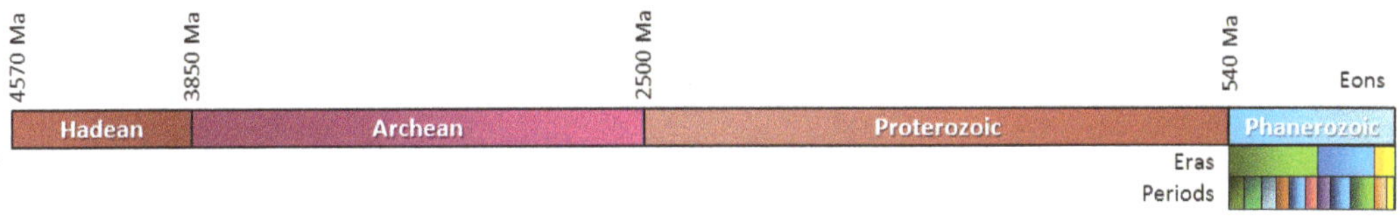

Figure 3: The eons of Earth's history [SE]

The Phanerozoic — the past 540 Ma of Earth's history — is divided into three eras: the Paleozoic ("early life"), the Mesozoic ("middle life"), and the Cenozoic ("new life"), and each of these is divided into a number of periods (Figure 4). Most of the organisms that we share Earth with evolved at various times during the Phanerozoic.

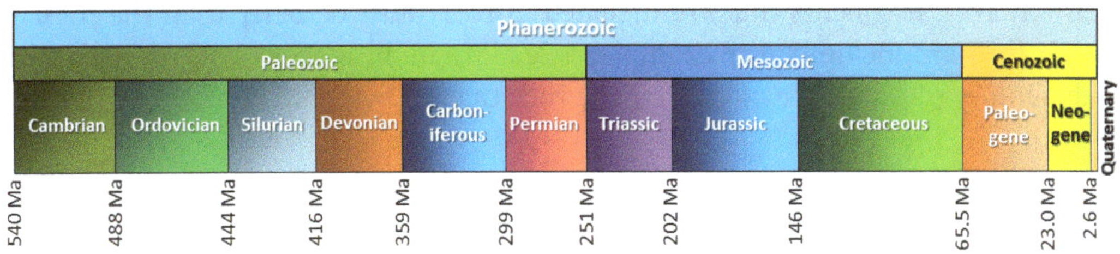

Figure 4: The eras (middle row) and periods (bottom row) of the Phanerozoic [SE]

The Cenozoic, which represents the past 65.5 Ma, is divided into three periods: Paleogene, Neogene, and Quaternary, and seven epochs (Figure 5). Dinosaurs became extinct at the start of the Cenozoic, after which birds and mammals radiated to fill the available habitats. Earth was very warm during the early Eocene and has steadily cooled ever since. Glaciers first appeared on Antarctica in the Oligocene and then on Greenland in the Miocene, and covered much of North America and Europe by the Pleistocene. The most recent of the Pleistocene glaciations ended around 11,700 years ago. The current epoch is known as the Holocene. Epochs are further divided into ages (a.k.a. stages), but we won't be going into that level of detail here.

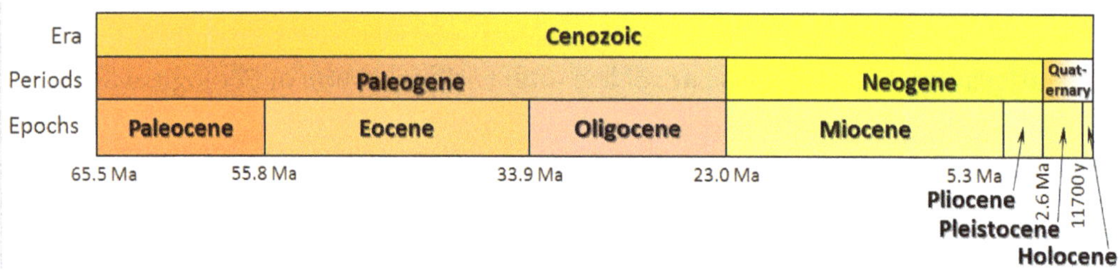

Figure 5: The periods (middle row) and epochs (bottom row) of the Cenozoic [SE]

Most of the boundaries between the periods and epochs of the geological time scale have been fixed on the basis of significant changes in the fossil record. For example, as already noted, the boundary between the Cretaceous and the Paleogene coincides exactly with the extinction of the dinosaurs. That's not a coincidence. Many other types of organisms went extinct at this time, and the boundary between the two periods marks the division between sedimentary rocks with Cretaceous organisms below, and Paleogene organisms above.

8.2 METHODS OF RELATIVE DATING

The simplest and most intuitive way of dating geological features is to look at the relationships between them. There are a few simple rules for doing this, some of which we've already looked at in Chapter 6. For example, the principle of superposition states that sedimentary layers are deposited in sequence, and, unless the entire sequence has been turned over by tectonic processes or disrupted by faulting, the layers at the bottom are older than those at the top. The principle of inclusions states that any rock fragments that are included in rock must be older than the rock in which they are included. For example, a xenolith in an igneous rock or a clast in sedimentary rock must be older than the rock that includes it (Figure 6).

Figure 6a: A xenolith of diorite incorporated into a basalt lava flow, Mauna Kea volcano, Hawaii. The lava flow took place some time after the diorite cooled, was uplifted, and then eroded. (Hammerhead for scale) [SE]

Figure 6b: Rip-up clasts of shale embedded in Gabriola Formation sandstone, Gabriola Island, B.C. The pieces of shale were eroded as the sandstone was deposited, so the shale is older than the sandstone. [SE]

The principle of cross-cutting relationships states that any geological feature that cuts across, or disrupts another feature must be younger than the feature that is disrupted. An example of this is given in Figure 7, which shows three different sedimentary layers. The lower sandstone layer is disrupted by two faults, so we can infer that the faults are younger than that layer. But the faults do not appear to continue into the coal seam, and they certainly do not continue into the upper sandstone. So we can infer that coal seam is younger than the faults (because it disrupts them), and of course the upper sandstone is youngest of all, because it lies on top of the coal seam.

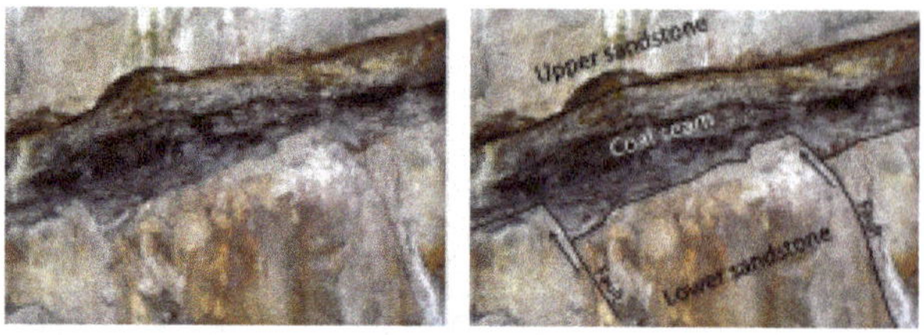

Figure 7: Superposition and cross-cutting relationships in Cretaceous Nanaimo Group rocks in Nanaimo, B.C. The coal seam is about 50 cm thick. [SE]

8.3 USING FOSSILS TO DATE ROCKS

Geologists get a wide range of information from fossils. They help us to understand evolution and life in general; they provide critical information for understanding depositional environments and changes in Earth's climate; and, of course, they can be used to date rocks.

Although the recognition of fossils goes back hundreds of years, the systematic cataloguing and assignment of relative ages to different organisms from the distant past — paleontology — only dates back to the earliest part of the 19th century. The oldest undisputed fossils are from rocks dated around 3.5 Ga, and although fossils this old are typically poorly preserved and are not useful for dating rocks, they can still provide important information about conditions at the time. The oldest well-understood fossils are from rocks dating back to around 600 Ma, and the sedimentary record from that time forward is rich in fossil remains that provide a detailed record of the history of life. However, as anyone who has gone hunting for fossils knows, that does not mean that all sedimentary rocks have visible fossils or that they are easy to find. Fossils alone cannot provide us with numerical ages of rocks, but over the past century geologists have acquired enough isotopic dates from rocks associated with fossil-bearing rocks (such as igneous dykes cutting through sedimentary layers) to be able to put specific time limits on most fossils.

A very selective history of life on Earth over the past 600 million years is provided in Figure 8.10. The major groups of organisms that we are familiar with evolved between the late Proterozoic and the Cambrian (~600 Ma to ~520 Ma). Plants, which evolved in the oceans as green algae, came onto land during the Ordovician (~450 Ma). Insects, which evolved from marine arthropods, came onto land during the Devonian (400 Ma), and amphibians (i.e., vertebrates) came onto land about 50 million years later. By the late Carboniferous, trees had evolved from earlier plants, and reptiles had evolved from amphibians. By the mid-Triassic, dinosaurs and mammals had evolved from very different branches of the reptiles; birds evolved from dinosaurs during the Jurassic. Flowering plants evolved in the late Jurassic or early Cretaceous. The earliest primates evolved from other mammals in the early Paleogene, and the genus *Homo* evolved during the late Neogene (~2.8 Ma).

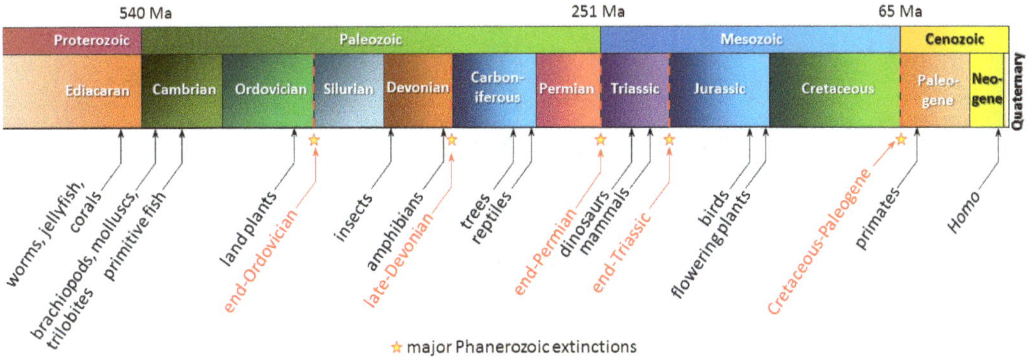

Figure 10: A summary of life on Earth during the late Proterozoic and the Phanerozoic. The top row shows geological eras, and the lower row shows the periods. [SE]

If we understand the sequence of evolution on Earth, we can apply knowledge to determining the relative ages of rocks. This is William Smith's principle of faunal succession, although of course it doesn't just apply to "fauna" (animals); it can also apply to fossils of plants and those of simple organisms.

The Phanerozoic has seen five major extinctions, as indicated in Figure 8.10. The most significant of these was at the end of the Permian, which saw the extinction of over 80% of all species and over 90% of all marine species. Most well-known types of organisms were decimated by this event, but only a few became completely extinct, including trilobites. The second most significant extinction was at the Cretaceous-Paleogene boundary (K-Pg, a.k.a. the K-T extinction). At that time, about 75% of marine species disappeared. Again, a few well-known types of organisms disappeared altogether, including dinosaurs (but not birds) and the pterosaurs. Other types were badly decimated but survived, and then flourished in the Paleogene. The K-Pg extinction is thought to have been caused by the impact of a large extraterrestrial body (10 km to 15 km across), but it is generally agreed that the other four Phanerozoic extinctions had other causes, although their exact nature is not clearly understood.

As already stated, it is no coincidence that the major extinctions all coincide with boundaries of geological periods and even eras. Paleontologists have placed most of the divisions of the geological time scale at points in the fossil record where there are major changes in the type of organisms observed.

If we can identify a fossil to the species level, or at least to the genus level, and we know the time period when the organism lived, we can assign a range of time to the rock. That range might be several million years because some organisms survived for a very long time. If the rock we are studying has several types of fossils in it, and we can assign time ranges to those fossils, we might be able to narrow the time range for the age of the rock considerably. An example of this is given in Figure 11.

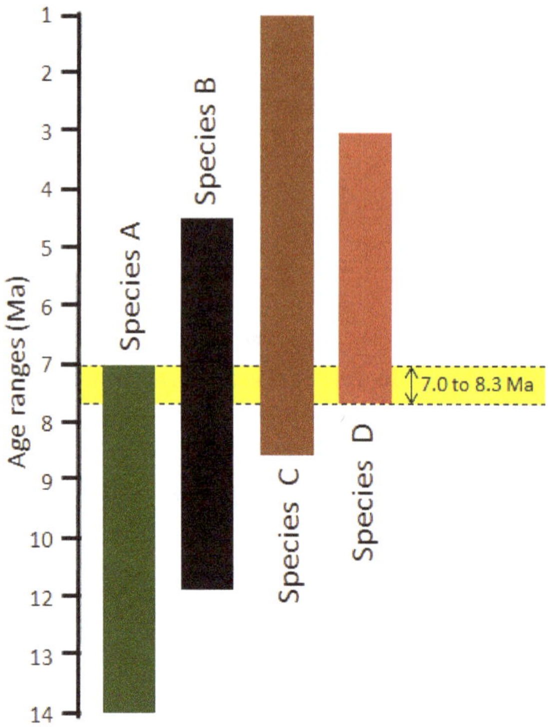

Figure 11: The application of bracketing to constrain the age of a rock based on several fossils. In this diagram, the coloured bar represents the time range during which each of the four species (A – D) existed on Earth. Although each species lived for several million years, we can narrow down the likely age of the rock to just 0.7 Ma during which all four species coexisted. [SE]

Some organisms survived for a very long time, and are not particularly useful for dating rocks. Sharks, for example, have been around for over 400 million years, and the great white shark has survived for 16 million years, so far. Organisms that lived for relatively short time periods are particularly useful for dating rocks, especially if they were distributed over a wide geographic area and so can be used to compare rocks from different regions. These are known as index fossils. There is no specific limit on how short the time span has to be to qualify as an index fossil. Some lived for millions of years, and others for much less than a million years.

Some well-studied groups of organisms qualify as biozone fossils because, although the genera and families lived over a long time, each species lived for a relatively short time and can be easily distinguished from others on the basis of specific features. For example, ammonites have a distinctive feature known as the suture line — where the internal shell layers that separate the individual chambers (septae) meet the outer shell wall, as shown in Figure 12. These suture lines are sufficiently variable to identify species that can be used to estimate the relative or absolute ages of the rocks in which they are found.

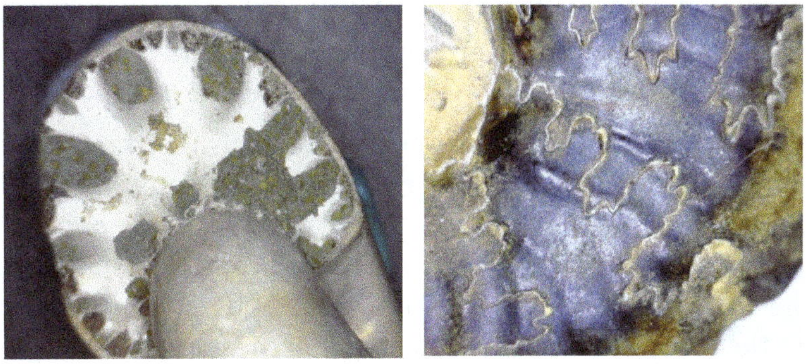

Figure 12: The septum of an ammonite (white part, left), and the suture lines where the septae meet the outer shell (right). [SE]

Foraminifera (small, carbonate-shelled marine organisms that originated during the Triassic and are still around today) are also useful biozone fossils. As shown in Figure 13, numerous different foraminifera lived during the Cretaceous. Some lasted for over 10 million years, but others for less than 1 million years. If the foraminifera in a rock can be identified to the species level, we can get a good idea of its age.

Time ranges of selected Cretaceous foraminifera species

Modern foraminifera from Belize (most are ~1 mm across)

Figure 13: Time ranges for Cretaceous foraminifera (left) and modern foraminifera from the Ambergris area of Belize (right) [left: SE, from data in Scott, R, 2014, A Cretaceous chronostratigraphic database: construction and applications, Carnets de Géologie, Vol. 14., right : SE]

8.4 METHODS FOR ISOTOPIC DATING

Originally fossils only provided us with relative ages because, although early paleontologists understood biological succession, they did not know the absolute ages of the different organisms. It was only in the early part of the 20th century, when isotopic dating methods were first applied, that it became possible to discover the absolute ages of the rocks containing fossils. In most cases, we cannot use isotopic techniques to directly date fossils or the sedimentary rocks they are found in, but we can constrain their ages by dating igneous rocks that cut across sedimentary rocks, or volcanic ash layers that lie within sedimentary layers.

Isotopic dating of rocks, or the minerals in them, is based on the fact that we know the decay rates of certain unstable isotopes of elements and that these rates have been constant over geological time. It is also based on the premise that when the atoms of an element decay within a mineral or a rock, they stay there and don't escape to the surrounding rock, water, or air. One of the isotope pairs widely used in geology is the decay of ^{40}K to ^{40}Ar (potassium-40 to argon-40). ^{40}K is a radioactive isotope of potassium that is present in very small amounts in all minerals that have potassium in them. It has a half-life of 1.3 billion years, meaning that over a period of 1.3 Ga one-half of the ^{40}K atoms in a mineral or rock will decay to ^{40}Ar, and over the next 1.3 Ga one-half of the remaining atoms will decay, and so on (Figure 14).

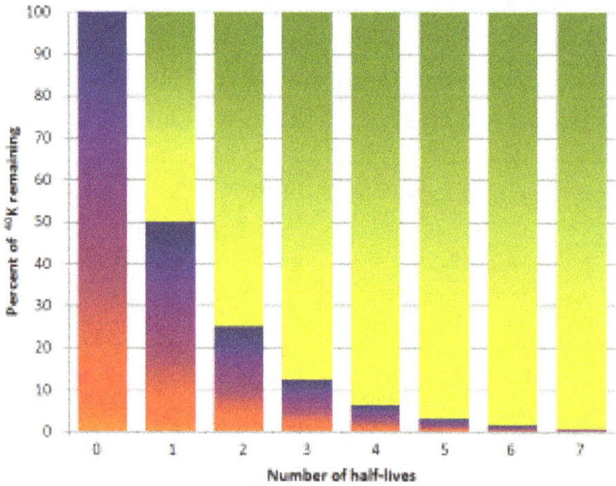

Figure 14: The decay of 40K over time. Each half-life is 1.3 billion years, so after 3.9 billion years (three half-lives) 12.5% of the original 40K will remain. The red-blue bars represent 40K and the green-yellow bars represent 40Ar. [SE]

In order to use the K-Ar dating technique, we need to have an igneous or metamorphic rock that includes a potassium-bearing mineral. One good example is granite, which normally has some potassium feldspar (Figure 15). Feldspar does not have any argon in it when it forms. Over time, the ^{40}K in the feldspar decays to ^{40}Ar. Argon is a gas and the atoms of ^{40}Ar remain embedded within the crystal, unless the rock is subjected to high temperatures after it forms. The sample must be analyzed using a very sensitive mass-spectrometer, which can detect the differences between the masses of atoms, and can therefore distinguish between ^{40}K and the much more abundant ^{39}K. Biotite and hornblende are also commonly used for K-Ar dating.

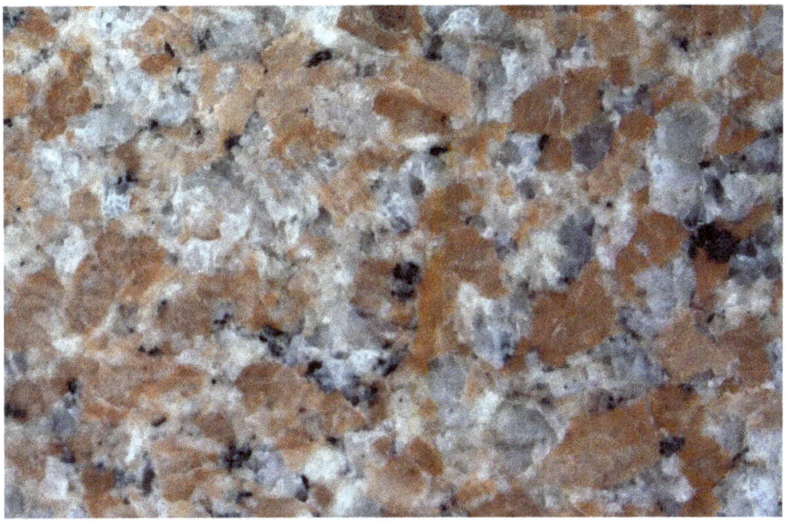

Figure 15: Crystals of potassium feldspar (pink) in a granitic rock are candidates for isotopic dating using the K-Ar method because they contained potassium and no argon when they formed. [SE]

Why can't we use isotopic dating techniques with sedimentary rocks?

An important assumption that we have to be able to make when using isotopic dating is that when the rock formed none of the daughter isotope was present (e.g., ^{40}Ar in the case of the K-Ar method). A clastic sedimentary rock is made up of older rock and mineral fragments, and when the rock forms it is almost certain that all of the fragments already have daughter isotopes in them. Furthermore, in almost all cases, the fragments have come from a range of source rocks that all formed at different times. If we dated a number of individual grains in the sedimentary rock, we would likely get a range of different dates, all older than the age of the rock. It might be possible to date some chemical sedimentary rocks isotopically, but there are no useful isotopes that can be used on old chemical sedimentary rocks. Radiocarbon dating can be used on sediments or sedimentary rocks that contain carbon, but it cannot be used on materials older than about 60 ka.

Exercise 8.2 Isotopic Dating

Assume that a feldspar crystal from the granite shown in Figure 15 was analyzed for ^{40}K and ^{40}Ar. The proportion of ^{40}K remaining is 0.91. Using the decay curve shown on this graph, estimate the age of the rock.

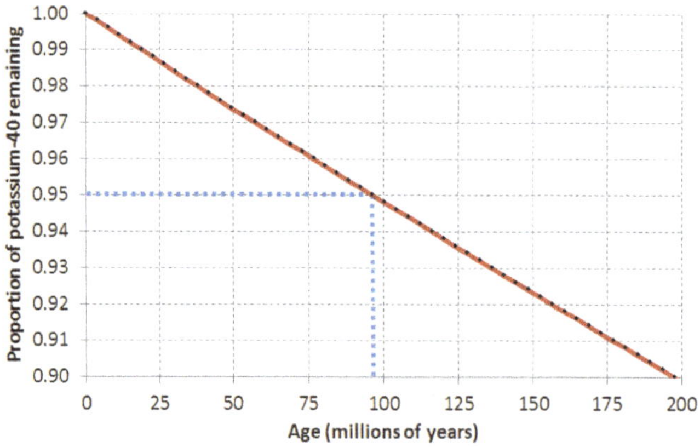

An example is provided (in blue) for a ^{40}K proportion of 0.95, which is equivalent to an age of approximately 96 Ma. This is determined by drawing a horizontal line from 0.95 to the decay curve line, and then a vertical line from there to the time axis. [SE]

K-Ar is just one of many isotope-pairs that are useful for dating geological materials. Some of the other important pairs are listed in Table 2, along with the age ranges that they apply to and some comments on their applications. When radiometric techniques are applied to metamorphic rocks, the results normally tell us the date of metamorphism, not the date when the parent rock formed.

Isotope System	Half-Life	Useful Range	Comments
Potassium-argon	1.3 Ga	10 Ka – 4.57 Ga	Widely applicable because most rocks have some potassium
Uranium-lead	4.5 Ga	1 Ma – 4.57 Ga	The rock must have uranium-bearing minerals
Rubidium-strontium	47 Ga	10 Ma – 4.57 Ga	Less precision than other methods at old dates
Carbon-nitrogen (a.k.a. radiocarbon dating)	5,730 y	100 y to 60,000 y	Sample must contain wood, bone, or carbonate minerals; can be applied to young sediments

Radiocarbon dating (using ^{14}C) can be applied to many geological materials, including sediments and sedimentary rocks, but the materials in question must be younger than 60 ka. Fragments of wood incorporated into young sediments are good candidates for carbon dating, and this technique has been used widely in studies involving late Pleistocene glaciers and glacial sediments. An example is shown in Figure 16; radiocarbon dates from wood fragments in glacial sediments have been used to estimate the time of the last glacial advance along the Strait of Georgia.

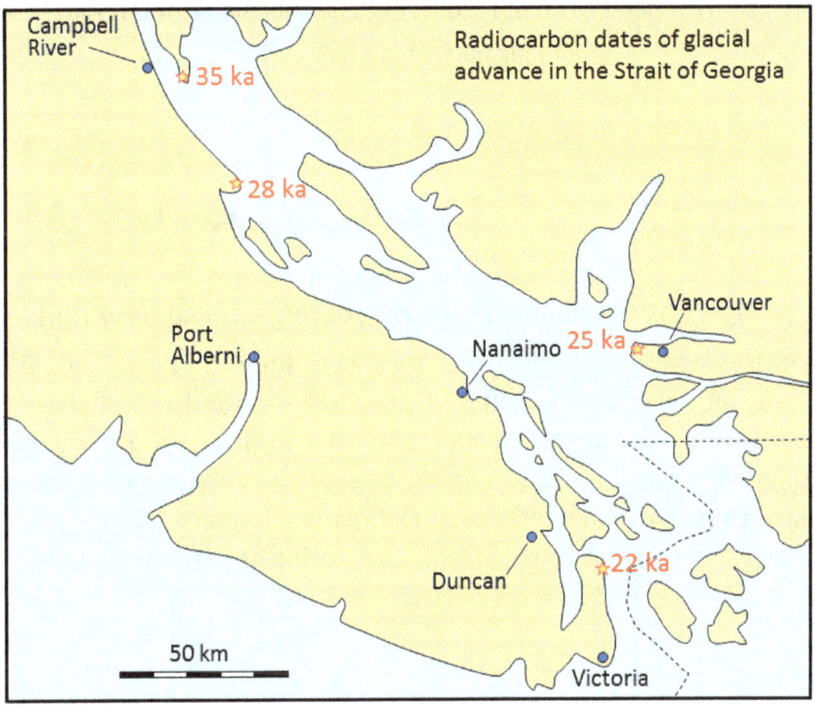

Figure 16: Radiocarbon dates on wood fragments in glacial sediments in the Strait of Georgia [SE after Clague, J, 1976, Quadra Sand and its relation to late Wisconsin glaciation of southeast British Columbia, Can. J. Earth Sciences, V. 13, p. 803-815]

8.5 OTHER DATING TECHNIQUES

There are numerous other techniques for dating geological materials, but we will examine just two of them here: tree-ring dating (i.e., dendrochronology) and dating based on the record of reversals of Earth's magnetic field.

Dendrochronology can be applied to dating very young geological materials based on reference records of tree-ring growth going back many millennia. The longest such records can take us back over 25 ka, to the height of the last glaciation. One of the advantages of dendrochronology is that, providing reliable reference records are available, the technique can be used to date events to the nearest year.

Dendrochronology has been used to date the last major subduction zone earthquake on the coast of B.C., Washington, and Oregon. When large earthquakes strike in this setting, there is a tendency for some coastal areas to subside by one or two metres. Seawater then rushes in, flooding coastal flats and killing trees and other vegetation within a few months. There are at least four locations along the coast of Washington that have such dead trees (and probably many more in other areas). Wood samples from these trees have been studied and the ring patterns have been compared with patterns from old living trees in the region (Figure 17).

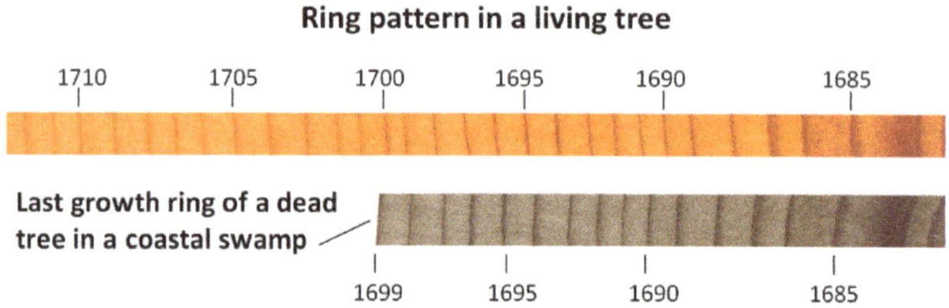

Figure 17: Example of tree-ring dating of dead trees [SE]

At all of the locations studied, the trees were found to have died either in the year 1699, or very shortly thereafter (Figure 18). On the basis of these results, it was concluded that a major earthquake took place in this region sometime between the end of growing season in 1699 and the beginning of the growing season in 1700. Evidence from a major tsunami that struck Japan on January 27, 1700, narrowed the timing of the earthquake to sometime in the evening of January 26, 1700.

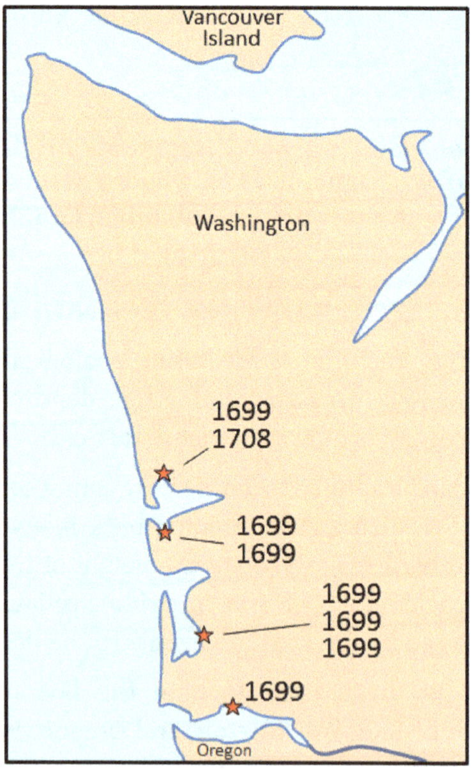

Figure 18: Sites in Washington where dead trees are present in coastal flats. The outermost wood of eight trees was dated using dendrochronology, and of these,

seven died during the year 1699, suggesting that the land was inundated by water at that time. [SE from data in Yamaguchi, D.K., B.F. Atwater, D.E. Bunker, B.E. Benson, and M.S. Reid. 1997. Tree-ring dating the 1700 Cascadia earthquake. Nature, Vol. 389, pp. 922 – 923, 30 October 1997.]

Changes in Earth's magnetic field can also be used to date events in geologic history. The magnetic field makes compasses point toward the North Pole, but, as we'll see in Chapter 10, this hasn't always been the case. At various times in the past, Earth's magnetic field has reversed itself completely, and during those times a compass would have pointed to the South Pole. By studying magnetism in volcanic rocks that have been dated isotopically, geologists have been able to delineate the chronology of magnetic field reversals going back for some 250 Ma. About 5 Ma of this record is shown in Figure 19, where the black bands represent periods of normal magnetism ("normal" meaning similar to the current magnetic field) and the white bands represent periods of reversed magnetism. These periods of consistent magnetic polarity are given names to make them easier to reference. The current normal magnetic field, known as Brunhes, has lasted for the past 780,000 years. Prior to that there was a short reversed period and then a short normal period known as Jaramillo.

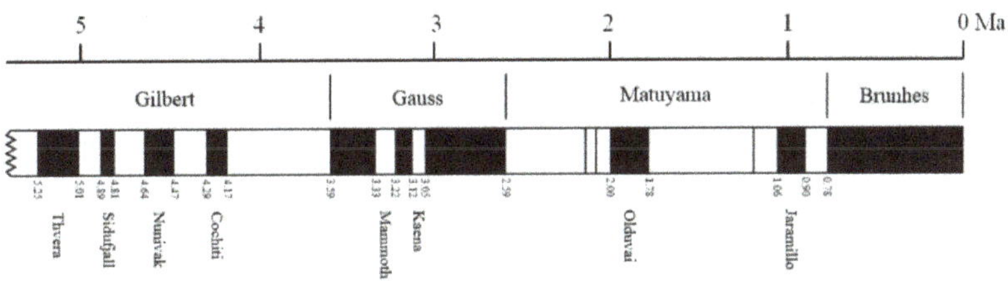

Figure 19: The last 5 Ma of magnetic field reversals. [SE after U.S. Geological Survey, http://upload.wikimedia.org/wikipedia/commons/1/13/Geomagnetic_polarity_late_Cenozoic.svg.]

Oceanic crust becomes magnetized by the magnetic field that exists as the crust forms from magma. As it cools, tiny crystals of magnetite that form within the magma become aligned with the existing magnetic field and then remain that way after all of the rock has hardened, as shown in Figure 20. Crust that is forming today is being magnetized in a "normal" sense, but crust that formed 780,000 to 900,000 years ago, in the interval between the Brunhes and Jaramillo normal periods, was magnetized in the "reversed" sense.

Chapter 9 has a discussion of Earth's magnetic field, including where and how it is generated and why its polarity changes periodically.

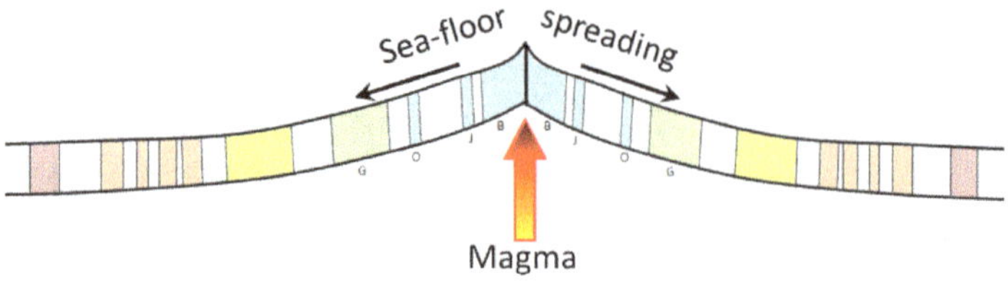

Figure 20: Depiction of the formation of magnetized oceanic crust at a spreading ridge. Coloured bars represent periods of normal magnetism, and the small capital letters denote the Brunhes, Jaramillio, Olduvai, and Gauss normal magnetic periods (see Figure 19). [SE]

Magnetic chronology can be used as a dating technique because we can measure the magnetic field of rocks using a magnetometer in a lab, or of entire regions by towing a magnetometer behind a ship or an airplane. For example, the Juan de Fuca Plate, which lies off of the west coast of B.C., Washington, and Oregon, is being and has been formed along the Juan de Fuca spreading ridge (Figure 21). The parts of the plate that are still close to the ridge have normal magnetism, while parts that are farther away (and formed much earlier) have either normal or reversed magnetism, depending on when the rock formed. By carefully matching the sea-floor magnetic stripes with the known magnetic chronology, we can determine the age at any point on the plate. We can see, for example, that the oldest part of the Juan de Fuca Plate that has not subducted (off the coast of Oregon) is just over 8 million years old, while the part that is subducting underneath Vancouver Island is between 0 and about 6 million years old.

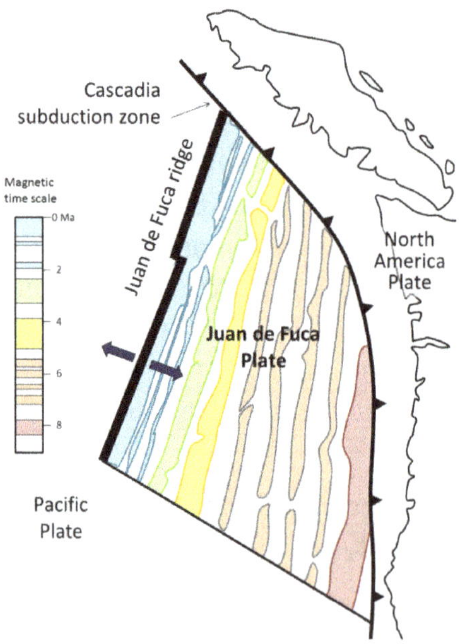

Figure 21: The pattern of magnetism within the area of the Juan de Fuca Plate, off the west coast of North America. The coloured shapes represent parts of the sea

floor that have normal magnetism, and the magnetic time scale is shown using the same colours. The blue bands represent Brunhes, Jaramillo, and Olduvai; the green represents Gauss; and so on. (Note that in this diagram, sea-floor magnetism is only shown for the Juan de Fuca Plate, although similar patterns exist on the Pacific Plate.) [SE]

8.6 UNDERSTANDING GEOLOGICAL TIME

It's one thing to know the facts about geological time — how long it is, how we measure it, how we divide it up, and what we call the various periods and epochs — but it is quite another to really understand geological time. The problem is that our lives are short and our memories are even shorter. Our experiences span only a few decades, so we really don't have a way of knowing what 11,700 years means. What's more, it's hard for us to understand how 11,700 years differs from 65.5 Ma, or even from 1.8 Ga. It's not that we can't comprehend what the numbers mean — we can all get that figured out with a bit of practice — but even if we do know the numerical meaning of 65.5 Ma, we can't really appreciate how long ago it was.

You may be wondering why it's so important to really "understand" geological time. There are some very good reasons. One is so that we can fully understand how geological processes that seem impossibly slow can produce anything of consequence. For example, we are familiar with the concept of driving from one major city to another: a journey of several hours at around 100 km/h. Continents move toward each other at rates of a fraction of a millimetre per day, or something in the order of 0.00000001 km/h, and yet, at this impossibly slow rate (try walking at that speed!), they can move thousands of kilometres. Sediments typically accumulate at even slower rates — less than a millimetre per year — but still they are thick enough to be thrust up into monumental mountains and carved into breathtaking canyons.

Another reason is that for our survival on this planet, we need to understand issues like extinction of endangered species and anthropogenic (human-caused) climate change. Some people, who don't understand geological time, are quick to say that the climate has changed in the past, and that what is happening now is no different. And it certainly has changed in the past. For example, from the Eocene (50 Ma) to the present day, Earth's climate cooled by about 12°C. That's a huge change that ranks up there with many of the important climate changes of the distant past, and yet the rate of change over that time was only 0.000024°C/century. Anthropogenic climate change has been 1.1°C over the past 100 years,Climate change data from NASA Goddard Institute for Space Studies: http://data.giss.nasa.gov/gistemp/tabledata_v3/GLB.Ts.txt and that is 45,800 times faster than the rate of natural climate change since the Eocene!

One way to wrap your mind around geological time is to put it into the perspective of single year, because we all know how long it is from one birthday to the next. At that rate, each hour of the year is equivalent to approximately 500,000 years, and each day is equivalent to 12.5 million years.

If all of geological time is compressed down to a single year, Earth formed on January 1, and the first life forms evolved in late March (~3,500 Ma). The first large life forms appeared on November 13 (~600 Ma), plants appeared on land around November 24, and amphibians on December 3. Reptiles evolved from amphibians during the first week of December and dinosaurs and early mammals evolved from reptiles by December 13, but the dinosaurs, which survived for 160 million years, were gone by Boxing Day (December 26). The Pleistocene Glaciation got started at around 6:30 p.m. on New Year's Eve, and the last glacial ice left southern Canada by 11:59 p.m.

It's worth repeating: on this time scale, the earliest ancestors of the animals and plants with which we are familiar did not appear on Earth until mid-November, the dinosaurs disappeared after Christmas, and most of Canada was periodically locked in ice from 6:30 to 11:59 p.m. on New Year's Eve. As for people, the first to inhabit B.C. got here about one minute before midnight, and the first Europeans arrived about two seconds before midnight.

It is common for the popular press to refer to distant past events as being "prehistoric." For example, dinosaurs are reported as being "prehistoric creatures," even by the esteemed National Geographic Society. The written records of our history date back to about 6,000 years ago, so anything prior to that is considered "prehistoric." But to call the dinosaurs prehistoric is equivalent to — and about as useful as — saying that Singapore is beyond the city limits of Kamloops! If we are going to become literate about geological time, we have to do better than calling dinosaurs, or early horses (54 Ma), or even early humans (2.8 Ma), "prehistoric."

AT A GLANCE

The Geological Time Scale	The work of William Smith was critical to the establishment of the first geological time scale early in the 19th century, but it wasn't until the 20th century that geologists were able to assign reliable dates to the various time periods. The geological time scale is now maintained by the International Commission on Stratigraphy. Geological time is divided into eons, eras, periods, and epochs.
Relative Dating Methods	We can determine the relative ages of different rocks by observing and interpreting relationships among them, such as superposition, cross-cutting, and inclusions. Gaps in the geological record are represented by various types of unconformities.
Dating Rocks Using Fossils	Fossils are useful for dating rocks date back to about 600 Ma. If we know the age range of a fossil, we can date the rock, but some organisms lived for many millions of years. Index fossils represent shorter geological times, and if a rock has several different fossils with known age ranges, we can normally narrow the time during which the rock formed.

Isotopic Dating Methods	Radioactive isotopes decay at predictable and known rates, and can be used to date igneous and metamorphic rocks. Some of the more useful isotope systems are potassium-argon, rubidium-strontium, uranium-lead, and carbon-nitrogen. Radiocarbon dating can be applied to sediments and sedimentary rocks, but only if they are younger than 60 ka.
Other Dating Methods	There are many other methods for dating geological materials. Two that are widely used are dendrochronology and magnetic chronology. Dendrochronology, based on studies of tree rings, is widely applied to dating glacial events. Magnetic chronology is based on the known record of Earth's magnetic field reversals.
Understanding Geological Time	While knowing about geological time is relatively easy, actually comprehending the significance of the vast amounts of geological time is a great challenge. To be able to solve important geological problems and critical societal challenges, like climate change, we need to really understand geological time.

CHAPTER

INTERIOR OF THE EARTH 9

Fortunately there are a few places where mantle rock is exposed on Earth's surface, and we have some samples of material from the insides of other planetary bodies, in the form of meteorites that have landed on Earth (Figure 1). We also have a great deal of seismic information that can help us understand the nature of Earth's interior.

Figure 1: Left: a fragment of the Tagish Lake meteorite, discovered in 2000 on the ice of Tagish Lake, B.C. It is a "stony" meteorite that is dominated by ferromagnesian silicate minerals, and is similar in composition to Earth's mantle. Right: part of the Elbogen meteorite discovered in Germany around 1400. It is an iron meteorite, similar in composition to Earth's core. Both samples are a few centimetres across.

Earth's interior is broadly divided by composition and depth into crust, mantle, and core (Figure 9.2). The crust is primarily (~95%) made up of igneous rock and metamorphic rock with an overall composition between intermediate and felsic. The remaining 5% is made up of sedimentary rock, which is dominated by mudstone.

The mantle includes several layers, all with the same overall ultramafic composition. The upper mantle is typically composed of peridotite, a rock dominated by olivine and pyroxene. The lower mantle has a similar chemical composition, but because of the extreme pressures, different minerals are present, including spinels and garnets. The properties of the mantle also vary with depth, as follows:

- Lithosphere: solid

- Asthenosphere: partially liquid
- Upper and lower mantle: solid but plastic (the difference between the two is in the type of minerals)
- "D" layer (the part of the mantle within 200 km of the core): partially liquid
- The core-mantle boundary (CMB) is at a depth of 2,900 km.

The core is primarily composed of iron, with lesser amounts of nickel (about 5%) and several percent oxygen. It is extremely hot (~3500° to 5000°C). The outer core is liquid while the inner core is solid — even though it is hotter — because the pressure is so much greater at that depth.

Although the CMB is a little less than half of the way to Earth's centre, the mantle, being on the outside, is by far the major component of Earth. The mantle makes up 82.5% of the volume, the core 16.1%, and the crust only 1.4%.

In the remainder of this chapter, we'll look first at how we know about Earth's interior structure, and then at the properties of the different layers and the processes that take place within them.

Figure 2: Earth's layers: crust is pink, mantle is green, core is blue [SE]

9.1 EARTH AND SEISMOLOGY

Seismology is the study of vibrations within Earth. These vibrations are caused by various events, including earthquakes, extra-terrestrial impacts, explosions, storm waves hitting the shore, and tidal effects. Of course, seismic techniques have been most widely applied to the detection and study of earthquakes, but there are many other applications, and arguably seismic waves provide the most important information that we have concerning Earth's interior. Before going any deeper into Earth, however, we need to take a look at the properties of seismic waves. The types of waves that are useful for understanding Earth's interior are called body waves, meaning that, unlike the surface waves on the ocean, they are transmitted through Earth materials.

Imagine hitting a large block of strong rock (e.g., granite) with a heavy sledgehammer (Figure 3). At the point where the hammer strikes it, a small part of the rock will be compressed by a fraction of a millimetre. That compression will transfer to the neighbouring part of the rock, and so on through to the far side of the rock, from where it will bounce back to the top — all in a fraction of a second. This is known as a compression wave, and it can be illustrated by holding a loose spring (like a Slinky) that is attached to something (or someone) at the other end. If you give it a sharp push so the coils are compressed, the compression propagates (travels) along the length of the spring and back (Figure 4). You can think of a compression wave as a "push" wave — it's called a P-wave (although the "P" stands for "primary" because P-waves arrive first at seismic stations).

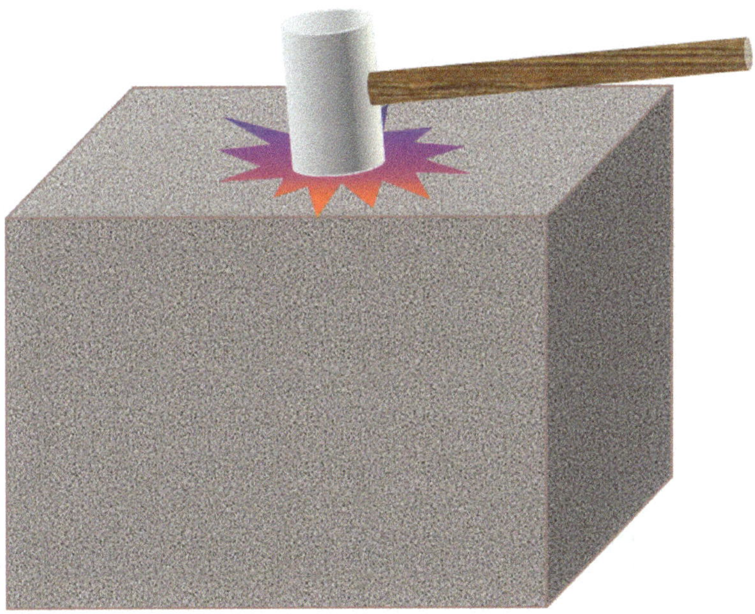

Figure 3: Hitting a large block of rock with a heavy hammer will create seismic waves within the rock. Please don't try this at home! [SE]

When we hit a rock with a hammer, we also create a different type of body wave, one that is characterized by back-and-forth vibrations (as opposed to compressions). This is known as a shear wave (S-wave, where the "S" stands for "secondary"), and an analogy would be what happens when you flick a length of rope with an up-and-down motion. As shown in Figure 4, a wave will form in the rope, which will travel to the end of the rope and back.

Compression waves and shear waves travel very quickly through geological materials. As shown in Figure 5, typical P-wave velocities are between 0.5 km/s and 2.5 km/s in unconsolidated sediments, and between 3.0 km/s and 6.5 km/s in solid crustal rocks. Of the common rocks of the crust, velocities are greatest in basalt and granite. S-waves are slower than P-waves, with velocities between 0.1 km/s and 0.8 km/s in soft sediments, and between 1.5 km/s and 3.8 km/s in solid rocks.

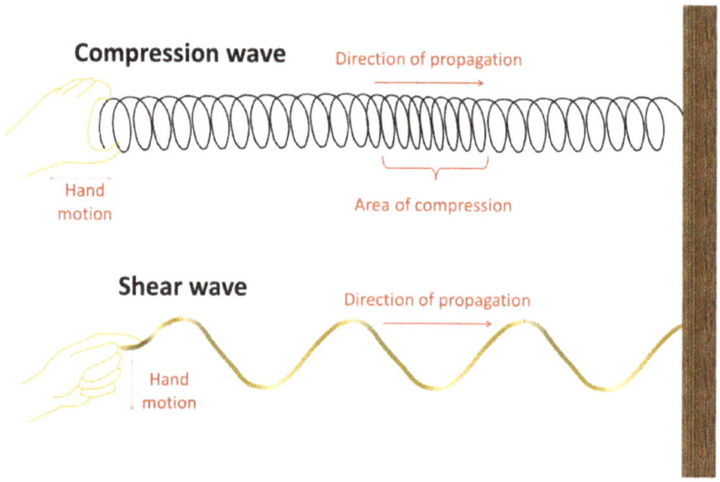

Figure 4: A compression wave can be illustrated by a spring (like a Slinky) that is given a sharp push at one end. A shear wave can be illustrated by a rope that is given a quick flick. [SE]

Exercise 1 How Soon Will Seismic Waves Get Here?

Imagine that a strong earthquake takes place on Vancouver Island within Strathcona Park (west of Courtenay). Assuming that the crustal average P-wave velocity is 5 km/s, how long will it take for the first seismic waves (P-waves) to reach you in the following places (distances from the epicentre are shown)?

Location/distance	Time (s)
Nanaimo (120 km)	
Surrey (200 km)	
Kamloops (390 km)	

Mantle rock is generally denser and stronger than crustal rock and both P- and S-waves travel faster through the mantle than they do through the crust. Moreover, seismic-wave velocities are related to how tightly compressed a rock is, and the level of compression increases dramatically with depth. Finally, seismic waves are affected by the phase state of rock. They are slowed if there is any degree of melting in the rock. If the material is completely liquid, P-waves are slowed dramatically and S-waves are stopped altogether.

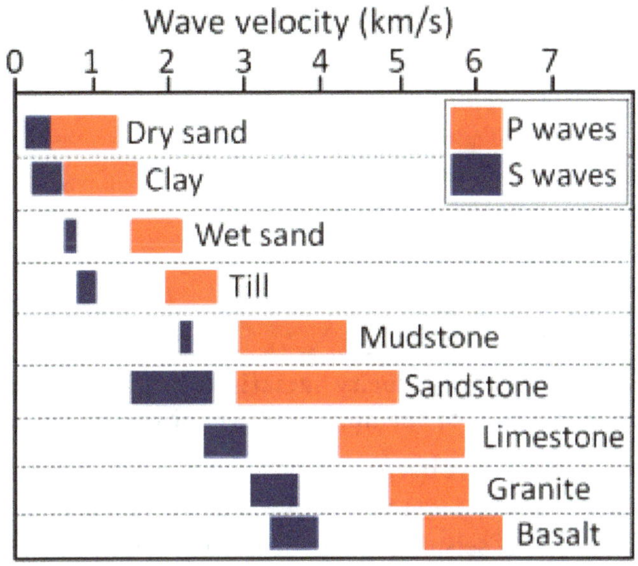

Figure 5: Typical velocities of P-waves (red) and S-waves (blue) in sediments and in solid crustal rocks [SE after: US Env. Prot. Agency http://www.epa.gov/esd/cmb/GeophysicsWebsite/ pages/reference/properties/Geomechanical_(Engineering)_ Properties.htm]

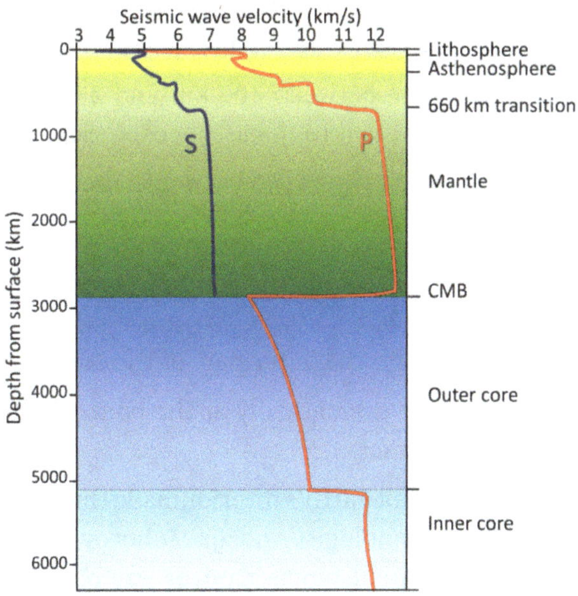

Figure 6a: P-wave and S-wave velocity variations with depth in Earth. [SE]

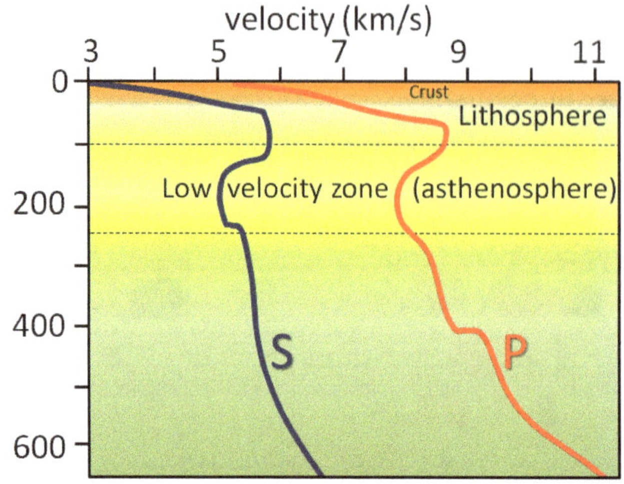

Figure 6b: P-wave and S-wave velocity variations in the upper mantle and crust (This is an expanded view of the upper 600 km of the curves in Figure 9.6a)

Accurate seismometers have been used for earthquake studies since the late 1800s, and systematic use of seismic data to understand Earth's interior started in the early 1900s. The rate of change of seismic waves with depth in Earth (as shown in Figure 6) has been determined over the past several decades by analyzing seismic signals from large earthquakes at seismic stations around the world. Small differences in arrival time of signals at different locations have been interpreted to show that:

- Velocities are greater in mantle rock than in the crust.
- Velocities generally increase with pressure, and therefore with depth.
- Velocities slow in the area between 100 km and 250 km depth (called the "low-velocity zone"; equivalent to the asthenosphere).
- Velocities increase dramatically at 660 km depth (because of a mineralogical transition).
- Velocities slow in the region just above the core-mantle boundary (the D" layer or "ultra-low-velocity zone").
- S-waves do not pass through the outer part of the core.
- P-wave velocities increase dramatically at the boundary between the liquid outer core and the solid inner core.

One of the first discoveries about Earth's interior made through seismology was in the early 1900s when Croatian seismologist Andrija Mohorovičić (pronounced Moho-ro-vi-chich) realized that at certain distances from an earthquake, two separate sets of seismic waves arrived at a seismic station within a few seconds of each other. He reasoned that the waves that went down into the mantle, travelled through the mantle, and then were bent upward back into the crust, reached the seismic station first because although they had farther to go, they travelled faster through mantle rock (as shown in Figure 7). The boundary between the crust and the mantle is known as the Mohorovičić discontinuity (or Moho). Its depth is between 60 km and 80 km beneath major mountain ranges, around 30 km to 50 km beneath most of the continental crust, and between 5 km and 10 km beneath the oceanic crust.

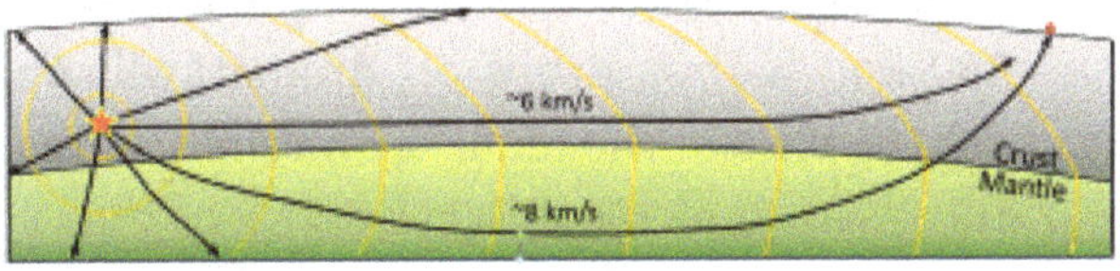

Figure 7: Depiction of seismic waves emanating from an earthquake (red star). Some waves travel through the crust to the seismic station (at about 6 km/s), while others go down into the mantle (where they travel at around 8 km/s) and are bent upward toward the surface, reaching the station before the ones that travelled only through the crust. [SE]

Our current understanding of the patterns of seismic wave transmission through Earth is summarized in Figure 9.8. Because of the gradual increase in density (and therefore rock strength) with depth, all waves are refracted (toward the lower density material) as they travel through homogenous parts of Earth and thus tend to curve outward toward the surface. Waves are also refracted at boundaries within Earth, such as at the Moho, at the core-mantle boundary (CMB), and at the outer-core/inner-core boundary.

S-waves do not travel through liquids — they are stopped at the CMB — and there is an S-wave shadow on the side of Earth opposite a seismic source. The angular distance from the seismic source to the shadow zone is 103° on either side, so the total angular distance of the shadow zone is 154°. We can use this information to infer the depth to the CMB.

P-waves do travel through liquids, so they can make it through the liquid part of the core. Because of the refraction that takes place at the CMB, waves that travel through the core are bent away from the surface, and this creates a P-wave shadow zone on either side, from 103° to 150°. This information can be used to discover the differences between the inner and outer parts of the core.

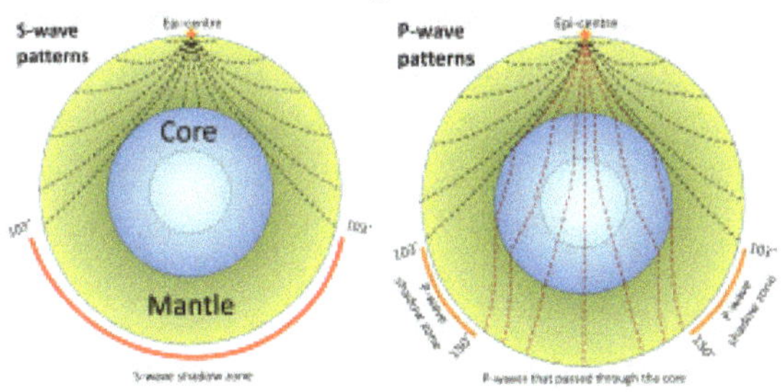

Figure 8: Patterns of seismic wave propagation through Earth's mantle and core. S-waves do not travel through the liquid outer core, so they leave a shadow on

Earth's far side. P-waves do travel through the core, but because the waves that enter the core are refracted, there are also P-wave shadow zones.

9.2 TEMPERATURE OF EARTH'S INTERIOR

Earth's internal temperature increases with depth. However, as shown in Figure 10, that rate of increase is not linear. The temperature gradient is around 15° to 30°C/km within the upper 100 km; it then drops off dramatically through the mantle, increases more quickly at the base of the mantle, and then increases slowly through the core. The temperature is around 1000°C at the base of the crust, around 3500°C at the base of the mantle, and around 5,000°C at Earth's centre. The temperature gradient within the lithosphere (upper 100 km) is quite variable depending on the tectonic setting. Gradients are lowest in the central parts of continents, higher in the vicinity of subduction zones, and higher still at divergent boundaries.

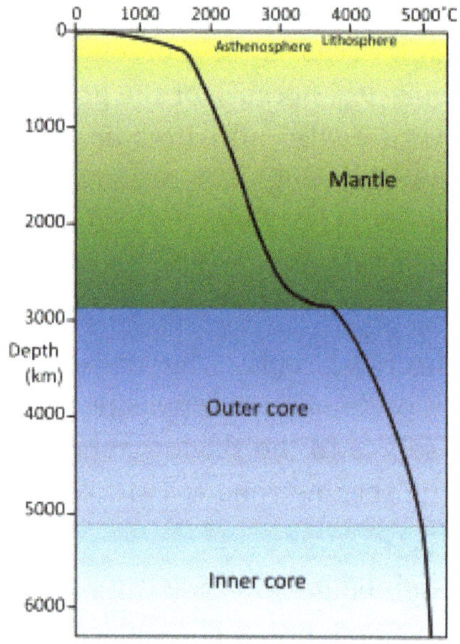

Figure 10: Generalized rate of temperature increase with depth within Earth. Temperature increases to the right, so the flatter the line, the steeper the temperature gradient. Our understanding of the temperature gradient comes from seismic wave information and knowledge of the melting points of Earth's materials. [

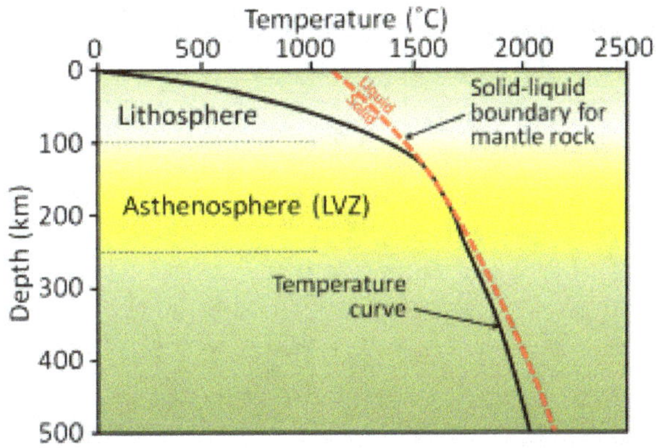

Figure 11: Rate of temperature increase with depth in Earth's upper 500 km, compared with the dry mantle rock melting curve (red dashed line). LVZ= low-velocity zone

Figure 11 shows a typical temperature curve for the upper 500 km of the mantle, in comparison with the melting curve for dry mantle rock. Within the depth interval between 100 and 250 km, the temperature curve comes very close to the melting boundary for dry mantle rock. At these depths, therefore, mantle rock is either very nearly melted or partially melted. In some situations, where extra heat is present and the temperature line crosses over the melting line, or where water is present, it may be completely molten. This region of the mantle is known as the low-velocity zone because seismic waves are slowed within rock that is near its melting point, and of course it is also known as the asthenosphere. Below 250 km, the temperature stays on the left side of the melting line; in other words, the mantle is solid from here all the way down to the core-mantle boundary.

The fact that the temperature gradient is much less in the main part of the mantle than in the lithosphere has been interpreted to indicate that the mantle is convecting, and therefore that heat from depth is being brought toward the surface faster than it would be with only heat conduction. As we'll see in Chapter 10, a convecting mantle is an essential feature of plate tectonics.

The convection of the mantle is a product of the transfer of heat from the core to the lower mantle. As in a pot of soup on a hot stove (Figure 12), the material near the heat source becomes hot and expands, making it lighter than the material above. The force of buoyancy causes it to rise, and cooler material flows in from the sides. The mantle convects in this way because the heat transfer from below is not perfectly even, and also because, even though mantle material is solid rock, it is sufficiently plastic to slowly flow (at rates of centimetres per year) as long as a steady force is applied to it.

As in the soup pot example, Earth's mantle will no longer convect once the core has cooled to the point where there is not enough heat transfer to overcome the strength of the rock. This has already happened on smaller planets like Mercury and Mars, as well as on Earth's Moon.

Figure 12: Convection in a pot of soup on a hot stove (left). As long as heat is being transferred from below, the liquid will convect. If the heat is turned off (right), the liquid remains hot for a while, but convection will cease. [SE]

The inside of Earth is hot

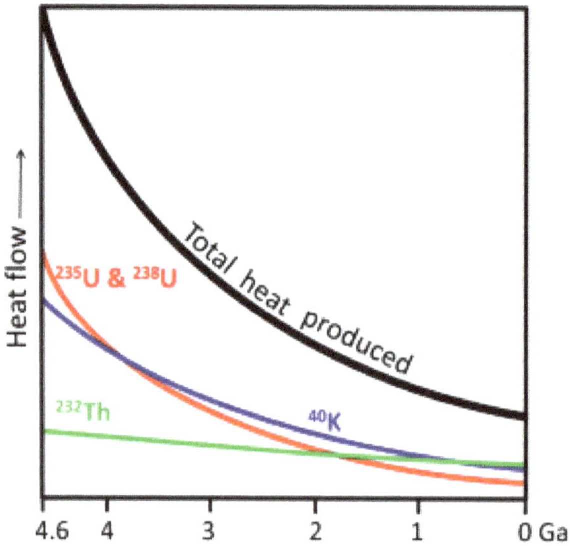

The heat of Earth's interior comes from two main sources, each contributing about 50% of the heat. One of those is the frictional heat left over from the collisions of large and small particles that created Earth in the first place, plus the subsequent frictional heat of redistribution of material within Earth by gravitational forces (e.g., sinking of iron to form the core).

The other source is radioactivity, specifically the spontaneous radioactive decay of ^{235}U, ^{238}U, ^{40}K, and ^{232}Th, which are primarily present in the mantle. As shown on this figure, the total heat produced that way has been decreasing over time (because these isotopes are getting used up), and is now roughly 25% of what it was when Earth formed. This means that Earth's interior is slowly becoming cooler.

9.3 EARTH'S MAGNETIC FIELD

Heat is also being transferred from the solid inner core to the liquid outer core, and this leads to convection of the liquid iron of the outer core. Because iron is a metal and conducts electricity (even when molten), its motion generates a magnetic field.

Earth's magnetic field is defined by the North and South Poles that align generally with the axis of rotation (Figure 13). The lines of magnetic force flow into Earth in the northern hemisphere and out of Earth in the southern hemisphere. Because of the shape of the field lines, the magnetic force trends at different angles to the surface in different locations (red arrows of Figure 13). At the North and South Poles, the force is vertical. Anywhere on the equator the force is horizontal, and everywhere in between, the magnetic force is at some intermediate angle to the surface. As we'll see in Chapter 10, the variations in these orientations provide a critical piece of evidence to the understanding of continental drift as an aspect of plate tectonics.

Earth's magnetic field is generated within the outer core by the convective movement of liquid iron, but as we discovered in Chapter 8, the magnetic field is not stable over geological time. For reasons that are not completely understood, the magnetic field decays periodically and then becomes re-established. When it does re-establish, it may be oriented the way it was before the decay, or it may be oriented with the reversed polarity. Over the past 250 Ma, there have a few hundred magnetic field reversals, and their timing has been anything but regular. The shortest ones that geologists have been able to define lasted only a few thousand years, and the longest one was more than 30 million years, during the Cretaceous (Figure 14).

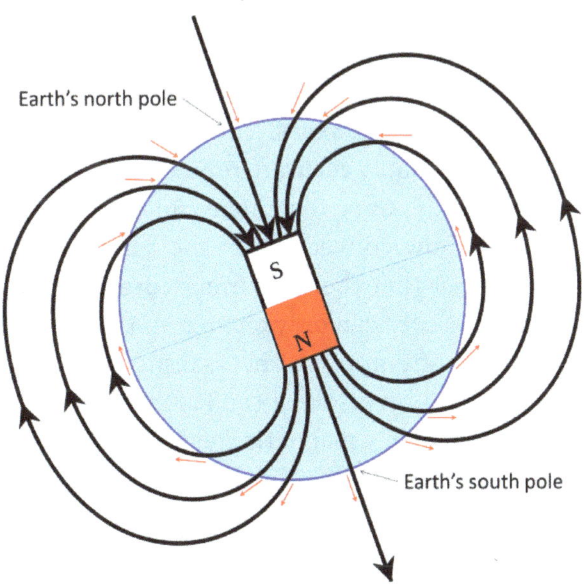

Figure 13: Depiction of Earth's magnetic field as a bar magnet coinciding with the core. The south pole of such a magnet points to Earth's North Pole. The red

arrows represent the orientation of the magnetic field at various locations on Earth's surface. [SE after: http://upload.wikimedia.org/wikipedia/commons/1/17/Earths_Magnetic_Field_ Confusion.svg]

Exercise 3What Does Your Magnetic Dip Meter Tell You?

Regular compasses point only to the north magnetic pole, but if you have a magnetic dip meter (or an iPhone with the appropriate app*), you could also measure the angle of the magnetic field at your location in the up-and-down sense. You don't need to get the app (or an iPhone) to do this exercise!

Using Figure 13 as a guide, describe where you'd be on Earth if the vertical angles are as follows:

Up at a shallow angle Parallel to the ground

Vertical orientation	General location	Vertical orientation	General location
Straight down			
Down at a steep angle			

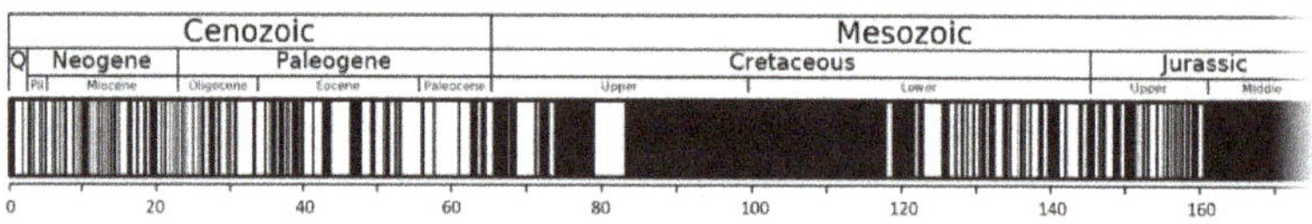

Figure 14: Magnetic field reversal chronology for the past 170 Ma. The first 5 Ma of the magnetic chronology are shown in more detail in Figure 9.15. [SE after: http://upload.wikimedia.org/wikipedia/en/c/c0/Geomagnetic_polarity_0-169_Ma.svg]

Changes in Earth's magnetic field have been studied using a mathematical model, and reversals have been shown to take place when the model was run to simulate a period of several hundred thousand years. The fact that field reversals took place shows that the model is a reasonably accurate representation of the Earth. According to the lead author of the study, Gary Glatzmaier, of University of California Santa Cruz: "Our solution shows how convection in the fluid outer core is continually trying to reverse the field but that the solid inner core inhibits magnetic reversals because the field in the inner core can only change on the much longer time scale of diffusion. Only once in many attempts is a reversal successful, which is probably the reason why the times between reversals of the Earth's field are long and randomly distributed." A depiction of Earth's magnetic field lines during a stable period and during a reversal is shown in Figure 15.

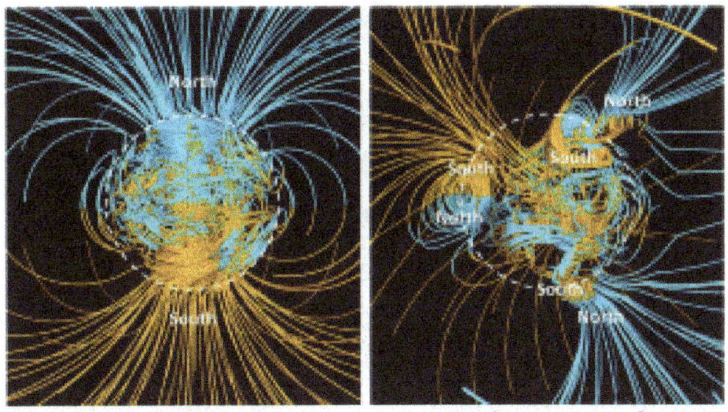

Figure 15: Depiction of Earth's magnetic field between reversals (left) and during a reversal (right). The lines represent magnetic field lines: blue where the field points toward Earth's centre and yellow where it points away. The rotation axis of Earth is vertical, and the outline of the core is shown as a dashed white circle.

9.4 ISOSTASY

Theory holds that the mantle is able to convect because of its plasticity, and this property also allows for another very important Earth process known as isostasy. The literal meaning of the word isostasy is "equal standstill," but the importance behind it is the principle that Earth's crust is floating on the mantle, like a raft floating in the water, rather than resting on the mantle like a raft sitting on the ground.

The relationship between the crust and the mantle is illustrated in Figure 16. On the right is an example of a non-isostatic relationship between a raft and solid concrete. It's possible to load the raft up with lots of people, and it still won't sink into the concrete. On the left, the relationship is an isostatic one between two different rafts and a swimming pool full of peanut butter. With only one person on board, the raft floats high in the peanut butter, but with three people, it sinks dangerously low. We're using peanut butter here, rather than water, because its viscosity more closely represents the relationship between the crust and the mantle. Although it has about the same density as water, peanut butter is much more viscous (stiff), and so although the three-person raft will sink into the peanut butter, it will do so quite slowly.

Figure 16: Illustration of a non-isostatic relationship between a raft and solid ground (right) and of isostatic relationships between rafts and peanut butter (left).

The relationship of Earth's crust to the mantle is similar to the relationship of the rafts to the peanut butter. The raft with one person on it floats comfortably high. Even with three people on it the raft is less dense than the peanut butter, so it floats, but it floats uncomfortably low for those three people. The crust, with an average density of around 2.6 grams per cubic centimetre (g/cm3), is less dense than the mantle (average density of approximately 3.4 g/cm3 near the surface, but more than that at depth), and so it is floating on the "plastic" mantle. When more weight is added to the crust, through the process of mountain building, it slowly sinks deeper into the mantle and the mantle material that was there is pushed aside (Figure 17, left). When that weight is removed by erosion over tens of millions of years, the crust rebounds and the mantle rock flows back (Figure 17, right).

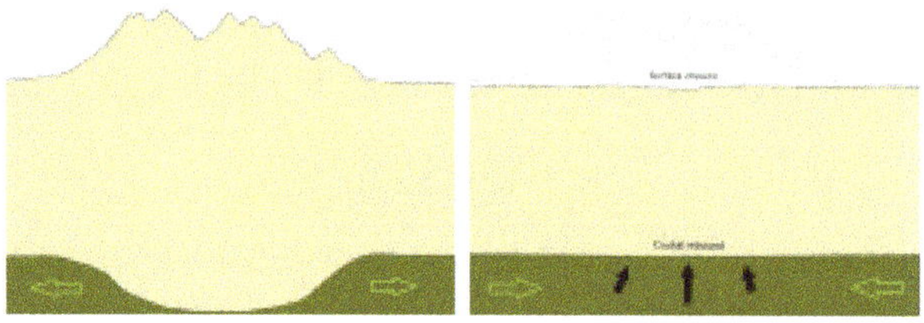

Figure 17: Illustration of the isostatic relationship between the crust and the mantle. Following a period of mountain building, mass has been added to a part of the crust, and the thickened crust has pushed down into the mantle (left). Over the following tens of millions of years, the mountain chain is eroded and the crust rebounds (right). The green arrows represent slow mantle flow.

The crust and mantle respond in a similar way to glaciation. Thick accumulations of glacial ice add weight to the crust, and as the mantle beneath is squeezed to the sides, the crust subsides. This process is illustrated for the current ice sheet on Greenland in Figure 18. The Greenland Ice Sheet at this location is over 2,500 m thick, and the crust beneath the thickest part has been depressed to the point where it is below sea level over a wide area. When the ice eventually melts, the crust and mantle will slowly rebound, but full rebound will likely take more than 10,000 years.

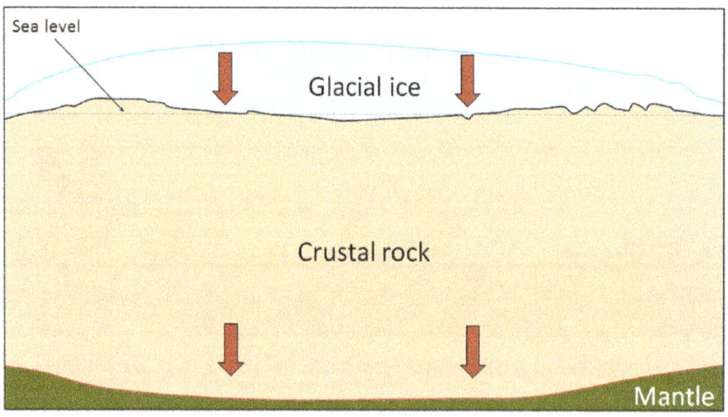

Figure 18a: A cross-section through the crust in the northern part of Greenland (The ice thickness is based on data from NASA and the Center for Remote

Sensing of Ice Sheets, but the crust thickness is less than it should be for the sake of illustration.) The maximum ice thickness is over 2,500 m. The red arrows represent downward pressure on the mantle because of the mass of the ice.

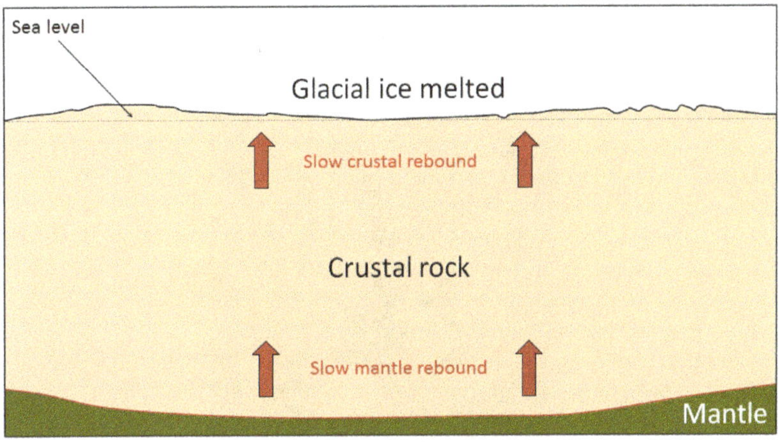

Figure 18b: Depiction of the situation after complete melting of the ice sheet, a process that could happen within 2,000 years if people and their governments continue to ignore climate change. The isostatic rebound of the mantle would not be able to keep up with this rate of melting, so for several thousand years the central part of Greenland would remain close to sea level, in some areas even below sea level.

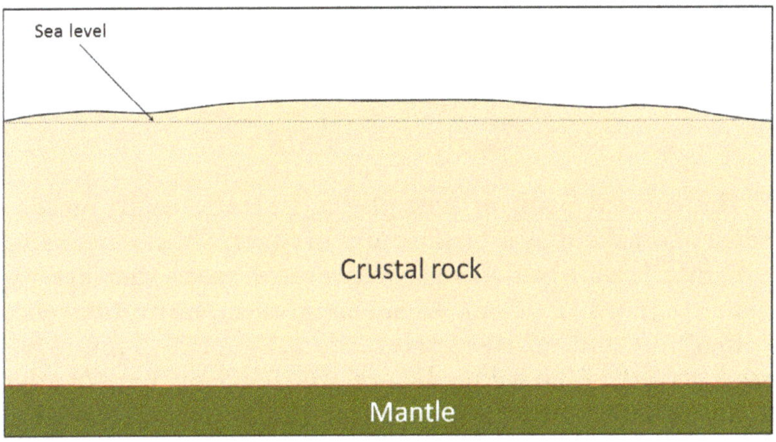

Figure 18c: It is likely that complete rebound of the mantle beneath Greenland would take more than 10,000 years.

How can the mantle be both solid and plastic?

You might be wondering how it is possible that Earth's mantle is rigid enough to break during an earthquake, and yet it convects and flows like a very viscous liquid. The explanation is that the mantle behaves as a non-Newtonian fluid, meaning that it responds differently to stresses depending on how quickly the stress is applied. A good example of this is the behaviour of the material known as Silly Putty, which can bounce and will break if you pull on it sharply, but will deform in a liquid manner if stress is applied slowly. In this photo, Silly Putty was placed over a hole in a glass tabletop, and in response to gravity, it slowly flowed into the hole. The mantle will flow when placed under the slow but steady stress of a growing (or melting) ice sheet.

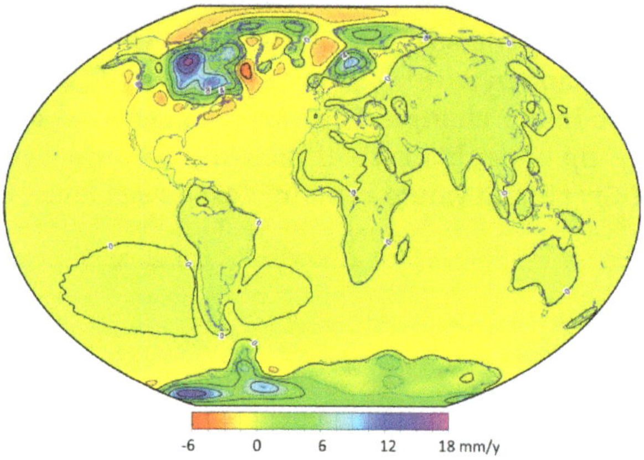

Figure 19: The current rates of post-glacial isostatic uplift (green, blue, and purple shades) and subsidence (yellow and orange). Subsidence is taking place where the mantle is slowly flowing back toward areas that are experiencing post-glacial uplift. [From: Paulson, A., S. Zhong, and J. Wahr. Inference of mantle viscosity from GRACE and relative sea level data, Geophys. J. Int. (2007) 171, 497–508. Accessed at: http://en.wikipedia.org/wiki/Hudson_Bay#/media/File:PGR_Paulson2007_Rate_of_Lithospheric_Uplift_due_to_PGR.png]

Large parts of Canada are still rebounding as a result of the loss of glacial ice over the past 12 ka, and as shown in Figure 19, other parts of the world are also experiencing isostatic rebound. The highest rate of uplift is in within a large area to the west of Hudson Bay, which is where the Laurentide Ice Sheet was the thickest (over 3,000 m). Ice finally left this region around 8,000 years ago, and the crust is currently rebounding at a rate of nearly 2 cm/year. Strong isostatic rebound is also occurring in northern Europe where the Fenno-Scandian Ice Sheet was thickest, and in the eastern part of Antarctica, which also experienced significant ice loss during the Holocene.

There are also extensive areas of subsidence surrounding the former Laurentide and Fenno-Scandian Ice Sheets. During glaciation, mantle rock flowed away from the areas beneath the main ice sheets, and this material is now slowly flowing back, as illustrated in Figure 18b.

AT A GLANCE

Understanding Earth through Seismology	Seismic waves that travel through Earth are either P-waves (compression, or "push" waves) or S-waves (shear waves). P-waves are faster than S-waves, and can pass through fluids. By studying seismic waves, we can discover the nature and temperature characteristics of the various parts of Earth's interior.
The Temperature of Earth's Interior	Earth's temperature increases with depth (to around 5000°C at the centre), but there are significant variations in the rate of temperature increase. These variations are related to differences in composition and the existence of convection in the mantle and liquid part of the core.
Earth's Magnetic Field	Because of outer-core convection, Earth has a magnetic field. The magnetic force directions are different at different latitudes. The polarity of the field is not constant, and has flipped from "normal" (as it is now) to reversed and back to normal hundreds of times in the past.
Isostasy	The "plastic" nature of the mantle, which allows for mantle convection, also determines the nature of the relationship between the crust and the mantle. The crust floats on the mantle in an isostatic relationship. Where the crust becomes thicker because of mountain building, it pushes farther down into the mantle. Oceanic crust, being heavier than continental crust, floats lower on the mantle.

CHAPTER

PLATE TECTONICS THEORY 10

Plate tectonics is the model or theory that we use to understand how our planet works. More specifically it is a model that explains the origins of continents and oceans, folded rocks and mountain ranges, igneous and metamorphic rocks, earthquakes and volcanoes, and continental drift. Plate tectonics was first proposed just over 100 years ago, but did not become an accepted part of geology until about 50 years ago. It took 50 years for this theory to become accepted for a few reasons. First, it was a true revolution in thinking about Earth, which was difficult for many established geologists. Second, there was a political gulf between the main proponent of the theory Alfred Wegener (from Germany) and the geological establishment of the day, which was mostly centred in Britain and the United States. Third, the evidence and understanding of Earth that would have supported plate tectonic theory simply didn't exist until the middle of the 20th century.

10.1 FATHER OF PLATE TECTONICS: ALFRED WEGENER

Alfred Wegener (1880-1930) (Figure 10.1) earned a PhD in astronomy at the University of Berlin in 1904, but he had always been interested in geophysics and meteorology and spent most of his academic career working in meteorology. In 1911 he happened on a scientific publication that included a description of the existence of matching Permian-aged terrestrial fossils in various parts of South America, Africa, India, Antarctica, and Australia (Figure 2).

Wegener concluded that this distribution of fossils could only exist if these continents were joined together during the Permian, and he coined the term Pangea ("all land") for the supercontinent that he thought included all of the present-day continents.

Figure 1: Alfred Wegener a few years before his death in 1930 [http://upload.wikimedia.org/wikipedia/ commons/6/65/Alfred_Wegener_ca.1924-30.jpg]

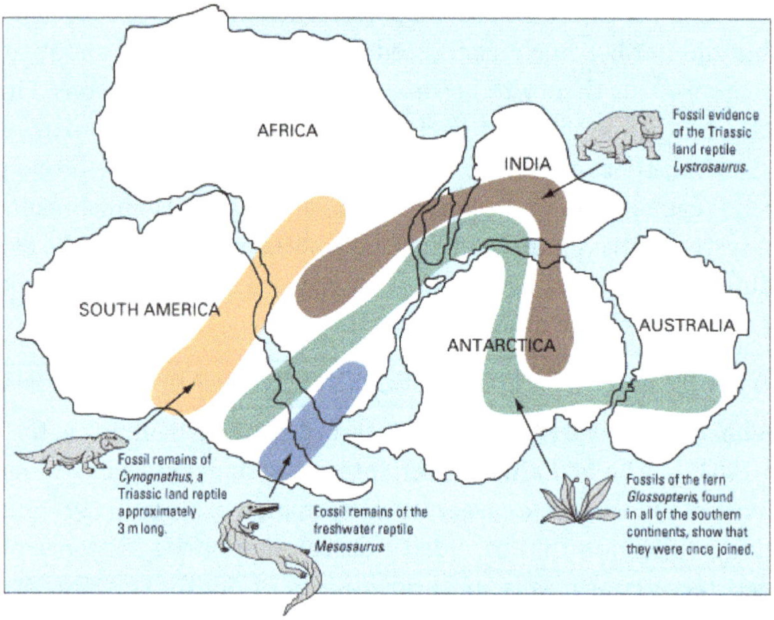

Figure 2: The distribution of several Permian terrestrial fossils that are present in various parts of continents that are now separated by oceans. During the Permian, the supercontinent Pangea included the supercontinent Gondwana, shown here, along with North America and Eurasia.

Wegener pursued his theory with determination — combing the libraries, consulting with colleagues, and making observations — looking for evidence to support it. He relied heavily on matching geological patterns across oceans, such as sedimentary strata in South America matching those in Africa (Figure 3), North American coalfields matching those in Europe, and the mountains of Atlantic Canada matching those of northern Britain — both in morphology and rock type. Wegener also referred to the evidence for the Carboniferous and Permian (~300 Ma) Karoo Glaciation in South America, Africa, India, Antarctica, and Australia (Figure 10.4). He argued that this could only have happened if these continents were once all connected as a single supercontinent. He also cited evidence (based on his own astronomical observations) that showed that the continents were moving with respect to each other, and determined a separation rate between Greenland and Scandinavia of 11 m per year, although he admitted that the measurements were not accurate. In fact they weren't even close — the separation rate is actually about 2.5 cm per year!

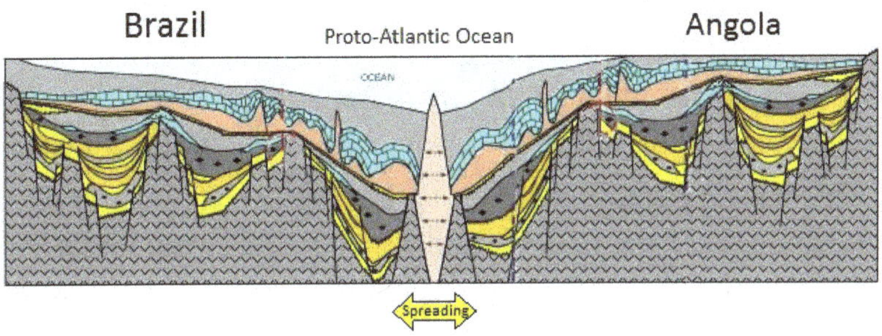

Figure 3: A cross-section showing the geological similarities between parts of Brazil on the left and Angola (Africa) on the right. The pink layer is a salt deposit, which is now known to be common in areas of continental rifting. [Source: U.S. Energy Information Administration (March 2015) http://www.eia.gov/countries/analysisbriefs/Angola/angola.pdf]

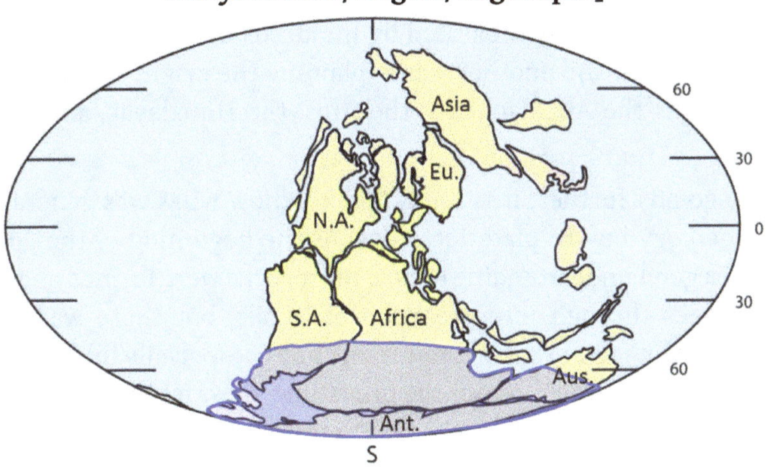

Figure 4: The distribution of the Carboniferous and Permian Karoo Glaciation (in blue) [SE, after http://upload.wikimedia.org/wikipedia/commons/9/96/Karoo_Glaciation.png]

Wegener first published his ideas in 1912 in a short book called *Die Entstehung der Kontinente* (*The Origin of Continents*), and then in 1915 in *Die Entstehung der Kontinente und Ozeane* (*The Origin of Continents and Oceans*). He revised this book several times up to 1929. It was translated into French, English, Spanish, and Russian in 1924.

In fact the continental fits were not perfect and the geological matchups were not always consistent, but the most serious problem of all was that Wegener could not conceive of a good mechanism for moving the continents around. It was understood by this time that the continents were primarily composed of sialic material (SIAL: silicon and aluminum dominated), and that the ocean floors were primarily simatic (SIMA: silicon and magnesium dominated). Wegener proposed that the continents were like icebergs floating on the heavier SIMA crust, but the only forces that he could invoke to propel continents around were poleflucht, the effect of Earth's rotation pushing objects toward the equator, and the lunar and solar tidal forces, which tend to push objects toward the west. It was quickly shown that these forces were far too weak to move continents, and without any reasonable mechanism to make it work, Wegener's theory was quickly dismissed by most geologists of the day.

Alfred Wegener died in Greenland in 1930 while carrying out studies related to glaciation and climate. At the time of his death, his ideas were tentatively accepted by only a small minority of geologists, and soundly rejected by most. However, within a few decades that was all to change.

10.2 GLOBAL GEOLOGICAL MODELS: THE EARLY 20TH CENTURY

The untimely death of Alfred Wegener didn't solve any problems for those who opposed his ideas because they still had some inconvenient geological truths to deal with. One of those was explaining the distribution of terrestrial species across five continents that are currently separated by hundreds or thousands of kilometres of ocean water (Figure 2), and another was explaining the origin of extensive fold-belt mountains, such as the Appalachians, the Alps, the Himalayas, and the Canadian Rockies.

Before we go any further, it is important to know what was generally believed about global geology before plate tectonics. At the beginning of the 20th century, geologists had a good understanding of how most rocks were formed and understood their relative ages through interpretation of fossils, but there was considerable controversy regarding the origin of mountain chains, especially fold-belt mountains. At the end of the 19th century, one of the prevailing views on the origin of mountains was the theory of contractionism — the idea that since Earth is slowly cooling, it must also be shrinking. In this scenario, mountain ranges had formed like the wrinkles on a dried-up apple, and the oceans had submerged parts of former continents. While this theory helped to address the dilemma of the terrestrial fossils, it came with its own set of problems, one being that the amount of cooling couldn't produce the necessary amount of shrinking, and the other being the principle of isostasy (which had already been around for several decades), which wouldn't allow continents to sink. (See Section 9.4 for a review of the important principle of isostasy.)

Another widely held view was permanentism, in which it was believed that the continents and oceans have always been generally as they are today. This view incorporated a mechanism for creation of mountain chains known as the geosyncline theory. A geosyncline is a thick deposit of sediments and sedimentary rocks, typically situated along the edge of a continent (Figure 5).

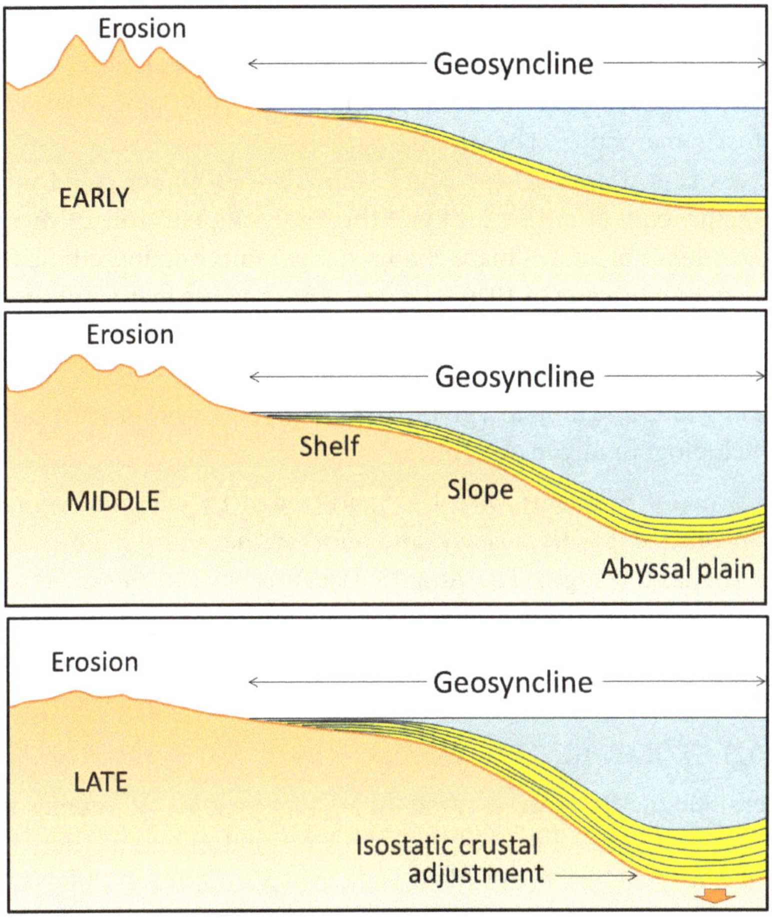

Figure 5: The development of a geosyncline along a continental margin. (Note that a geosyncline is not related to a syncline, which is a downward fold in sedimentary rocks.) [SE]

The idea of geosynclines developing into fold-belt mountains originated in the middle of the 19th century, proposed first by James Hall and later elaborated by Dwight Dana, both of whom worked extensively in the Appalachian Mountains of the eastern United States. The process of converting a geosyncline into a mountain belt was never really adequately explained, although it was widely believed that mountain belts formed when geosynclines were compressed by forces pushing from either side. The problem is that, without the lateral forces related to plate tectonics, no one was able to adequately describe what would do the pushing. The sediments that accumulate within a geosyncline are derived from erosion of the adjacent continent. Geosynclinal sediments — which eventually turn into sedimentary rocks — may be many thousands of metres thick. As they accumulate, they push down the pre-existing crustal rocks. Extensive geosynclinal deposits exist around much of the coastline of most of the continents; there is a large geosyncline along the eastern edge of North America.

Proponents of the geosyncline theory of mountain formation, and there were many well into the 1960s, also had the problem of explaining the intercontinental terrestrial fossil matchups. The simple explanation was that there were "land bridges" across the Atlantic along which animals and plants could migrate back and forth. One proponent of this idea was the American naturalist Ernest Ingersoll. Referring to evidence of past climate changes, Ingersoll contributed the following to the Encyclopedia Americana in 1920: "The most interesting feature of these changes, however, is that by which, now and again, the Old World was connected with the New by necks or spaces of land, known as "land-bridges"; especially as these permitted an interchange of plants and animals, giving to us many new ones from the other side of the ocean, including, finally, man himself

There are many problems with the land-bridge theory, one being that it is completely inconsistent with isostasy, and another that there is no evidence of the remnants of the land bridges. The Atlantic Ocean is several thousand metres deep over wide areas, and so the underwater slopes leading up to a land bridge would have to have been at least tens of kilometres wide in most places, and many times that in others. A land bridge of that size would certainly have left some trace.

10.3 GEOLOGICAL RENAISSANCE: MID-20TH CENTURY

As the mineral magnetite (Fe3O4) crystallizes from magma, it becomes magnetized with an orientation parallel to that of Earth's magnetic field at that time. This is called remnant magnetism. Rocks like basalt, which cool from a high temperature and commonly have relatively high levels of magnetite, are particularly susceptible to being magnetized in this way, but even sediments and sedimentary rocks, as long as they have small amounts of magnetite, will take on remnant magnetism because the magnetite grains gradually become reoriented following deposition. By studying both the horizontal and vertical components of the remnant magnetism, one can tell not only the direction to magnetic north at the time of the rock's formation, but also the latitude where the rock formed relative to magnetic north.

In the early 1950s, a group of geologists from Cambridge University, including Keith Runcorn, Ted Irving,Ted Irving later set up a paleomagnetic lab at the Geological Survey of Canada in Sidney, B.C., and did a great deal of important work on understanding the geology of western North America. and several others, started looking at the remnant magnetism of Phanerozoic British and European volcanic rocks, and collecting paleomagnetic data. They found that rocks of different ages sampled from generally the same area showed quite different apparent magnetic pole positions (Figure 6). They initially assumed that this meant that Earth's magnetic field had, over time, departed significantly from its present position – which is close to the rotational pole.

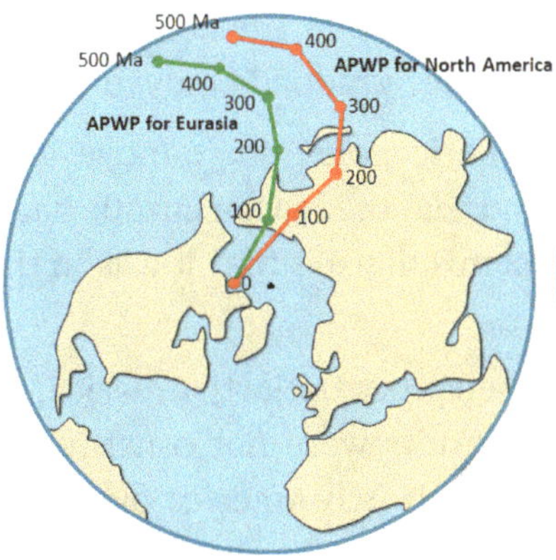

Figure 6: Apparent polar-wandering paths (APWP) for Eurasia and North America. The view is from the North Pole (black dot) looking down. The outer circle is the equator. In the diagram to the right the curve locations have been corrected taking continental drift into account. [SE]

The curve defined by the paleomagnetic data was called a polar wandering path because Runcorn and his students initially thought that their data represented actual movement of the magnetic poles (since geophysical models of the time suggested that the magnetic poles did not need to be aligned with the rotational poles). We now know that the magnetic data define movement of continents, and not of the magnetic poles, so we call it an apparent polar wandering path (APWP).

Polar wandering path

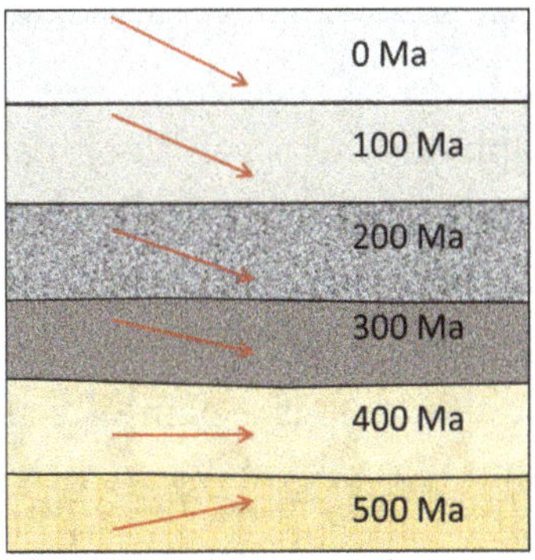

The red arrows show magnetic dip

At around 500 Ma, what we now call Europe was south of the equator, and so European rocks formed then would have acquired an upward-pointing magnetic field orientation (see Figure 9.13 and the figure shown here). Between then and now, Europe gradually moved north, and the rocks forming at various times acquired steeper and steeper *downward-pointing* magnetic orientations.

When researchers evaluated magnetic data in this way in the 1950s, they plotted where the North Pole would have appeared to be based on the magnetic data and assumed that the continent was always where it is now. That means that the 500 Ma "apparent" north pole would have been somewhere in the South Pacific, and that over the following 500 million years it would have gradually moved north.

Of course we now know that the magnetic poles don't move around much (although polarity reversals do take place) and that the reason Europe had a magnetic orientation characteristic of the southern hemisphere is that it was in the southern hemisphere at 500 Ma.

Runcorn and colleagues soon extended their work to North America, and this also showed apparent polar wandering, but the results were not consistent with those from Europe. For example, the 200 Ma pole for North America plotted somewhere in China, while the 200 Ma pole for Europe plotted in the Pacific Ocean. Since there could only have been one pole position at 200 Ma, this evidence strongly supported the idea that North America and Europe had moved relative to each other since 200 Ma. Subsequent paleomagnetic work showed that South America, Africa, India, and Australia also have unique polar wandering curves. In 1956, Runcorn changed his mind and became a proponent of continental drift.

This paleomagnetic work of the 1950s was the first new evidence in favour of continental drift, and it led a number of geologists to start thinking that the idea might have some merit. Nevertheless, for a majority of geologists working on global geology at the time, this type of evidence was not sufficiently convincing to get them to change their views.

During the 20th century, our knowledge and understanding of the ocean basins and their geology increased dramatically. Before 1900, we knew virtually nothing about the bathymetry and geology of the oceans. By the end of the 1960s, we had detailed maps of the topography of the ocean floors, a clear picture of the geology of ocean floor sediments and the solid rocks underneath them, and almost as much information about the geophysical nature of ocean rocks as of continental rocks.

Up until about the 1920s, ocean depths were measured using weighted lines dropped overboard. In deep water this is a painfully slow process and the number of soundings in the deep oceans was probably fewer than 1,000. That is roughly one depth sounding for every 350,000 square kilometres of the ocean. To put that in perspective, it would be like trying to describe the topography of British Columbia with elevation data for only a half a dozen points! The voyage of the Challenger in

1872 and the laying of trans-Atlantic cables had shown that there were mountains beneath the seas, but most geologists and oceanographers still believed that the oceans were essentially vast basins with flat bottoms, filled with thousands of metres of sediments.

Following development of acoustic depth sounders in the 1920s (Figure 7), the number of depth readings increased by many orders of magnitude, and by the 1930s, it had become apparent that there were major mountain chains in all of the world's oceans. During and after World War II, there was a well-organized campaign to study the oceans, and by 1959, sufficient bathymetric data had been collected to produce detailed maps of all the oceans (Figure 8).

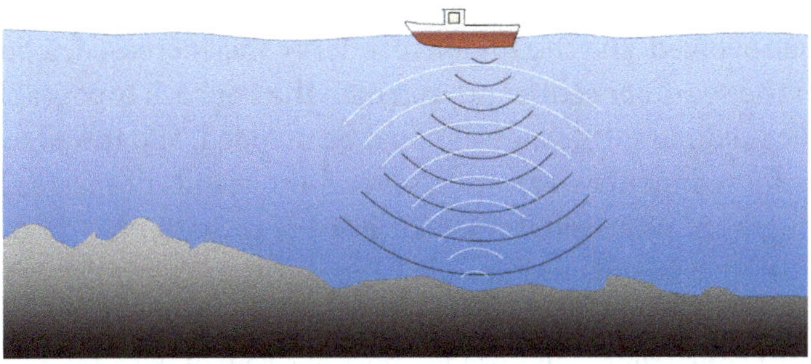

Figure 7: Depiction of a ship-borne acoustic depth sounder. The instrument emits a sound (black arcs) that bounces off the sea floor and returns to the surface (white arcs). The travel time is proportional to the water depth. [SE]

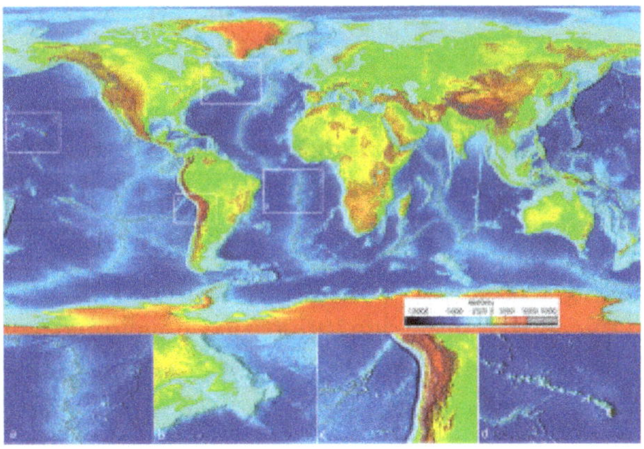

Figure 8: Ocean floor bathymetry (and continental topography). Inset (a): the mid-Atlantic ridge, (b): the Newfoundland continental shelf, (c): the Nazca trench adjacent to South America, and (d): the Hawaiian Island chain. [SE after NOAA, http://upload.wikimedia.org/wikipedia/commons/9/93/Elevation.jpg]

The important physical features of the ocean floor are:

- Extensive linear ridges (commonly in the central parts of the oceans) with water depths in the order of 2,000 to 3,000 m (Figure 10.8, inset a)

- Fracture zones perpendicular to the ridges (inset a)
- Deep-ocean plains at depths of 5,000 to 6,000 m (insets a and d)
- Relatively flat and shallow continental shelves with depths under 500 m (inset b)
- Deep trenches (up to 11,000 m deep), most near the continents (inset c)
- Seamounts and chains of seamounts (inset d)

Seismic reflection sounding involves transmitting high-energy sound bursts and then measuring the echos with a series of geophones towed behind a ship. The technique is related to acoustic sounding as described above; however, much more energy is transmitted and the sophistication of the data processing is much greater. As the technique evolved, and the amount of energy was increased, it became possible to see through the sea-floor sediments and map the bedrock topography and crustal thickness. Hence sediment thicknesses could be mapped, and it was soon discovered that although the sediments were up to several thousands of metres thick near the continents, they were relatively thin — or even non-existent — in the ocean ridge areas (Figure 9). The seismic studies also showed that the crust is relatively thin under the oceans (5 km to 6 km) compared to the continents (30 km to 60 km) and geologically very consistent, composed almost entirely of basalt.

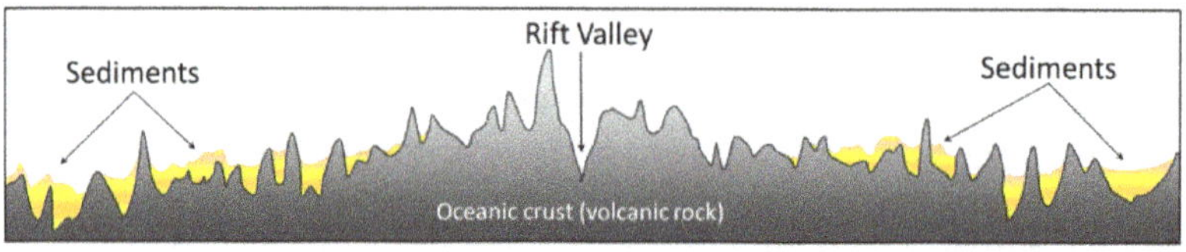

Figure 9: Topographic section at an ocean ridge based on reflection seismic data. Sediments are not thick enough to be detectable near the ridge, but get thicker on either side. The diagram represents approximately 50 km width, and has a 10x vertical exaggeration. [SE]

In the early 1950s, Edward Bullard, who spent time at the University of Toronto but is mostly associated with Cambridge University, developed a probe for measuring the flow of heat from the ocean floor. Bullard and colleagues found the rate to be higher than average along the ridges, and lower than average in the trench areas. Although Bullard was a plate-tectonics sceptic, these features were interpreted to indicate that there is convection within the mantle — the areas of high heat flow being correlated with upward convection of hot mantle material, and the areas of low heat flow being correlated with downward convection.

With developments of networks of seismographic stations in the 1950s, it became possible to plot the locations and depths of both major and minor earthquakes with great accuracy. It was found that there is a remarkable correspondence between earthquakes and both the mid-ocean ridges and the deep ocean trenches. In 1954

Gutenberg and Richter showed that the ocean-ridge earthquakes were all relatively shallow, and confirmed what had first been shown by Benioff in the 1930s — that earthquakes in the vicinity of ocean trenches were both shallow and deep, but that the deeper ones were situated progressively farther inland from the trenches (Figure 10).

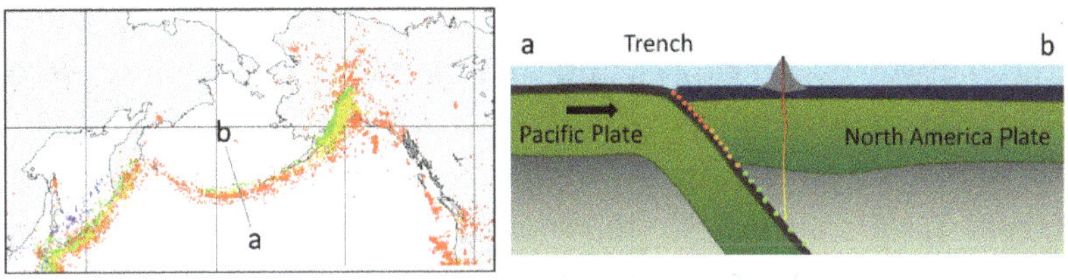

Figure 10: Cross-section through the Aleutian subduction zone with a depiction of the increasing depth of earthquakes "inshore" from the trench. [SE]

In the 1950s, scientists from the Scripps Oceanographic Institute in California persuaded the U.S. Coast Guard to include magnetometer readings on one of their expeditions to study ocean floor topography. The first comprehensive magnetic data set was compiled in 1958 for an area off the coast of B.C. and Washington State. This survey revealed a bewildering pattern of low and high magnetic intensity in sea-floor rocks (Figure 10.11). When the data were first plotted on a map in 1961, nobody understood them — not even the scientists who collected them. Although the patterns made even less sense than the stripes on a zebra, many thousands of kilometres of magnetic surveys were conducted over the next several years.

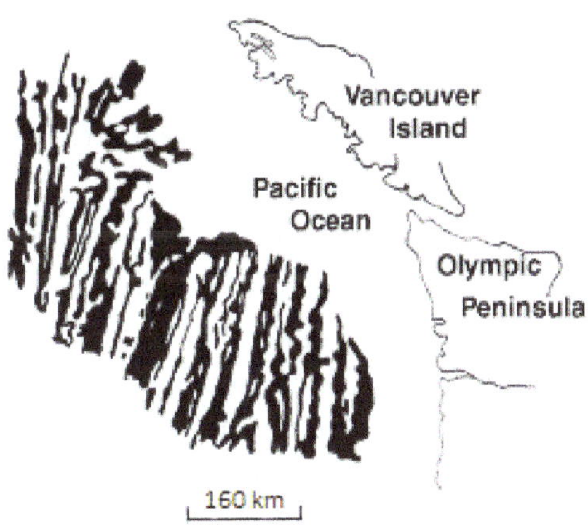

Figure 11: Pattern of sea-floor magnetism off of the west coast of British Columbia and Washington [SE after http://geomaps.wr.usgs.gov/parks/noca/nocageol4c.html, adapted from: Raff, A and Mason, R, 1961, Magnetic survey off the west coast of North America, 40° N to 52° N latitude, Geol. Soc. America Bulletin, V. 72, p. 267-270.]

The wealth of new data from the oceans began to significantly influence geological thinking in the 1960s. In 1960, Harold Hess, a widely respected geologist from Princeton University, advanced a theory with many of the elements that we now accept as plate tectonics. He maintained some uncertainty about his proposal however, and in order to deflect criticism from mainstream geologists, he labelled it geopoetry. In fact, until 1962, Hess didn't even put his ideas in writing – except internally to the U.S. Navy (which funded his research) – but presented them mostly in lectures and seminars. Hess proposed that new sea floor was generated from mantle material at the ocean ridges, and that old sea floor was dragged down at the ocean trenches and re-incorporated into the mantle. He suggested that the process was driven by mantle convection currents, rising at the ridges and descending at the trenches (Figure 12). He also suggested that the less-dense continental crust did not descend with oceanic crust into trenches, but that colliding land masses were thrust up to form mountains. Hess's theory formed the basis for our ideas on sea-floor spreading and continental drift, but it did not deal with the concept that the crust is made up of specific plates. Although the Hess model was not roundly criticized, it was not widely accepted (especially in the U.S.), partly because it was not well supported by hard evidence.

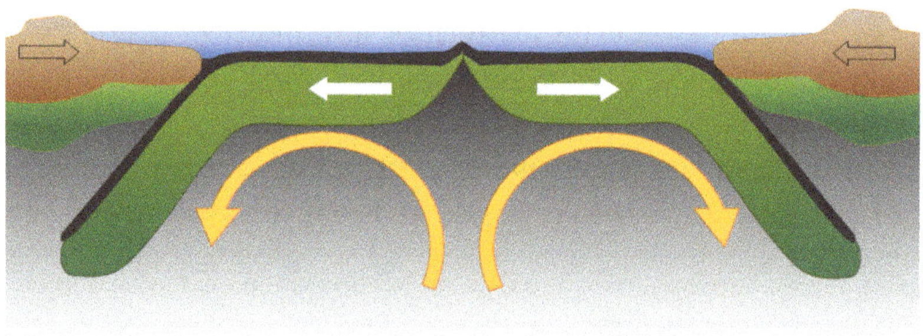

Figure 12: A representation of Harold Hess's model for sea-floor spreading and subduction

Collection of magnetic data from the oceans continued in the early 1960s, but still nobody could explain the origin of the zebra-like patterns. Most assumed that they were related to variations in the composition of the rocks – such as variations in the amount of magnetite – as this is a common explanation for magnetic variations in rocks of the continental crust. The first real understanding of the significance of the striped anomalies was the interpretation by Fred Vine, a Cambridge graduate student. Vine was examining magnetic data from the Indian Ocean and, like others before, he noted the symmetry of the magnetic patterns with respect to the oceanic ridge.

At the same time, other researchers, led by groups in California and New Zealand, were studying the phenomenon of reversals in Earth's magnetic field. They were trying to determine when such reversals had taken place over the past several million years by analyzing the magnetic characteristics of hundreds of samples from basaltic flows. As discussed in Chapter 9, it is evident that Earth's magnetic field becomes weakened periodically and then virtually non-existent, before becoming re-established with the reverse polarity. During periods of reversed polarity, a compass would point south instead of north.

The time scale of magnetic reversals is irregular. For example, the present "normal" event, known as the Bruhnes magnetic chron, has persisted for about 780,000 years. This was preceded by a 190,000-year reversed event; a 50,000-year normal event known as Jaramillo; and then a 700,000-year reversed event .

In a paper published in September 1963, Vine and his PhD supervisor Drummond Matthews proposed that the patterns associated with ridges were related to the magnetic reversals, and that oceanic crust created from cooling basalt during a normal event would have polarity aligned with the present magnetic field, and thus would produce a positive anomaly (a black stripe on the sea-floor magnetic map), whereas oceanic crust created during a reversed event would have polarity opposite to the present field and thus would produce a negative magnetic anomaly (a white stripe). The same idea had been put forward a few months earlier by Lawrence Morley, of the Geological Survey of Canada; however, his papers submitted earlier in 1963 to Nature and The Journal of Geophysical Research were rejected. Many people refer to the idea as the Vine-Matthews-Morley (VMM) hypothesis.

Vine, Matthews, and Morley were the first to show this type of correspondence between the relative widths of the stripes and the periods of the magnetic reversals. The VMM hypothesis was confirmed within a few years when magnetic data were compiled from spreading ridges around the world. It was shown that the same general magnetic patterns were present straddling each ridge, although the widths of the anomalies varied according to the spreading rates characteristic of the different ridges. It was also shown that the patterns corresponded with the chronology of Earth's magnetic field reversals. This global consistency provided strong support for the VMM hypothesis and led to rejection of the other explanations for the magnetic anomalies.

In 1963, J. Tuzo Wilson of the University of Toronto proposed the idea of a mantle plume or hot spot — a place where hot mantle material rises in a stationary and semi-permanent plume, and affects the overlying crust. He based this hypothesis partly on the distribution of the Hawaiian and Emperor Seamount island chains in the Pacific Ocean (Figure 13). The volcanic rock making up these islands gets progressively younger toward the southeast, culminating with the island of Hawaii itself, which consists of rock that is almost all younger than 1 Ma. Wilson suggested that a stationary plume of hot upwelling mantle material is the source of the Hawaiian volcanism, and that the ocean crust of the Pacific Plate is moving toward the northwest over this hot spot. Near the Midway Islands, the chain takes a pronounced change in direction, from northwest-southeast for the Hawaiian Islands and to nearly north-south for the Emperor Seamounts. This change is widely ascribed to a change in direction of the Pacific Plate moving over the stationary mantle plume, but a more plausible explanation is that the Hawaiian mantle plume has not actually been stationary throughout its history, and in fact moved at least 2,000 km south over the period between 81 and 45 Ma.J. A. Tarduno et al., 2003, The Emperor Seamounts: Southward Motion of the Hawaiian Hotspot Plume in Earth's Mantle, Science 301 (5636): 1064–1069.

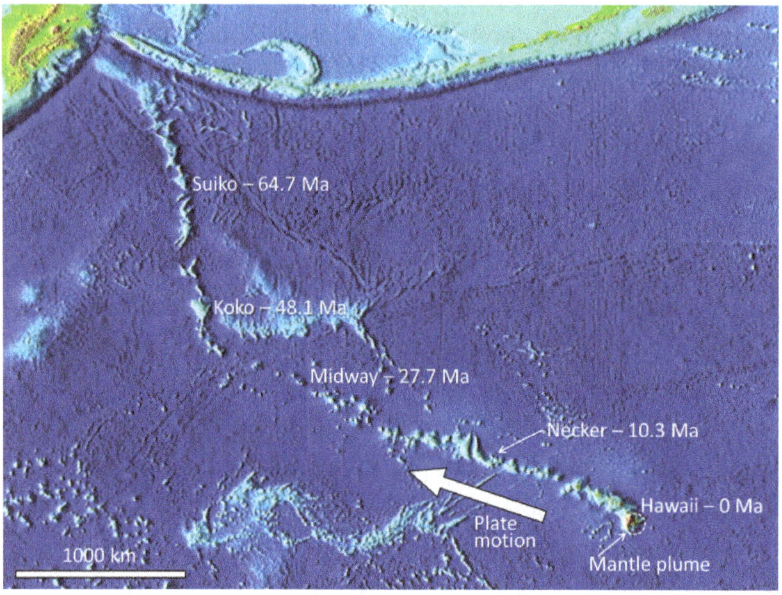

Figure 13: The ages of the Hawaiian Islands and the Emperor Seamounts in relation to the location of the Hawaiian mantle plume [SE. Basemap from the National Geophysical Data Centre, accessed at: http://en.wikipedia.org/wiki/Hotspot_(geology)#/ media/File:Hawaii_hotspot.jpg.]

10.4 PLATES, PLATE MOTIONS, AND PLATE-BOUNDARY PROCESSES

Continental drift and sea-floor spreading became widely accepted around 1965 as more and more geologists started thinking in these terms. By the end of 1967, Earth's surface had been mapped into a series of plates (Figure 16). The major plates are Eurasia, Pacific, India, Australia, North America, South America, Africa, and Antarctic. There are also numerous small plates (e.g., Juan de Fuca, Nazca, Scotia, Philippine, Caribbean), and many very small plates or sub-plates. For example the Juan de Fuca Plate is actually three separate plates (Gorda, Juan de Fuca, and Explorer) that all move in the same general direction but at slightly different rates.

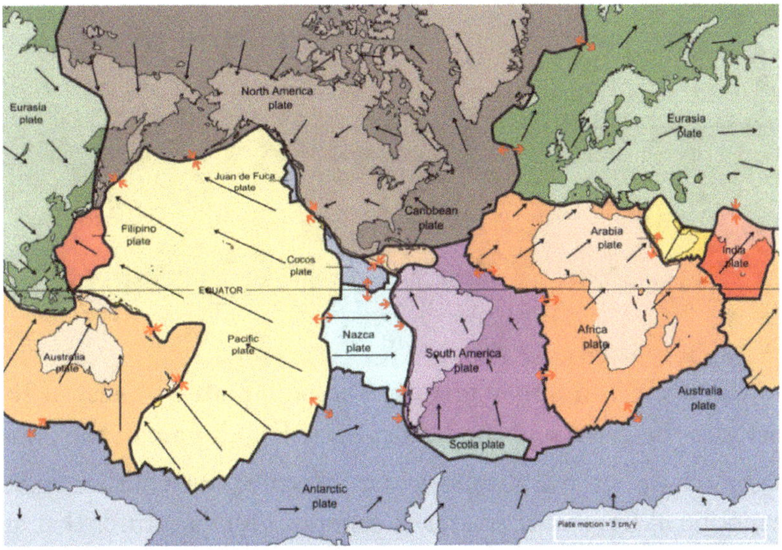

Figure 16: A map showing 15 of the Earth's tectonic plates and the approximate rates and directions of plate motions. [SE after USGS, http://en.wikipedia.org/wiki/Plate_tectonics#/media/File:Plates_tect2_en.svg

Rates of motions of the major plates range from less than 1 cm/y to over 10 cm/y. The Pacific Plate is the fastest at over 10 cm/y in some areas, followed by the Australian and Nazca Plates. The North American Plate is one of the slowest, averaging around 1 cm/y in the south up to almost 4 cm/y in the north.

Plates move as rigid bodies, so it may seem surprising that the North American Plate can be moving at different rates in different places. The explanation is that plates move in a rotational manner. The North American Plate, for example, rotates counter-clockwise; the Eurasian Plate rotates clockwise.

Boundaries between the plates are of three types: divergent (i.e., moving apart), convergent (i.e., moving together), and transform (moving side by side). Before we talk about processes at plate boundaries, it's important to point out that there are never gaps between plates. The plates are made up of crust and the lithospheric part of the mantle (Figure 17), and even though they are moving all the time, and in different directions, there is never a significant amount of space between them. Plates are thought to move along the lithosphere-asthenosphere boundary, as the asthenosphere is the zone of partial melting. It is assumed that the relative lack of strength of the partial melting zone facilitates the sliding of the lithospheric plates.

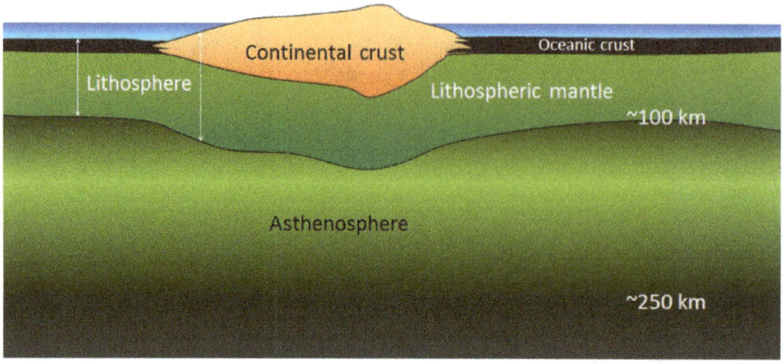

Figure 17: The crust and upper mantle. Tectonic plates consist of lithosphere, which includes the crust and the lithospheric (rigid) part of the mantle. [SE]

At spreading centres, the lithospheric mantle may be very thin because the upward convective motion of hot mantle material generates temperatures that are too high for the existence of a significant thickness of rigid lithosphere (Figure 12). The fact that the plates include both crustal material and lithospheric mantle material makes it possible for a single plate to be made up of both oceanic and continental crust. For example, the North American Plate includes most of North America, plus half of the northern Atlantic Ocean. Similarly the South American Plate extends across the western part of the southern Atlantic Ocean, while the European and African plates each include part of the eastern Atlantic Ocean. The Pacific Plate is almost entirely oceanic, but it does include the part of California west of the San Andreas Fault.

Divergent Boundaries

Divergent boundaries are spreading boundaries, where new oceanic crust is created from magma derived from partial melting of the mantle caused by decompression as hot mantle rock from depth is moved toward the surface (Figure 18). The triangular zone of partial melting near the ridge crest is approximately 60 km thick and the proportion of magma is about 10% of the rock volume, thus producing crust that is about 6 km thick. Most divergent boundaries are located at the oceanic ridges (although some are on land), and the crustal material created at a spreading boundary is always oceanic in character; in other words, it is mafic igneous rock (e.g., basalt or gabbro, rich in ferromagnesian minerals). Spreading rates vary considerably, from 1 cm/y to 3 cm/y in the Atlantic, to between 6 cm/y and 10 cm/y in the Pacific. Some of the processes taking place in this setting include:

- Magma from the mantle pushing up to fill the voids left by divergence of the two plates
- Pillow lavas forming where magma is pushed out into seawater (Figure 19)
- Vertical sheeted dykes intruding into cracks resulting from the spreading

- Magma cooling more slowly in the lower part of the new crust and forming gabbro bodies

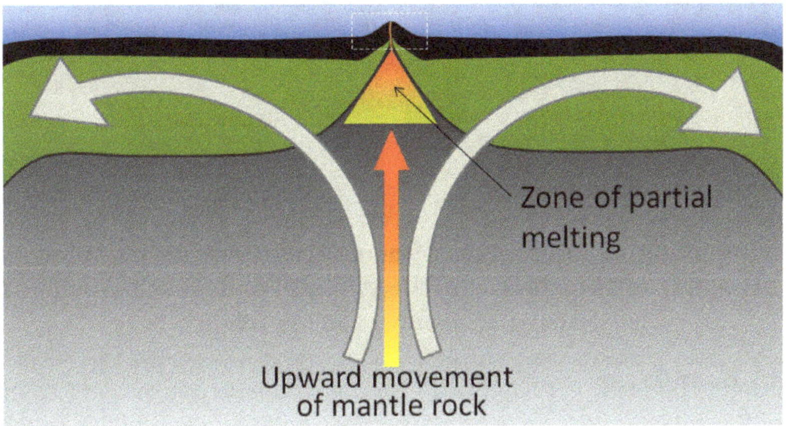

Figure 18: The general processes that take place at a divergent boundary. The area within the dashed white rectangle is shown in Figure 19.

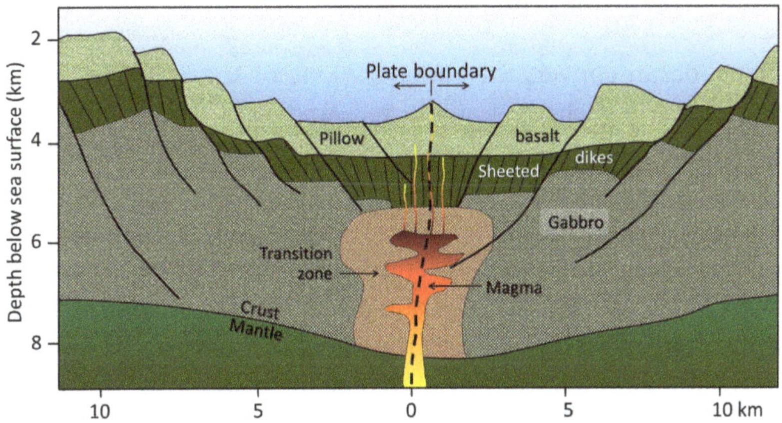

Figure 19: Depiction of the processes and materials formed at a divergent boundary

Spreading is hypothesized to start within a continental area with up-warping or doming related to an underlying mantle plume or series of mantle plumes. The buoyancy of the mantle plume material creates a dome within the crust, causing it to fracture in a radial pattern, with three arms spaced at approximately 120° (Figure 20). When a series of mantle plumes exists beneath a large continent, the resulting rifts may align and lead to the formation of a rift valley (such as the present-day Great Rift Valley in eastern Africa). It is suggested that this type of valley eventually develops into a linear sea (such as the present-day Red Sea), and finally into an ocean (such as the Atlantic). It is likely that as many as 20 mantle plumes, many of which still exist, were responsible for the initiation of the rifting of Pangea along what is now the mid-Atlantic ridge (see Figure 14).

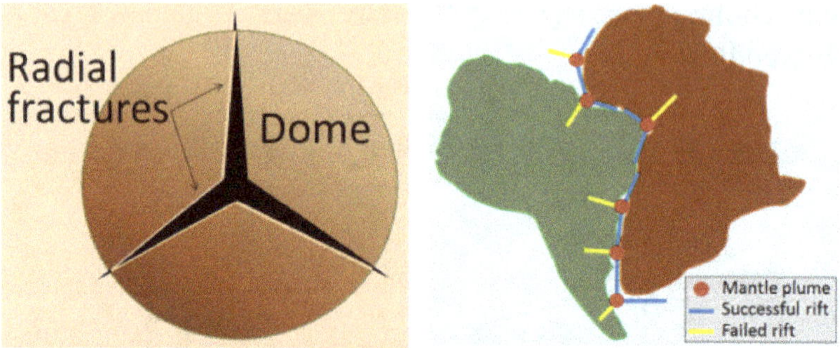

Figure 20: Depiction of the process of dome and three-part rift formation (left) and of continental rifting between the African and South American parts of Pangea at around 200 Ma (right)

Convergent Boundaries

Convergent boundaries, where two plates are moving toward each other, are of three types, depending on the type of crust present on either side of the boundary — oceanic or continental. The types are ocean-ocean, ocean-continent, and continent-continent.

At an ocean-ocean convergent boundary, one of the plates (oceanic crust and lithospheric mantle) is pushed, or subducted**,** under the other. Often it is the older and colder plate that is denser and subducts beneath the younger and hotter plate. There is commonly an ocean trench along the boundary. The subducted lithosphere descends into the hot mantle at a relatively shallow angle close to the subduction zone, but at steeper angles farther down (up to about 45°). As discussed in the context of subduction-related volcanism in Chapter 4, the significant volume of water within the subducting material is released as the subducting crust is heated. This water is mostly derived from alteration of pyroxene and olivine to serpentine near the spreading ridge shortly after the rock's formation. It mixes with the overlying mantle, and the addition of water to the hot mantle lowers the crust's melting point and leads to the formation of magma (flux melting). The magma, which is lighter than the surrounding mantle material, rises through the mantle and the overlying oceanic crust to the ocean floor where it creates a chain of volcanic islands known as an island arc. A mature island arc develops into a chain of relatively large islands (such as Japan or Indonesia) as more and more volcanic material is extruded and sedimentary rocks accumulate around the islands.

As described above in the context of Benioff zones (Figure 10), earthquakes take place close to the boundary between the subducting crust and the overriding crust. The largest earthquakes occur near the surface where the subducting plate is still cold and strong.

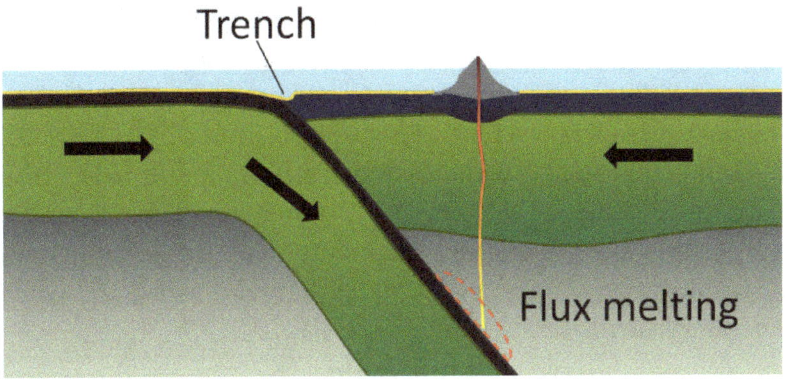

Figure 21: Configuration and processes of an ocean-ocean convergent boundary [SE]

Examples of ocean-ocean convergent zones are subduction of the Pacific Plate south of Alaska (Aleutian Islands) and west of the Philippines, subduction of the India Plate south of Indonesia, and subduction of the Atlantic Plate beneath the Caribbean Plate (Figure 21).

At an ocean-continent convergent boundary, the oceanic plate is pushed under the continental plate in the same manner as at an ocean-ocean boundary. Sediment that has accumulated on the continental slope is thrust up into an accretionary wedge, and compression leads to thrusting within the continental plate (Figure 22). The mafic magma produced adjacent to the subduction zone rises to the base of the continental crust and leads to partial melting of the crustal rock. The resulting magma ascends through the crust, producing a mountain chain with many volcanoes.

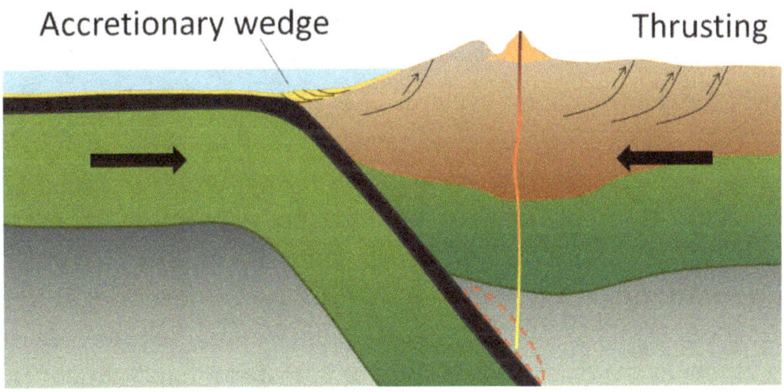

Figure 22: Configuration and processes of an ocean-continent convergent boundary [SE]

Examples of ocean-continent convergent boundaries are subduction of the Nazca Plate under South America (which has created the Andes Range) and subduction of the Juan de Fuca Plate under North America (creating the mountains Garibaldi, Baker, St. Helens, Rainier, Hood, and Shasta, collectively known as the Cascade Range).

A continent-continent collision occurs when a continent or large island that has been moved along with subducting oceanic crust collides with another continent (Figure 23). The colliding continental material will not be subducted because it is too light (i.e., because it is composed largely of light continental rocks [SIAL]), but the root of the oceanic plate will eventually break off and sink into the mantle. There is tremendous deformation of the pre-existing continental rocks, and creation of mountains from that rock, from any sediments that had accumulated along the shores (i.e., within geosynclines) of both continental masses, and commonly also from some ocean crust and upper mantle material.

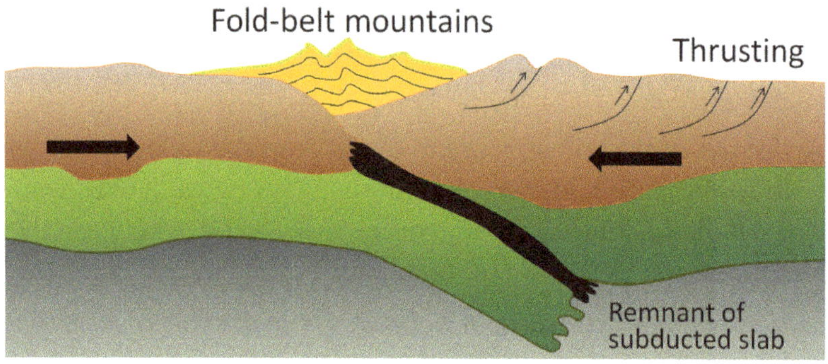

Figure 23: Configuration and processes of a continent-continent convergent boundary [SE]

Examples of continent-continent convergent boundaries are the collision of the India Plate with the Eurasian Plate, creating the Himalaya Mountains, and the collision of the African Plate with the Eurasian Plate, creating the series of ranges extending from the Alps in Europe to the Zagros Mountains in Iran. The Rocky Mountains in B.C. and Alberta are also a result of continent-continent collisions.

Transform boundaries exist where one plate slides past another without production or destruction of crustal material. As explained above, most transform faults connect segments of mid-ocean ridges and are thus ocean-ocean plate boundaries (Figure 15). Some transform faults connect continental parts of plates. An example is the San Andreas Fault, which connects the southern end of the Juan de Fuca Ridge with the northern end of the East Pacific Rise (ridge) in the Gulf of California (Figures 24 an 25). The part of California west of the San Andreas Fault and all of Baja California are on the Pacific Plate. Transform faults do not just connect divergent boundaries. For example, the Queen Charlotte Fault connects the north end of the Juan de Fuca Ridge, starting at the north end of Vancouver Island, to the Aleutian subduction zone.

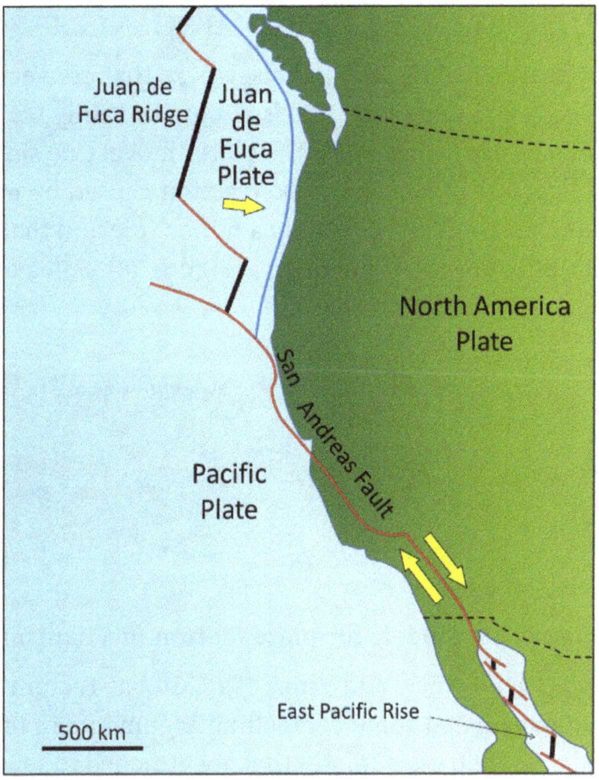

Figure 24: The San Andreas Fault extends from the north end of the East Pacific Rise in the Gulf of California to the southern end of the Juan de Fuca Ridge. All of the red lines on this map are transform faults. [SE]

Figure 25: The San Andreas Fault at Parkfield in central California. The person with the orange shirt is standing on the Pacific Plate and the person at the far side of the bridge is on the North American Plate. The bridge is designed to slide on its foundation. [SE]

10.5 PLATE MOTION MECHANISMS

It has been often repeated in this text and elsewhere that convection of the mantle is critical to plate tectonics, and while this is almost certainly so, there is still some debate about the actual forces that make the plates move. One side in the argument holds that the plates are only moved by the traction caused by mantle convection. The other side holds that traction plays only a minor role and that two other forces, ridge-push and slab-pull, are more important (Figure 28). Some argue that the real answer lies somewhere in between.

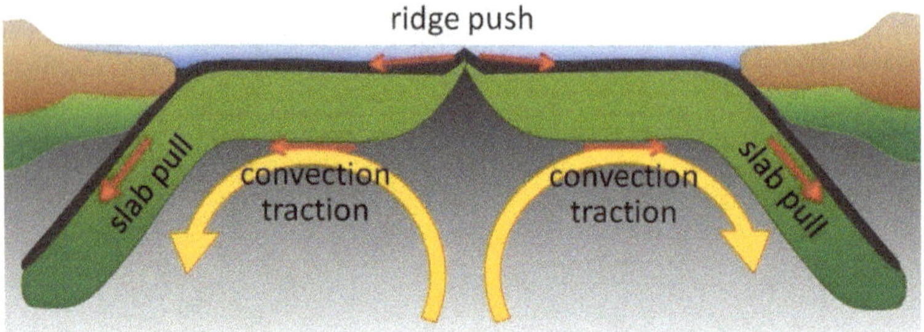

Figure 28: Models for plate motion mechanisms

Kearey and Vine (1996)Kearey and Vine , 1996, Global Tectonics (2ed), Blackwell Science Ltd., Oxford have listed some compelling arguments in favour of the ridge-push/slab-pull model, as follows: (a) plates that are attached to subducting slabs (e.g., Pacific, Australian, and Nazca Plates) move the fastest, and plates that are not (e.g., North American, South American, Eurasian, and African Plates) move significantly slower; (b) in order for the traction model to apply, the mantle would have to be moving about five times faster than the plates are moving (because the coupling between the partially liquid asthenosphere and the plates is not strong), and such high rates of convection are not supported by geophysical models; and (c) although large plates have potential for much higher convection traction, plate velocity is not related to plate area.

In the ridge-push/slab-pull model, which is the one that has been adopted by most geologists working on plate-tectonic problems, the lithosphere is the upper surface of the convection cells, as is illustrated in Figure 29.

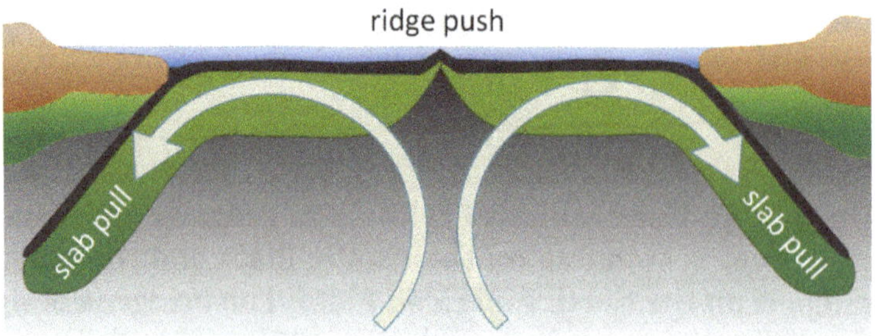

Figure 29 The ridge-push/slab-pull model for plate motion, in which the lithosphere is the upper surface of the convective systems. [SE]

Although ridge-push/slab-pull is the favoured mechanism for plate motion, it's important not to underestimate the role of mantle convection. Without convection, there would be no ridges to push from because upward convection brings hot buoyant rock to surface. Furthermore, many plates, including our own North American Plate, move along nicely — albeit slowly — without any slab-pull happening.

AT A GLANCE

Alfred Wegener – the Father of Plate Tectonics	The evidence for continental drift in the early 20th century included the matching of continental shapes on either side of the Atlantic and the geological and fossil matchups between continents that are now thousands of kilometres apart.
Global Geological Models of the Early 20th Century	The established theories of global geology were permanentism and contractionism, but neither of these theories was able to explain some of the evidence that supported the idea of continental drift.
Geological Renaissance of the Mid-20th Century	Giant strides were made in understanding Earth during the middle decades of the 20th century, including discovering magnetic evidence of continental drift, mapping the topography of the ocean floor, describing the depth relationships of earthquakes along ocean trenches, measuring heat flow differences in various parts of the ocean floor, and mapping magnetic reversals on the sea floor. By the mid-1960s, the fundamentals of the theory of plate tectonics were in place.
Plates, Plate Motions, and Plate-Boundary Processes	Earth's lithosphere is made up of over 20 plates that are moving in different directions at rates of between 1 cm/y and 10 cm/y. The three types of plate boundaries are divergent (plates moving apart and new crust forming), convergent (plates moving together and one being subducted), and transform (plates moving side by side). Divergent boundaries form where existing plates are rifted apart, and it is hypothesized that this is caused by a series of mantle plumes. Subduction zones are assumed to form where accumulation of sediment at a passive margin leads to separation of oceanic and continental lithosphere. Supercontinents form and break up through these processes.
Mechanisms for Plate Motion	It is widely believed that ridge-push and slab-pull are the main mechanisms for plate motion, as opposed to traction by mantle convection. Mantle convection is a key factor for producing the conditions necessary for ridge-push and slab-pull.

CHAPTER 11
EARTHQUAKES

Earthquakes scare people ... a lot! That's not surprising because time and time again earthquakes have caused massive damage and many, many casualties. Anyone who has lived through a damaging earthquake cannot forget the experience (Figure 1). But geoscientists and engineers are getting better at understanding earthquakes, minimizing the amount of damage they cause, and reducing the number of people affected. People living in western Canada don't need to be frightened by earthquakes, but they do need to be prepared.

Figure 1: A schoolroom in Courtenay damaged by the 1946 Vancouver Island earthquake. If the earthquake had not happened on a Sunday, the casualties would have been much greater.

11.1 INTRODUCTION TO EARTHQUAKE?

An earthquake is the shaking caused by the rupture (breaking) and subsequent displacement of rocks (one body of rock moving with respect to another) beneath

Earth's surface.

A body of rock that is under stress becomes deformed. When the rock can no longer withstand the deformation, it breaks and the two sides slide past each other. Most earthquakes take place near plate boundaries, but not necessarily right on a boundary, and not necessarily even on a pre-existing fault.

The engineering principle of elastic deformation, which can be used to understand earthquakes, is illustrated in Figure 2. The stress applied to a rock — typically because of ongoing plate movement — results in strain or deformation of the rock (Figure 2b). Because most rock is strong (unlike loose sand, for example), it can withstand a significant amount of deformation without breaking. But every rock has a deformation limit and will rupture (break) once that limit is reached. At that point, in the case of rocks within the crust, the rock breaks and there is displacement along the rupture surface (Figure 2c). The magnitude of the earthquake depends on the extent of the area that breaks (the area of the rupture surface) and the average amount of displacement (sliding).

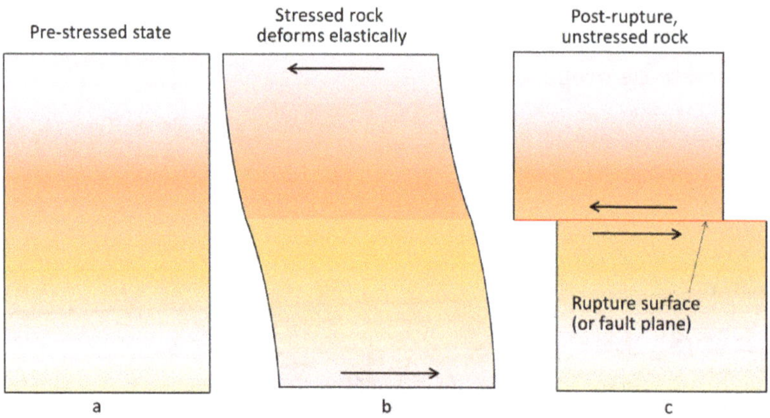

Figure 2: Depiction of the concept of elastic deformation and rupture, looking down. [SE]

The concept of a rupture surface, which is critical to understanding earthquakes, is illustrated in Figure 3. An earthquake does not happen at a point, it happens over an area within a plane, although not necessarily a flat plane. Within the area of the rupture surface, the amount of displacement is variable (Figure 3), and, by definition, it decreases to zero at the edges of the rupture surface because the rock beyond that point isn't displaced at all. The extent of a rupture surface and the amount of displacement will depend on a number of factors, including the type and strength of the rock, and the degree to which it was stressed beforehand.

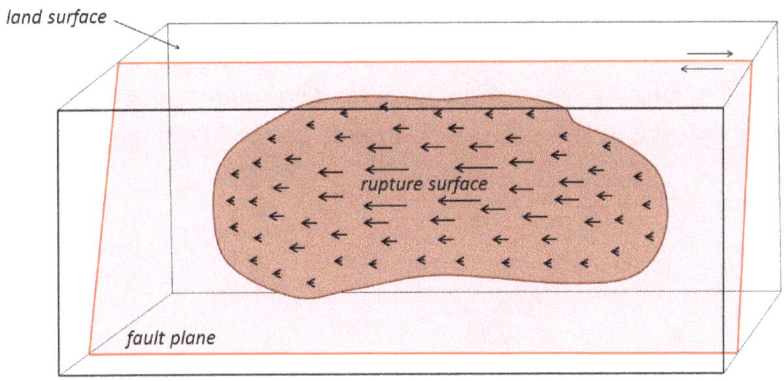

Figure 3: A rupture surface (dark pink), on a steeply dipping fault plane (light pink). The diagram represents a part of the crust that may be up to tens or hundreds of kilometres long. The rupture surface is the part of the fault plane along which displacement occurred. In this example, the near side of the fault is moving to the left, and the lengths of the arrows within the rupture surface represent relative amounts of displacement.

Earthquake rupture doesn't happen all at once; it starts at a single point and spreads rapidly from there. Depending on the extent of the rupture surface, the propagation of failures out from the point of initiation is typically completed within seconds to several tens of seconds (Figure 4). The initiation point isn't necessarily in the centre of the rupture surface; it may be close to one end, near the top, or near the bottom.

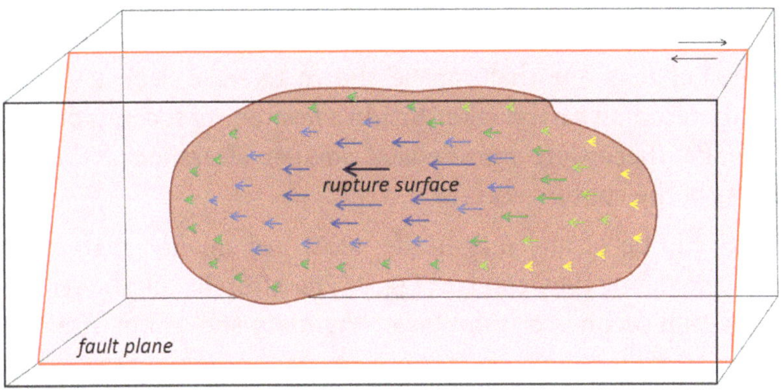

Figure 4: Propagation of failure on a rupture surface. In this case, the failure starts at the dark blue heavy arrow and propagates outward, reaching the left side first (green arrows) and the right side last (yellow arrows).

Figure 5 shows the distribution of immediate aftershocks associated with the 1989 Loma Prieta earthquake. Panel (b) is a section along the San Andreas Fault; this view is equivalent to what is shown in Figures 3 and 4. The area of red dots is the rupture surface; each red dot is a specific aftershock that was recorded on a seismometer. The hexagon labelled "main earthquake" represents the first or main shock. When that happened, the rock at that location broke and was displaced. That released the stress on that particular part of the fault, but it resulted in an increase of the stress on other nearby parts of the fault, and contributed to a cascade of smaller ruptures (aftershocks), in this case, over an area about 60 km long and 15 km wide.

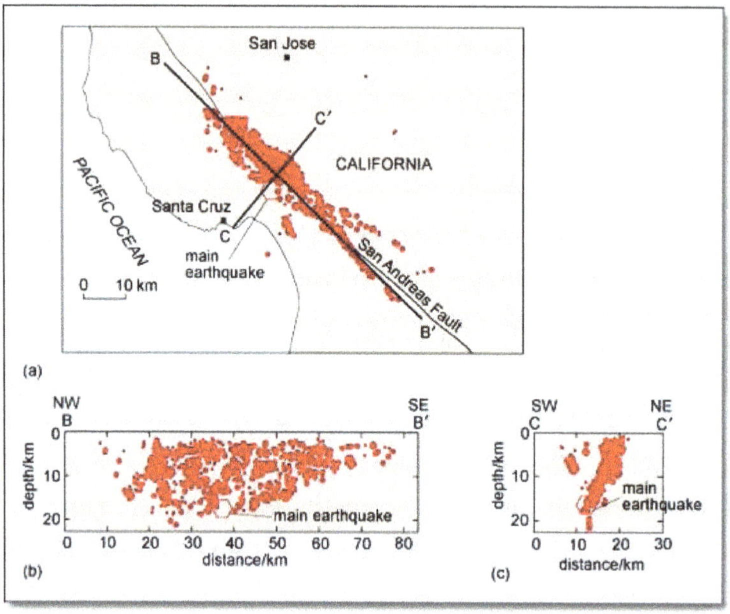

Figure 5: Distribution of the aftershocks of the 1989 M 6.9 Loma Prieta earthquake (a: plan view, b: section along the fault, c: section across the fault.)

So, what exactly is an aftershock then? An aftershock is an earthquake just like any other, but it is one that can be shown to have been triggered by stress transfer from a preceding earthquake. Within a few tens of seconds of the main Loma Prieta earthquake, there were hundreds of smaller aftershocks; their distribution defines the area of the rupture surface.

Aftershocks can be of any magnitude. Most are smaller than the earthquake that triggered them, but they can be bigger. The aftershocks shown in Figure 11.5 all happened within seconds or minutes of the main shock, but aftershocks can be delayed for hours, days, weeks, or even years. As already noted, aftershocks are related to stress transfer. For example, the main shock of the Loma Prieta earthquake triggered aftershocks in the immediate area, which triggered more in the surrounding area, eventually extending for 30 km along the fault in each direction and for 15 km toward the surface. But the earthquake as a whole also changed the stress on adjacent parts of the San Andreas Fault. This effect, which has been modelled for numerous earthquakes and active faults around the world, is depicted in Figure 6. Stress was reduced in the area of the rupture (blue), but was increased at either end of the rupture surface (red and yellow).

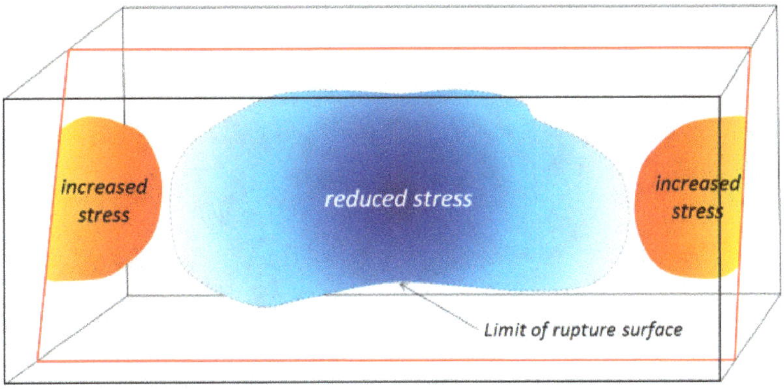

Figure 6: Depiction of stress changes related to an earthquake. Stress decreases in the area of the rupture surface, but increases on adjacent parts of the fault. [by SE based on data from 2010 Laguna Salada earthquake by Stein and Toda at: http://supersites.earthobservations.org/Baja_stress.png]

Stress transfer isn't necessarily restricted to the fault along which an earthquake happened. It will affect the rocks in general around the site of the earthquake and may lead to increased stress on other faults in the region. The effects of stress transfer don't necessarily show up right away. Segments of faults are typically in some state of stress, and the transfer of stress from another area is only rarely enough to push a fault segment beyond its limits to the point of rupture. The stress that is added by stress transfer accumulates along with the ongoing buildup of stress from plate motion and eventually leads to another earthquake.

Episodic tremor and slip (ETS)

Episodic tremor and slip (ETS) is periodic slow sliding along part of a subduction boundary. It does not produce recognizable earthquakes, but does produce seismic tremor (rapid seismic vibrations on a seismometer). It was first discovered on the Vancouver Island part of the Cascadia subduction zone by Geological Survey of Canada geologists Herb Dragert and Gary Rogers.*

The boundary between the subducting Juan de Fuca Plate and the North America Plate can be divided into three segments, as shown below. The cold upper part of the boundary is locked. The plates are stuck and don't move, except with very large earthquakes that happen approximately every 500 years (the last one was M8.5+ in January 26, 1700). The warm lower part of the boundary is sliding continuously because the warm rock is weaker. The central part of the boundary isn't cold enough to be stuck, but isn't warm enough to slide continuously. Instead it slips episodically, approximately every 14 months for about 2 weeks, moving a few centimetres each time.

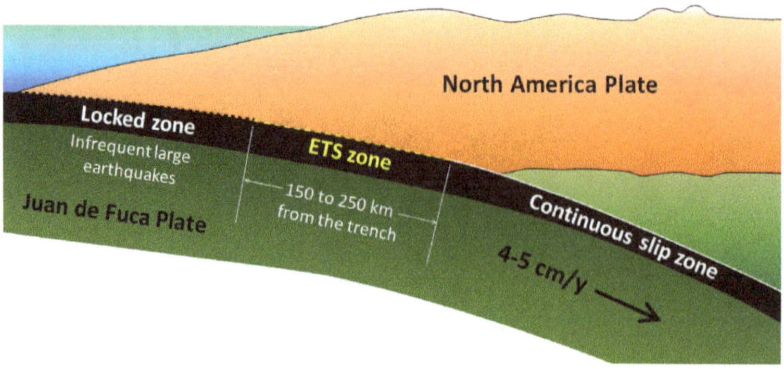

You might be inclined to think that it's a good thing that there is periodic slip on this part of the plate because it releases some of the tension and reduces the risk of a large earthquake. In fact, the opposite is likely the case. The movement along the ETS part of the plate boundary acts like a medium-sized earthquake and leads to stress transfer to the adjacent locked part of the plate. Approximately every 14 months, during the two-week ETS period, there is a transfer of stress to the shallow locked part of the Cascadia subduction zone, and therefore an increased chance of a large earthquake.

11.2 EARTHQUAKES AND PLATE TECTONICS

The distribution of earthquakes across the globe is shown in Figure 7. It is relatively easy to see the relationships between earthquakes and the plate boundaries. Along divergent boundaries like the mid-Atlantic ridge and the East Pacific Rise, earthquakes are common, but restricted to a narrow zone close to the ridge, and consistently at less than 30 km depth. Shallow earthquakes are also common along transform faults, such as the San Andreas Fault. Along subduction zones, as we saw in Chapter 10, earthquakes are very abundant, and they are increasingly deep on the landward side of the subduction zone.

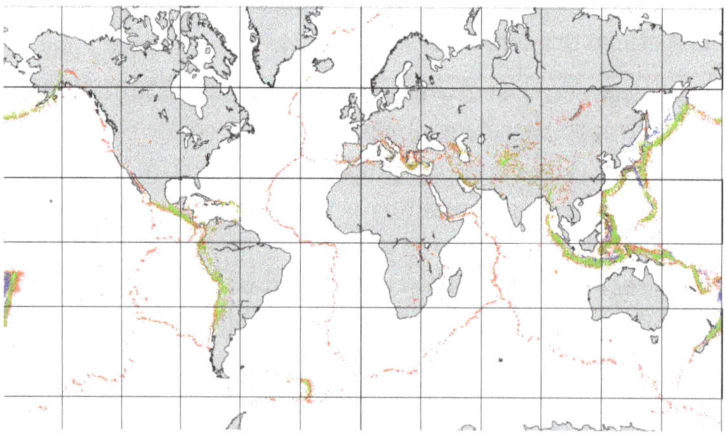

Figure 7: General distribution of global earthquakes of magnitude 4 and greater from 2004 to 2011, colour coded by depth (red: 0-33 km, orange 33-70 km,

green: 70-300 km, blue: 300-700 km) [from Dale Sawyer, Rice University, http://plateboundary.rice.edu ,used with permission]

Earthquakes are also relatively common at a few intraplate locations. Some are related to the buildup of stress due to continental rifting or the transfer of stress from other regions, and some are not well understood. Examples of intraplate earthquake regions include the Great Rift Valley area of Africa, the Tibet region of China, and the Lake Baikal area of Russia.

Earthquakes at Divergent and Transform Boundaries

Figure 8 provides a closer look at magnitude (M) 4 and larger earthquakes in an area of divergent boundaries in the mid-Atlantic region near the equator. Here, as we saw in Chapter 10, the segments of the mid-Atlantic ridge are offset by some long transform faults. Most of the earthquakes are located along the transform faults, rather than along the spreading segments, although there are clusters of earthquakes at some of the ridge-transform boundaries. Some earthquakes do occur on spreading ridges, but they tend to be small and infrequent because of the relatively high rock temperatures in the areas where spreading is taking place.

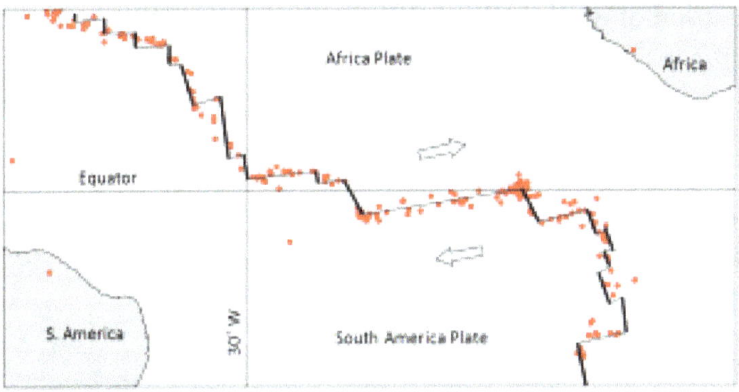

Figure 8: Distribution of earthquakes of M4 and greater in the area of the mid-Atlantic ridge near the equator from 1990 to 1996. All are at a depth of 0 to 33 km [SE after Dale Sawyer, Rice University, http://plateboundary.rice.edu]

Earthquakes at Convergent Boundaries

The distribution and depths of earthquakes in the Caribbean and Central America area are shown in Figure 9. In this region, the Cocos Plate is subducting beneath the North America and Caribbean Plates (ocean-continent convergence), and the South and North America Plates are subducting beneath the Caribbean Plate (ocean-ocean convergence). In both cases, the earthquakes get deeper with distance from the trench. In Figure 9, the South America Plate is shown as being subducted beneath the Caribbean Plate in the area north of Colombia, but since there is almost no earthquake activity along this zone, it is questionable whether subduction is actually taking place.

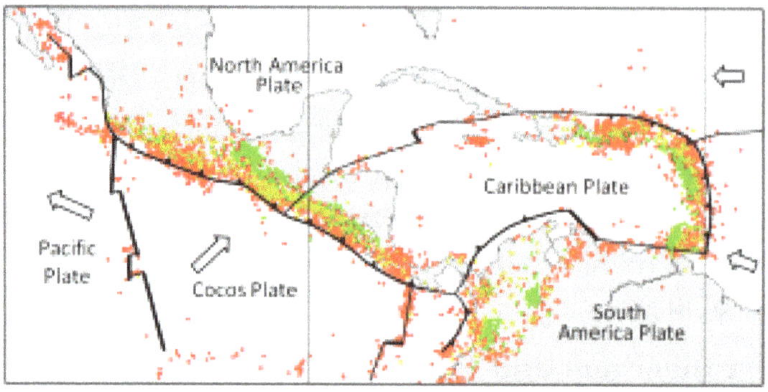

Figure 9 Distribution of earthquakes of M4 and greater in the Central America region from 1990 to 1996 (red: 0-33 km, orange: 33-70 km, green: 70-300 km, blue: 300-700 km) (Spreading ridges are heavy lines, subduction zones are toothed lines, and transform faults are light lines.) [SE after Dale Sawyer, Rice University, http://plateboundary.rice.edu]

There are also various divergent and transform boundaries in the area shown in Figure 9, and as we've seen in the mid-Atlantic area, most of these earthquakes occur along the transform faults.

The distribution of earthquakes with depth in the Kuril Islands of Russia in the northwest Pacific is shown in Figure 10. This is an ocean-ocean convergent boundary. The small red and yellow dots show background seismicity over a number of years, while the larger white dots are individual shocks associated with a M6.9 earthquake in April 2009. The relatively large earthquake took place on the upper part of the plate boundary between 60 km and 140 km inland from the trench. As we saw for the Cascadia subduction zone, this is where large subduction earthquakes are expected to occur.

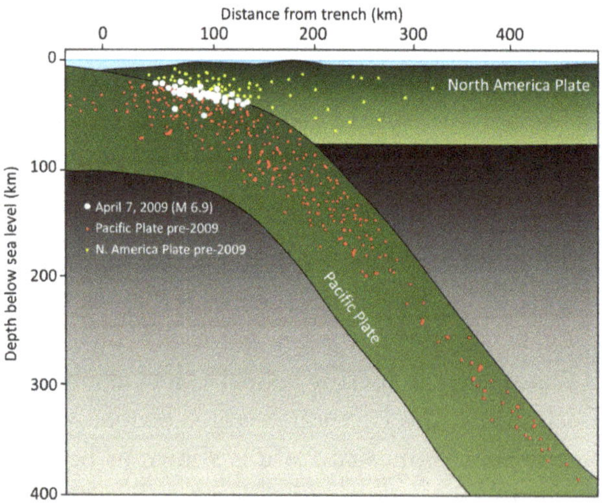

Figure 10: Distribution of earthquakes in the area of the Kuril Islands, Russia (just north of Japan) (White dots represent the April 2009 M6.9 earthquake. Red

and yellow dots are from background seismicity over several years prior to 2009.) [SE after Gavin Hayes, from data at http://earthquake.usgs.gov/earthquakes/eqarchives/subduction_zone/us2009fdak/szgc/ku6_trench.pdf]

In fact, all of the very large earthquakes — M9 or higher — take place at subduction boundaries because there is the potential for a greater width of rupture zone on a gently dipping boundary than on a steep transform boundary. The largest earthquakes on transform boundaries are in the order of M8.

The background seismicity at this convergent boundary, and on other similar ones, is predominantly near the upper side of the subducting plate. The frequency of earthquakes is greatest near the surface and especially around the area where large subduction quakes happen, but it extends to at least 400 km depth. There is also significant seismic activity in the overriding North America Plate, again most commonly near the region of large quakes, but also extending for a few hundred kilometres away from the plate boundary.

The distribution of earthquakes in the area of the India-Eurasia plate boundary is shown in Figure 11. This is a continent-continent convergent boundary, and it is generally assumed that although the India Plate continues to move north toward the Asia Plate, there is no actual subduction taking place. There are transform faults on either side of the India Plate in this area.

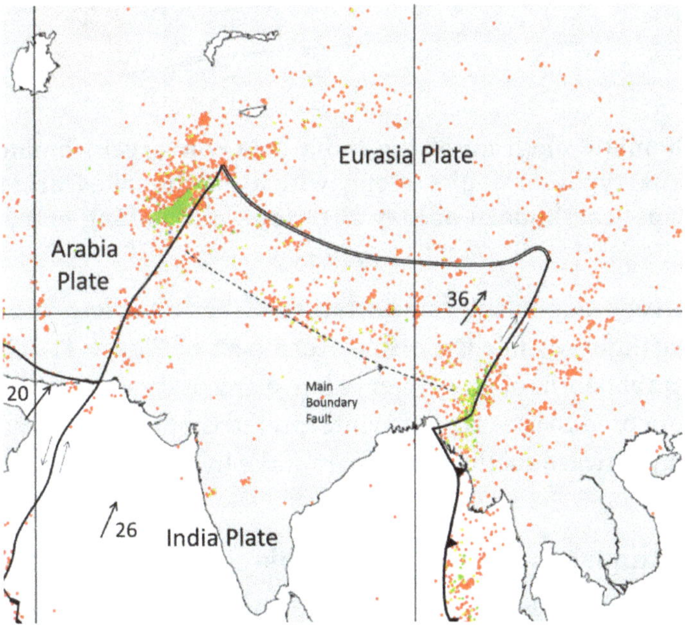

Figure 11: Distribution of earthquakes in the area where the India Plate is converging with the Asia Plate (data from 1990 to 1996, red: 0-33 km, orange: 33-70 km, green: 70-300 km). (Spreading ridges are heavy lines, subduction zones are toothed lines, and transform faults are light lines. The double line along the northern edge of the India Plate indicates convergence, but not subduction.

Plate motions are shown in mm/y.) [SE after Dale Sawyer, Rice University, http://plateboundary.rice.edu]

The entire northern India and southern Asia region is very seismically active. Earthquakes are common in northern India, Nepal, Bhutan, Bangladesh and adjacent parts of China, and throughout Pakistan and Afghanistan. Many of the earthquakes are related to the transform faults on either side of the India Plate, and most of the others are related to the significant tectonic squeezing caused by the continued convergence of the India and Asia Plates. That squeezing has caused the Asia Plate to be thrust over top of the India Plate, building the Himalayas and the Tibet Plateau to enormous heights. Most of the earthquakes of Figure 11 are related to the thrust faults shown in Figure 12 (and to hundreds of other similar ones that cannot be shown at this scale). The southernmost thrust fault in Figure 12 is equivalent to the Main Boundary Fault in Figure 11.

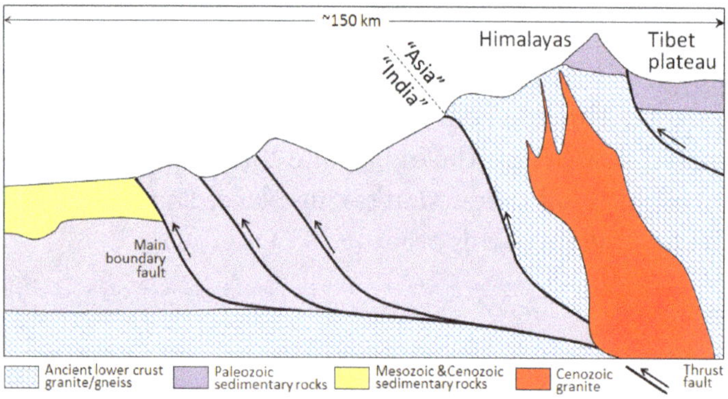

Figure 12 Schematic diagram of the India-Asia convergent boundary, showing examples of the types of faults along which earthquakes are focussed. The devastating Nepal earthquake of May 2015 took place along one of these thrust faults.

There is a very significant concentration of both shallow and deep (greater than 70 km) earthquakes in the northwestern part of Figure 11. This is northern Afghanistan, and at depths of more than 70 km, many of these earthquakes are within the mantle as opposed to the crust. It is interpreted that these deep earthquakes are caused by northwestward subduction of part of the India Plate beneath the Asia Plate in this area.

Exercise 1 Earthquakes in British Columbia

This map shows the incidence and magnitude of earthquakes in British Columbia over a one-month period in March and April 2015.

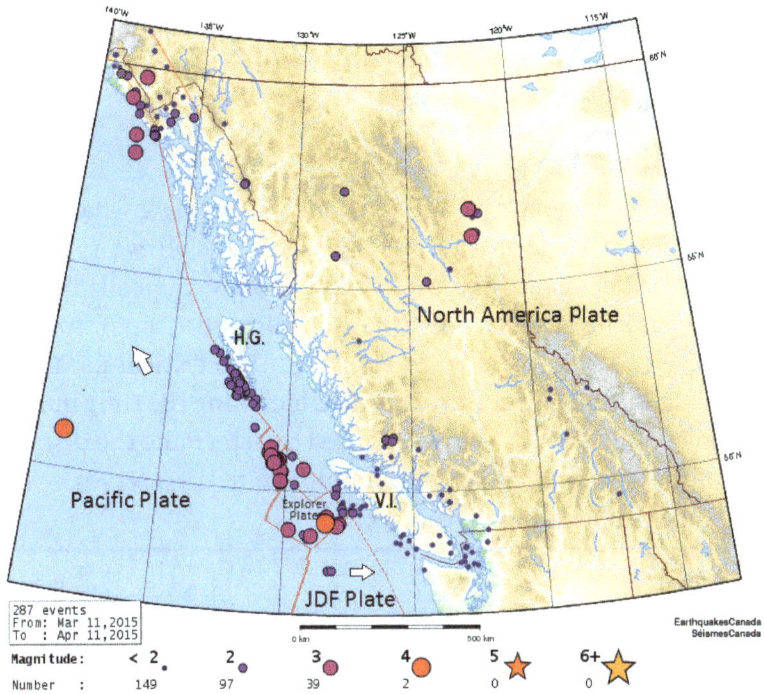

1. What is the likely origin of the earthquakes between the Juan de Fuca (JDF) and Explorer Plates?
2. The string of small earthquakes adjacent to Haida Gwaii (H.G.) coincides closely with the rupture surface of the 2012 M7.7 earthquake in that area. How might these earthquakes be related to that one?
3. Most of the earthquakes around Vancouver Island (V.I.) are relatively shallow. What is their likely origin?
4. Some of the earthquakes in B.C. are interpreted as being caused by natural gas extraction (including fracking). Which of the earthquakes here could fall into this category?

11.3 EARTHQUAKE MEASUREMENT

There are two main ways to measure earthquakes. The first of these is an estimate of the energy released, and the value is referred to as magnitude. This is the number that is typically used by the press when a big earthquake happens. It is often referred to as "Richter magnitude," but that is a misnomer, and it should be just "magnitude." There are many ways to measure magnitude — including Charles Richter's method developed in 1935 — but they are all ways to estimate the same number: the amount of energy released.

The other way of assessing the impact of an earthquake is to assess what people felt and how much damage was done. This is known as intensity. Intensity values are assigned to locations, rather than to the earthquake itself, and therefore intensity can vary widely, depending on the proximity to the earthquake and the types of materials and conditions of the subsurface.

Magnitude of Earthquakes

Before we look more closely at magnitude we need to review what we know about body waves, and look at surface waves. Body waves are of two types, P-waves, or primary or compression waves (like the compression of the coils of a spring), and S-waves, or secondary or shear waves (like the flick of a rope). An example of P and S seismic wave records is shown in Figure 13. The critical parameters for the measurement of Richter magnitude are labelled, including the time interval between the arrival of the P- and S-waves — which is used to determine the distance from the earthquake to the seismic station, and the amplitude of the S waves — which is used to estimate the magnitude of the earthquake.

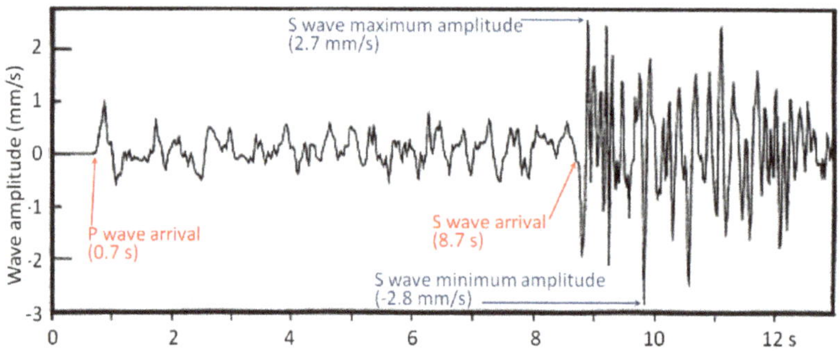

Figure 13: P-waves and S-waves from a small (M4) earthquake that took place near Vancouver Island in 1997. [SE]

When body waves (P or S) reach Earth's surface, some of their energy is transformed into surface waves, of which there are two main types, as illustrated in Figure 14. Rayleigh waves are characterized by vertical motion of the ground surface, like waves on water, while Love waves are characterized by horizontal motion. Both Rayleigh and Love waves are about 10% slower than S-waves (so they arrive later at a seismic station). Surface waves typically have greater amplitudes than body waves, and they do more damage.

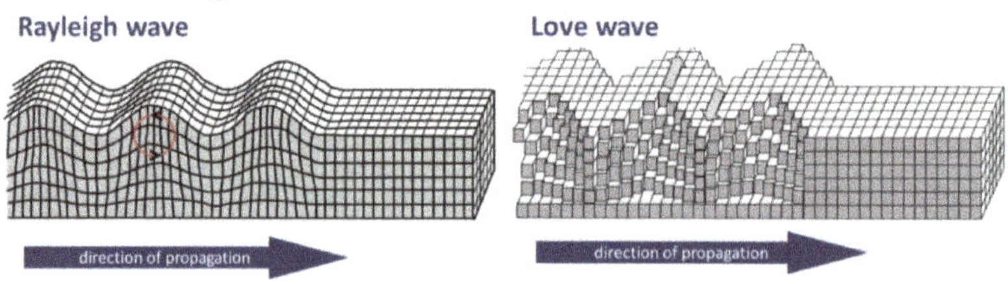

Figure 14: Depiction of seismic surface waves [SE after: https://en.wikipedia.org/wiki/Rayleigh_wave#/media/File:Rayleigh_wave.jpg and https://en.wikipedia.org/wiki/Love_wave#/media/File:Love_wave.jpg]

Other important terms for describing earthquakes are hypocentre (or focus) and epicentre. The hypocentre is the actual location of an individual earthquake shock at depth in the ground, and the epicentre is the point on the land surface directly above the hypocentre (Figure 15).

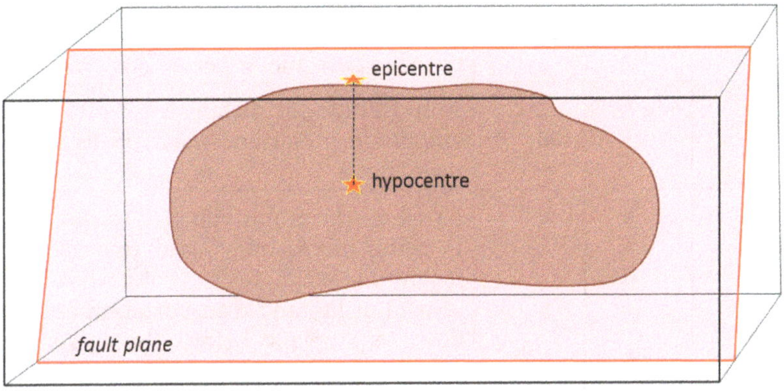

Figure 15: Epicentre and hypocentre [SE]

A number of methods for estimating magnitude are listed in Table 1. Local magnitude (ML) was widely used until late in the 20th century, but moment magnitude (MW) is now more commonly used because it gives more accurate estimates (especially with larger earthquakes) and can be applied to earthquakes at any distance from a seismometer. Surface-wave magnitudes can also be applied to measure distant large earthquakes.

Because of the increasing size of cities in earthquake-prone areas (e.g., China, Japan, California) and the increasing sophistication of infrastructure, it is becoming important to have very rapid warnings and magnitude estimates of earthquakes that have already happened. This can be achieved by using P-wave data to determine magnitude because P-waves arrive first at seismic stations, in many cases several seconds ahead of the more damaging S-waves and surface waves. Operators of electrical grids, pipelines, trains, and other infrastructure can use the information to automatically shut down systems so that damage and casualties can be limited.

Table 11.1 A summary of some of the different methods for estimating earthquake magnitude.

Type	M Range	Dist. Range	Comments
Local or Richter (ML)	2 to 6	0400 km	The original magnitude relationship defined in 1935 by Richter and Gutenberg. It is based on the maximum amplitude of S-waves recorded on a WoodAnderson torsion seismograph. ML values can be calculated using data from modern instruments. L stands for local because it only applies to earthquakes relatively close to the seismic station.

Type	M Range	Dist. Range	Comments
Moment (MW)	> 3.5	All	Based on the **seismic moment** of the earthquake, which is equal to the average amount of displacement on the fault times the fault area that slipped. It can also be estimated from seismic data if the seismometer is tuned to detect long-period body waves.
Surface wave (MS)	5 to 8	20 to 180°	A magnitude for distant earthquakes based on the amplitude of surface waves measured at a period near 20 s.
P-wave	2 to 8	Local	Based on the amplitude of P-waves. This technique is being increasingly used to provide very rapid magnitude estimates so that early warnings can be sent to utility and transportation operators to shut down equipment before the larger (but slower) S-waves and surface waves arrive.

Exercise 2 Moment Magnitude Estimates from Earthquake Parameters

You can use it to estimate the moment magnitude based on the approximate length, width, and displacement values provided in the following table:

Length (km)	Width (km)	Displacement (m)	Comments	MW?
60	15	4	The 1946 Vancouver Island earthquake	
0.4	0.2	.5	The small Vancouver Island earthquake shown in Figure 13	
20	8	4	The 2001 Nisqually earthquake described in Exercise 3	
1,100	120	10	The 2004 Indian Ocean earthquake	
30	11	4	The 2010 Haiti earthquake	

The largest recorded earthquake had a magnitude of 9.5. Could there be a 10? You can answer that question using this tool. See what numbers are needed to make MW = 10. Are they reasonable?

The magnitude scale is logarithmic; in fact, the amount of energy released by an earthquake of M4 is 32 times higher than that released by one of M3, and this ratio applies to all intervals in the scale. If we assign an arbitrary energy level of 1 unit to a M1 earthquake the energy for quakes up to M8 will be as shown on the following chart:

Magnitude	Energy
1	1
2	32
3	1,024

Magnitude	Energy
4	32,768
5	1,048,576
6	33,554,432
7	1,073,741,824
8	34,359,738,368

In any given year, when there is a large earthquake on Earth (M8 or M9), the amount of energy released by that one event will likely exceed the energy released by all smaller earthquake events combined.

Intensity of Earthquakes

The intensity of earthquake shaking at any location is determined not only by the magnitude of the earthquake and its distance, but also by the type of underlying rock or unconsolidated materials. If buildings are present, the size and type of buildings (and their inherent natural vibrations) are also important.

Intensity scales were first used in the late 19th century, and then adapted in the early 20th century by Giuseppe Mercalli and modified later by others to form what we know call the modified Mercalli intensity scale (Table 2). Intensity estimates are important because they allow us to characterize parts of any region into areas that are especially prone to strong shaking versus those that are not. The key factor in this regard is the nature of the underlying geological materials, and the weaker those are, the more likely it is that there will be strong shaking. Areas underlain by strong solid bedrock tend to experience much less shaking than those underlain by unconsolidated river or lake sediments.

Table 2: The modified Mercalli intensity scale.

I Not felt	Not felt except by a very few under especially favourable conditions
II Weak	Felt only by a few persons at rest, especially on upper floors of buildings
III Weak	Felt quite noticeably by persons indoors, especially on upper floors of buildings; many people do not recognize it as an earthquake; standing motor cars may rock slightly; vibrations similar to the passing of a truck; duration estimated
IV Light	Felt indoors by many, outdoors by few during the day; at night, some awakened; dishes, windows, doors disturbed; walls make cracking sound; sensation like heavy truck striking building; standing motor cars rocked noticeably
V Moderate	Felt by nearly everyone; many awakened; some dishes, windows broken; unstable objects overturned; pendulum clocks may stop
VI Strong	Felt by all, many frightened; some heavy furniture moved; a few instances of fallen plaster; damage slight

VII Very Strong	Damage negligible in buildings of good design and construction; slight to moderate in well-built ordinary structures; considerable damage in poorly built or badly designed structures; some chimneys broken
VIII Severe	Damage slight in specially designed structures; considerable damage in ordinary substantial buildings with partial collapse; damage great in poorly built structures; fall of chimneys, factory stacks, columns, monuments, walls; heavy furniture overturned
IX Violent	Damage considerable in specially designed structures; well-designed frame structures thrown out of plumb; damage great in substantial buildings, with partial collapse; buildings shifted off foundations
X Extreme	Some well-built wooden structures destroyed; most masonry and frame structures destroyed with foundations; rails bent
XI Extreme	Few, if any (masonry), structures remain standing; bridges destroyed; broad fissures in ground; underground pipelines completely out of service; earth slumps and land slips in soft ground; rails bent greatly
XII Extreme	Damage total; waves seen on ground surfaces; lines of sight and level distorted; objects thrown upward into the air

An example of this effect is the 1985 M8 earthquake that struck the Michoacán region of western Mexico, southwest of Mexico City. There was relatively little damage in the area around the epicentre, but there was tremendous damage and about 5,000 deaths in heavily populated Mexico City some 350 km from the epicentre. The key reason for this is that Mexico City was built largely on the unconsolidated and water-saturated sediment of former Lake Texcoco. These sediments resonate at a frequency of about two seconds, which was similar to the frequency of the body waves that reached the city. For the same reason that a powerful opera singer can break a wine glass by singing the right note, the amplitude of the seismic waves was amplified by the lake sediments. Survivors of the disaster recounted that the ground in some areas moved up and down by about 20 cm every two seconds for over two minutes. Damage was greatest to buildings between 5 and 15 storeys tall, because they also resonated at around two seconds, which amplified the shaking.

11.4 COMMON IMPACTS OF EARTHQUAKES

Some of the common impacts of earthquakes include structural damage to buildings, fires, damage to bridges and highways, initiation of slope failures, liquefaction, and tsunami. The types of impacts depend to a large degree on where the earthquake is located: whether it is predominantly urban or rural, densely or sparsely populated, highly developed or underdeveloped, and of course on the ability of the infrastructure to withstand shaking.

As we've seen from the example of the 1985 Mexico earthquake, the geological foundations on which structures are built can have a significant impact on earthquake shaking. When an earthquake happens, the seismic waves produced have a wide range of frequencies. The energy of the higher frequency waves tends to be absorbed by solid rock, while the lower frequency waves (with periods slower than one second) pass through the solid rock without being absorbed, but are eventually absorbed and amplified by soft sediments. It is therefore very common to see much worse earthquake damage in areas underlain by soft sediments than in areas of solid rock. A good example of this is in the Oakland area near San Francisco, where parts of a two-layer highway built on soft sediments collapsed during the 1989 Loma Prieta earthquake (Figure 17).

Figure 17: A part of the Cypress Freeway in Oakland California that collapsed during the 1989 Loma Prieta earthquake. [from: http://upload.wikimedia.org/wikipedia/ commons/9/91/Cypress_collapsed.jpg]

Building damage is also greatest in areas of soft sediments, and multi-storey buildings tend to be more seriously damaged than smaller ones. Buildings can be designed to withstand most earthquakes, and this practice is increasingly applied in earthquake-prone regions. Turkey is one such region, and even though Turkey had a relatively strong building code in the 1990s, adherence to the code was poor, as builders did whatever they could to save costs, including using inappropriate materials in concrete and reducing the amount of steel reinforcing. The result was that there were over 17,000 deaths in the 1999 M7.6 Izmit earthquake (Figure 18). After two devastating earthquakes that year, Turkish authorities strengthened the building code further, but the new code has been applied only in a few regions, and enforcement of the code is still weak, as revealed by the amount of damage from a M7.1 earthquake in eastern Turkey in 2011.

Figure 18: Buildings damaged by the 1999 earthquake in the Izmit area, Turkey. [from U.S. Geological Survey at: http://gallery.usgs.gov/sets/1999_Izmit,_Turkey_Earthquake/thumb/_/1]

Fires are commonly associated with earthquakes because fuel pipelines rupture and electrical lines are damaged when the ground shakes (Figure 19). Most of the damage in the great 1906 San Francisco earthquake was caused by massive fires in the downtown area of the city (Figure 20). Some 25,000 buildings were destroyed by those fires, which were fuelled by broken gas pipes. Fighting the fires was difficult because water mains had also ruptured. The risk of fires can be reduced through P-wave early warning systems if utility operators can reduce pipeline pressure and close electrical circuits.

Figure 19: Some of the effects of the 2011 Tohoku earthquake in the Sendai area of Japan. An oil refinery is on fire, and a vast area has been flooded by a tsunami. [from: http://en.wikipedia.org/wiki/2011_T%C5%8Dhoku_earthquake_and_tsunami#/media/File:SH-60B_helicopter_flies_over_Sendai.jpg]

Figure 20: Fires in San Francisco following the 1906 earthquake. [from: http://upload.wikimedia.org/wikipedia/commons/3/3e/San_francisco_fire_1906.jpg]

Earthquakes are important triggers for failures on slopes that are already weak. An example is the Las Colinas slide in the city of Santa Tecla, El Salvador, which was triggered by a M7.6 offshore earthquake in January 2001 (Figure 21).

Figure 21: The Las Colinas debris flow at Santa Tecla (a suburb of the capital San Salvador) triggered by the January 2001 El Salvador earthquake. This is just one

of many hundreds of slope failures that resulted from that earthquake. Over 500 people died in the area affected by this slide. [from: http://landslides.usgs.gov/learning/ images/foreign/ElSalvadorslide.jpg]

Ground shaking during an earthquake can be enough to weaken rock and unconsolidated materials to the point of failure, but in many cases the shaking also contributes to a process known as liquefaction, in which an otherwise solid body of sediment is transformed into a liquid mass that can flow. When water-saturated sediments are shaken, the grains become rearranged to the point where they are no longer supporting one another. Instead, the water between the grains is holding them apart and the material can flow. Liquefaction can lead to the collapse of buildings and other structures that might be otherwise undamaged. A good example is the collapse of apartment buildings during the 1964 Niigata earthquake (M7.6) in Japan (Figure 22). Liquefaction can also contribute to slope failures and to fountains of sandy mud (sand volcanoes) in areas where there is loose saturated sand beneath a layer of more cohesive clay.

Figure 22: Collapsed apartment buildings in the Niigata area of Japan. The material beneath the buildings was liquefied to varying degrees by the 1964 earthquake.

Parts of the Fraser River delta are prone to liquefaction-related damage because the region is characterized by a 2 m to 3 m thick layer of fluvial silt and clay over top of at least 10 m of water-saturated fluvial sand (Figure 23). Under these conditions, it can be expected that seismic shaking will be amplified and, the sandy sediments will liquefy. This could lead to subsidence and tilting of buildings, and to failure and sliding of the silt and clay layer. Current building-code regulations in the Fraser delta area require that measures be taken to strengthen the ground underneath multi-storey buildings prior to construction.

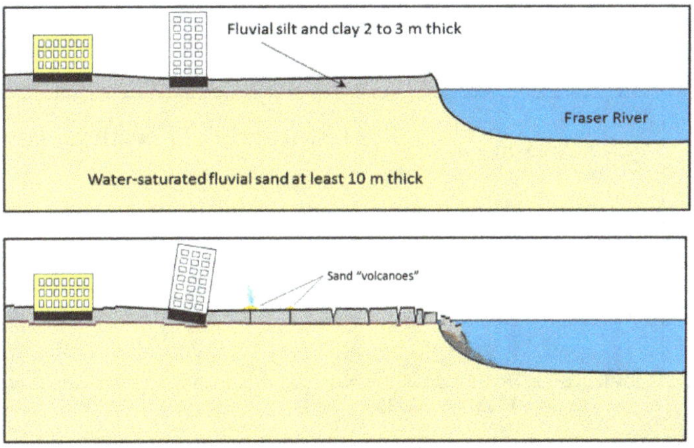

Figure 23: Recent unconsolidated sedimentary layers in the Fraser River delta area (top) and the potential consequences in the event of a damaging earthquake.

11.5 FORECASTING EARTHQUAKES: MINIMIZING DAMAGE AND CASUALTIES

It has long been a dream of seismologists, geologists, and public safety officials to be able to accurately predict the location, magnitude, and timing of earthquakes on time scales that would be useful for minimizing danger to the public and damage to infrastructure (e.g., weeks, days, hours). Many different avenues of prediction have been explored, such as using observations of warning foreshocks, changes in magnetic fields, seismic tremor, changing groundwater levels, strange animal behaviour, observed periodicity, stress transfer considerations, and others. So far, none of the research into earthquake prediction has provided a reliable method. Although there are some reports of successful earthquake predictions, they are rare, and many are surrounded by doubtful circumstances.

The problem with earthquake predictions, as with any other type of prediction, is that they have to be accurate *most* of the time, not just *some* of the time. We have come to rely on weather predictions because they are generally (and increasingly) accurate. But if we try to predict earthquakes and are only accurate 10% of the time (and even that isn't possible with the current state of knowledge), the public will lose faith in the process very quickly, and then will ignore all of the predictions. Efforts are currently focused on forecasting earthquake probabilities, rather than predicting their occurrence.

There was great hope for earthquake predictions late in the 1980s when attention was focused on part of the San Andreas Fault at Parkfield, about 200 km south of San Francisco. Between 1881 and 1965 there were five earthquakes at Parkfield, most spaced at approximately 20-year intervals, all confined to the same 20 km-long segment of the fault, and all very close to M6 (Figure 26). Both the 1934 and 1966 earthquakes were preceded by small foreshocks exactly 17 minutes before the main quake.

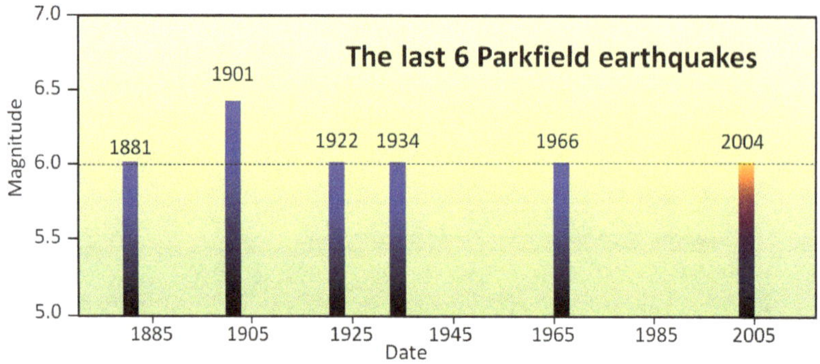

Figure 26: Earthquakes on the Parkfield segment of the San Andreas Fault between 1881 and 2004.

The U.S. Geological Survey recognized this as an excellent opportunity to understand earthquakes and earthquake prediction, so they armed the Parkfield area with a huge array of geophysical instruments and waited for the next quake, which was expected to happen around 1987. Nothing happened! The "1987 Parkfield earthquake" finally struck in September 2004. Fortunately all of the equipment was still there, but it was no help from the perspective of earthquake prediction. There were no significant precursors to the 2004 Parkfield earthquake in any of the parameters measured, including seismicity, harmonic tremor, strain (rock deformation), magnetic field, the conductivity of the rock, or creep, and there was no foreshock. In other words, even though every available technique was used to monitor it, the 2004 earthquake came as a complete surprise, with no warning whatsoever.

The hope for earthquake prediction is not dead, but it was hit hard by the Parkfield experiment. The current focus in earthquake-prone regions is to provide forecasts of the probability of an earthquake of a certain magnitude within a certain time period — typically a number of decades — while officials focus on ensuring that the population is educated about earthquake risks and that buildings and other infrastructure are as safe as can be. An example of this approach for the San Francisco Bay region of California is shown in Figure 27. Based on a wide range of information, including past earthquake history, accumulated stress from plate movement, and known stress transfer, seismologists and geologists have predicted the likelihood of a M6.7 or greater earthquake on each of eight major faults that cut through the region. The greatest probabilities are on the San Andreas, Rogers Creek, and Hayward Faults. As shown in Figure 27, there is a 63% chance that a major and damaging earthquake will take place somewhere in the region prior to 2036.

Figure 27: Probabilities of a M6.7 or larger earthquake over the period 2007 to 2036 on various faults in the San Francisco Bay region of California.

As we've discussed already, it's not sufficient to have strong building codes, they have to be enforced. Building code compliance is quite robust in most developed countries, but is sadly inadequate in many developing countries.

It's also not enough just to focus on new buildings; we have to make sure that existing buildings – especially schools and hospitals – and other structures such as bridges and dams, are as safe as they can be. An example of how this is applied to schools in B.C. is described in Box 2.

Making the seismic upgrade in B.C.'s schools

British Columbia is in the middle of a multi-billion-dollar program to make schools safer for students. The program is focused on older schools, because, according to the government, those built since 1992 already comply with modern seismic codes. Some schools would require too much work to make upgrading economically feasible and they are replaced. Where upgrading is feasible, the school is assessed carefully before any upgrade work is initiated.

An example is Sangster Elementary in Colwood on southern Vancouver Island. The school was originally built in 1957, with a major addition in 1973. Ironically, the newer part of the school, built of concrete blocks, required strengthening with the addition of a steel framework, while the 1957 part, which is a wood-frame building, did not require seismic upgrading. The work was completed in 2014.

[Sangster Elementary image from Google Maps – street view]

As of January 2015, upgrades had been completed at 145 B.C. schools, 11 were underway, and an additional 57 were ready to proceed with funding identified.* Another 129 schools were listed as needing upgrades. In May 2015, the provincial government announced that the target date for completion of the upgrades, originally set for 2020, had been delayed to 2030.

AT GLANCE

What Is an Earthquake?	An earthquake is the shaking that results when a body of rock that has been deformed breaks and the two sides quickly slide past each other. The rupture is initiated at a point but quickly spreads across an area of a fault, via a series of aftershocks initiated by stress transfer. Episodic tremor and slip is a periodic slow movement, accompanied by harmonic tremors, along the middle part of a subduction zone boundary.
Earthquakes and Plate Tectonics	Most earthquakes take place at or near plate boundaries, especially at transform boundaries (where most quakes are at less than 30 km depth) and at convergent boundaries (where they can be at well over 100 km depth). The largest earthquakes happen at subduction zones, typically in the upper section where the rock is relatively cool.

Measuring Earthquakes	Magnitude is a measure of the amount of energy released by an earthquake, and it is proportional to the area of the rupture surface and to the amount of displacement. Although any earthquake has only one magnitude value, it can be estimated in various ways, mostly involving seismic data. Intensity is a measure of the amount of shaking experienced and damage done at a particular location around the earthquake. Intensity will vary depending on the distance to the epicentre, the depth of the earthquake, and the geological nature of the material below surface.
The Impacts of Earthquakes	Damage to buildings is the most serious consequence of most large earthquakes. The amount of damage is related to the type and size of buildings, how they are constructed, and the nature of the material on which they are built. Other important consequences are fires, damage to bridges and highways, slope failures, liquefaction, and tsunami. Tsunami, which are almost all related to large subduction earthquakes, can be devastating.
Forecasting Earthquakes and Minimizing Damage and Casualties	There is no reliable technology for predicting earthquakes, but the probability of one happening within a certain time period can be forecast. We can minimize earthquake impacts by ensuring that citizens are aware of the risk, that building codes are enforced, that existing buildings like schools and hospitals are seismically sound, and that both public and personal emergency plans are in place.

CHAPTER 12
UNDERSTANDING OF GEOLOGICAL STRUCTURES

Observing and understanding geological structures helps us to determine the kinds of stresses that have existed within Earth in the past. This type of information is critical to our understanding of plate tectonics, earthquakes, the formation of mountains, metamorphism, and Earth resources. Some of the types of geological structures that are important to study include fractures, faults, and folds. Structural geologists make careful observations of the orientations of these structures and the amount and direction of offset along faults.

Figure 1: Folds in sedimentary rocks near Golden and the Kickinghorse River, B.C.

12.1 STRESS AND STRAIN

Rocks are subject to stress—mostly related to plate tectonics but also to the weight of overlying rocks—and their response to that stress is strain (deformation). In regions close to where plates are converging stress is typically compressive—the rocks are being squeezed. Where plates are diverging the stress is extensive—rocks are being pulled apart. At transform plate boundaries, where plates are moving side by side there is sideways or shear stress—meaning that there are forces in opposite directions parallel to a plane. Rocks have highly varying strain responses to stress because of their different compositions and physical properties, and because temperature is a big factor and rock temperatures within the crust can vary greatly.

We can describe the stress applied to a rock by breaking it down into three dimensions—all at right angles to one-another (Figure 2). If the rock is subject only to the pressure of burial, the stresses in all three directions will likely be the same. If it is subject to both burial and tectonic forces, the pressures will be different in different directions.

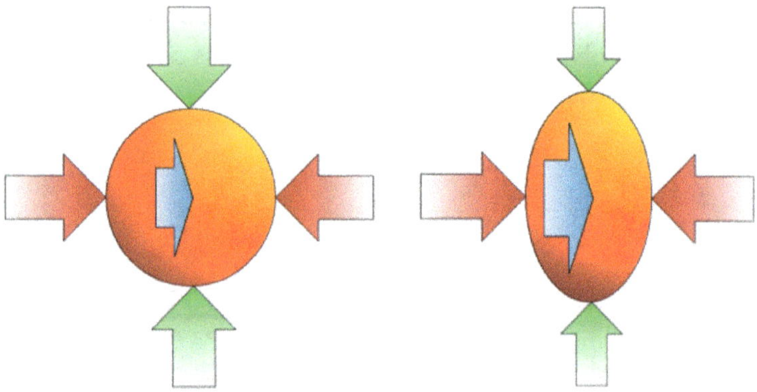

Figure 2: Depiction of the stress applied to rocks within the crust. The stress can be broken down into three components. Assuming that we're looking down in this case, the green arrows represent north-south stress, the red arrows represent east-west stress, and the blue arrows (the one underneath is not visible) represent up-down stress. On the left, all of the stress components are the same. On the right, the north-south stress is least and the up-down stress is greatest.

Rock can respond to stress in three ways: it can deform elastically, it can deform plastically, and it can break or fracture. Elastic strain is reversible; if the stress is removed, the rock will return to its original shape just like a rubber band that is stretched and released. Plastic strain is not reversible. As already noted, different rocks at different temperatures will behave in different ways to stress. Higher temperatures lead to more plastic behaviour. Some rocks or sediments are also more plastic when they are wet. Another factor is the rate at which the stress is applied. If the stress is applied quickly (for example, because of an extraterrestrial impact or an earthquake), there will be an increased tendency for the rock to fracture. Some different types of strain response are illustrated in Figure 3.

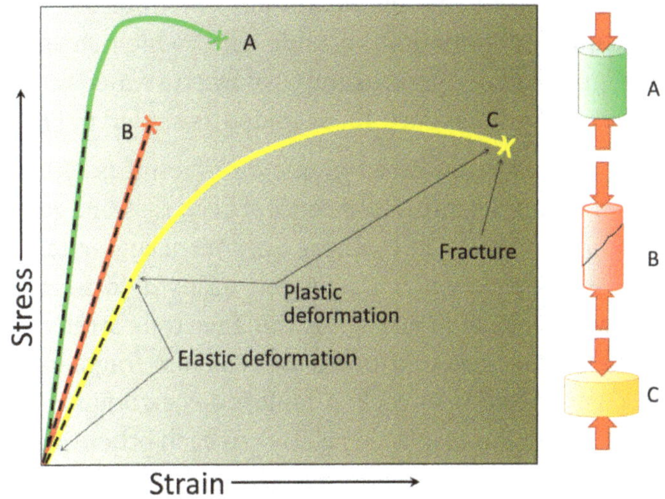

Figure 3: The varying types of response of geological materials to stress. The straight dashed parts are elastic strain and the curved parts are plastic strain. In each case the X marks where the material fractures. A, the strongest material, deforms relatively little and breaks at a high stress level. B, strong but brittle, shows no plastic deformation and breaks after relatively little elastic deformation. C, the most deformable, breaks only after significant elastic and plastic strain. The three deformation diagrams on the right show A and C before breaking and B after breaking.

Figure 4: Rock structures caused by various types of strain within rocks that have been stressed [all by SE]

The outcomes of placing rock under stress are highly variable, but they include fracturing, tilting and folding, stretching and squeezing, and faulting. A fracture is a simple break that does not involve significant movement of the rock on either side. Fracturing is particularly common in volcanic rock, which shrinks as it cools. The basalt columns in Figure 4a are a good example of fracture. Beds are sometimes tilted by tectonic forces, as shown in Figure 4b, or folded as shown in Figure 1.

When a body of rock is compressed in one direction it is typically extended (or stretched) in another. This is an important concept because some geological structures only form under compression, while others only form under tension. Most of the rock in Figure 4c is limestone, which is relatively easily deformed when heated. The dark rock is chert, which remains brittle. As the limestone stretched (parallel to the hammer handle) the brittle chert was forced to break into fragments to accommodate the change in shape of the body of rock. A fault is a rock boundary along which the rocks on either side have been displaced relative to each other (Figure 4d).

12.2 FOLDING

When a body of rock, especially sedimentary rock, is squeezed from the sides by tectonic forces, it is likely to fracture and/or become faulted if it is cold and brittle, or become folded if it is warm enough to behave in a plastic manner.

The nomenclature and geometry of folds are summarized on Figure 5. An upward fold is called an anticline, while a downward fold is called a syncline. In many areas it's common to find a series of anticlines and synclines (as in Figure 5), although some sequences of rocks are folded into a single anticline or syncline. A plane drawn through the crest of a fold in a series of beds is called the axial plane of the fold. The sloping beds on either side of an axial plane are limbs. An anticline or syncline is described as symmetrical if the angles between each of limb and the axial plane are generally similar, and asymmetrical if they are not. If the axial plane is sufficiently tilted that the beds on one side have been tilted past vertical, the fold is known as an overturned anticline or syncline.

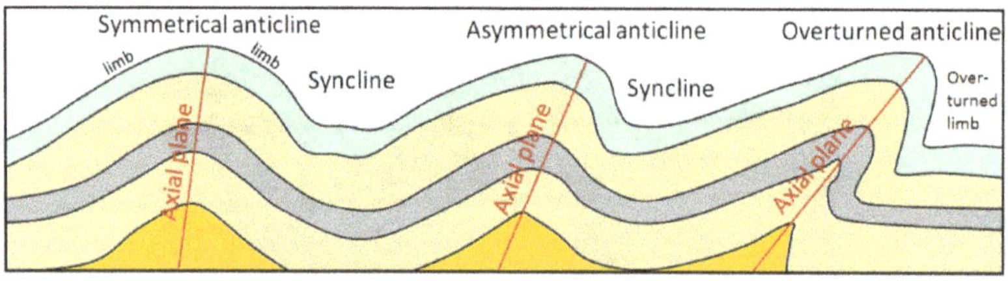

Figure 5: Examples of different types of folds and fold nomenclature. Axial planes are only shown for the anticlines, but synclines also have axial planes.

A very tight fold, in which the limbs are parallel or nearly parallel to one another is called an isoclinal fold (Figure 6). Isoclinal folds that have been overturned to the extent that their limbs are nearly horizontal are called recumbent folds.

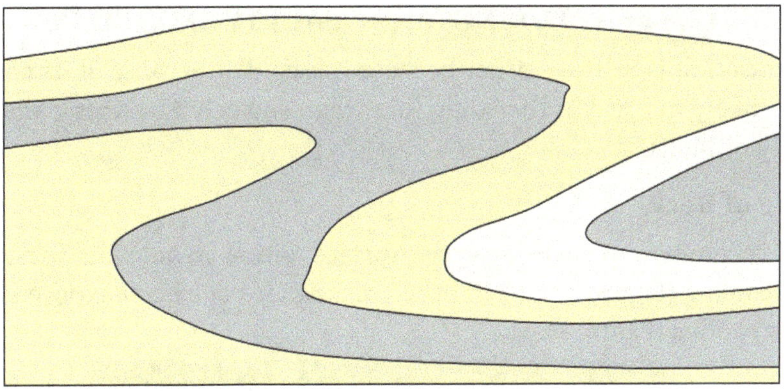

Figure 6: An isoclinal recumbent fold

Folds can be of any size, and it's very common to have smaller folds within larger folds (Figure 7). Large folds can have wavelengths of tens of kilometres, and very small ones might be visible only under a microscope. Anticlines are not necessarily, or even typically, expressed as ridges in the terrain, nor synclines as valleys. Folded rocks get eroded just like all other rocks and the topography that results is typically controlled mostly by the resistance of different layers to erosion (Figure 8).

Figure 7: Folded limestone (grey) and chert (rust-coloured) in Triassic Quatsino Formation rocks on Quadra Island, B.C. The image is about 1 m across.

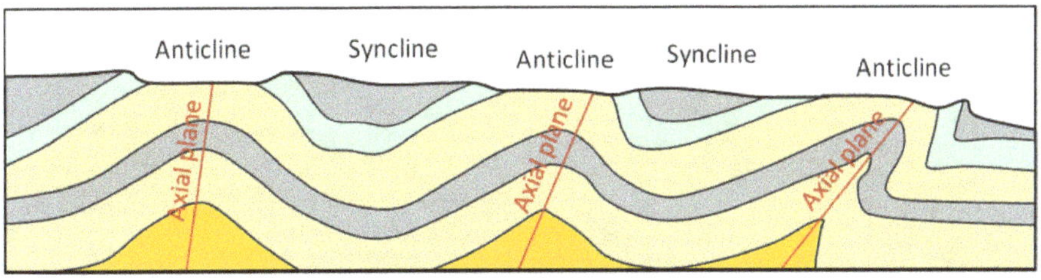

Figure 8 Example of the topography in an area of folded rocks that has been eroded. In this case the green and grey rocks are most resistant to erosion, and are represented by hills.

12.3 FRACTURING AND FAULTING OF ROCK

A body of rock that is brittle—either because it is cold or because of its composition, or both— is likely to break rather than fold when subjected to stress, and the result is fracturing or faulting.

Fracturing of Rock

Fracturing is common in rocks near the surface, either in volcanic rocks that have shrunk on cooling (Figure 4a), or in other rocks that have been exposed by erosion and have expanded (Figure 9).

Figure 9: Granite in the Coquihalla Creek area, B.C. (left) and sandstone at Nanoose,

A fracture in a rock is also called a joint. There is no side-to-side movement of the rock on either side of a joint. Most joints form where a body of rock is expanding because of reduced pressure, as shown by the two examples in Figure 9, or where the rock itself is contracting but the body of rock remains the same size (the cooling volcanic rock in Figure 4a). In all of these cases, the pressure regime is one of tension as opposed to compression. Joints can also develop where rock is being folded because, while folding typically happens during compression, there may be some parts of the fold that are in tension (Figure 10).

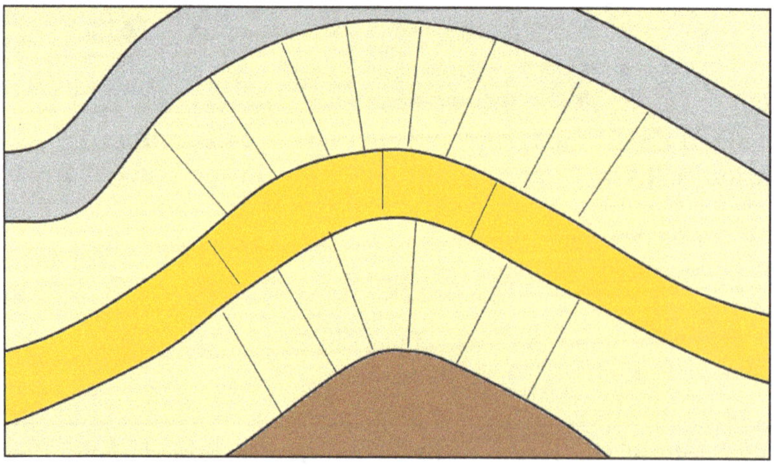

Figure 10: A depiction of joints developed in the hinge area of folded rocks. Note that in this situation some rock types are more likely to fracture than others.

Finally joints can also develop when rock is under compression as shown on Figure 11, where there is differential stress on the rock, and joint sets develop at angles to the compression directions.

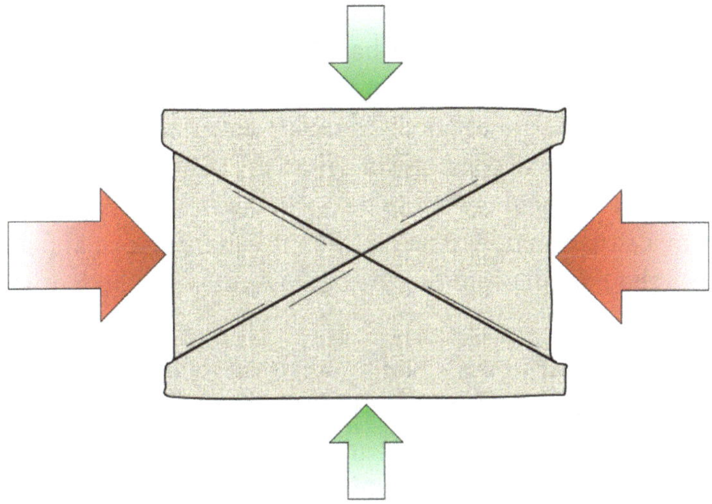

Figure 11: A depiction of joints developed in a rock that is under stress.

Faulting of Rock

A fault is boundary between two bodies of rock along which there has been relative motion (Figure 4d). As we discussed in Chapter 11, an earthquake involves the sliding of one body of rock past another. Earthquakes don't necessarily happen on existing faults, but once an earthquake takes place a fault will exist in the rock at that location. Some large faults, like the San Andreas Fault in California or the Tintina Fault, which extends from northern B.C. through central Yukon and into Alaska, show evidence of hundreds of kilometres of motion, while others show less than a millimetre. In order to estimate the amount of motion on a fault, we need to find some geological feature that shows up on both sides and has been offset (Figure 12).

Figure 12: A fault (white dashed line) in intrusive rocks on Quadra Island, B.C. The pink dyke has been offset by the fault and the extent of the offset is

shown by the white arrow (approximately 10 cm). Because the far side of the fault has moved to the right, this is a *right-lateral fault*. If the photo were taken from the other side, the fault would still appear to have a right-lateral offset.

There are several kinds of faults, as illustrated on Figure 13, and they develop under different stress conditions. The terms hanging wall and footwall in the diagrams apply to situations where the fault is not vertical. The body of rock above the fault is called the hanging wall, and the body of rock below it is called the footwall. If the fault develops in a situation of compression, then it will be a reverse fault because the compression causes the hanging wall to be pushed up relative to the footwall. If the fault develops in a situation of extension, then it will be a normal fault, because the extension allows the hanging wall to slide down relative to the footwall in response to gravity.

The third situation is where the bodies of rock are sliding sideways with respect to each other, as is the case along a transform fault (see Chapter 10). This is known as a strike-slip fault because the displacement is along the "strike" or the length of the fault. On strike-slip faults the motion is typically only horizontal, or with a very small vertical component, and as discussed above the sense of motion can be right lateral (the far side moves to the right), as in Figures 12 and 13, or it can be left lateral (the far side moves to the left). Transform faults are strike-slip faults.

Figure 13 Depiction of reverse, normal, and strike-slip faults. Reverse faults happen during compression while normal faults happen during extension. Most strike-slip faults are related to transform boundaries. [SE after: http://www.nature.nps.gov/geology/education/images/GRAPHICS/fault_types_2.jpg]

In areas that are characterized by extensional tectonics, it is not uncommon for a part of the upper crust to subside with respect to neighbouring parts. This is typical along areas of continental rifting, such as the Great Rift Valley of East Africa or in parts of Iceland, but it is also seen elsewhere. In such situations a down-dropped block is known as a graben (German for ditch), while an adjacent block that doesn't subside is called a horst (German for heap) (Figure 14).There are many horsts and grabens in the Basin and Range area of the western United States, especially in Nevada. Part of the Fraser Valley region of B.C., in the area around Sumas Prairie is a graben.

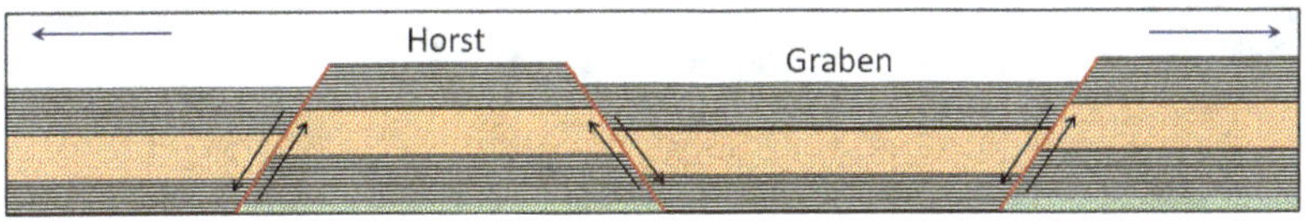

Figure 14: Depiction of graben and horst structures that form in extensional situations. All of the faults are normal faults.

A special type of reverse fault, with a very low-angle fault plane, is known as a thrust fault. Thrust faults are relatively common in areas where fold-belt mountains have been created during continent-continent collision. Some represent tens of kilometres of thrusting, where thick sheets of sedimentary rock have been pushed up and over top of other rock (Figure 15).

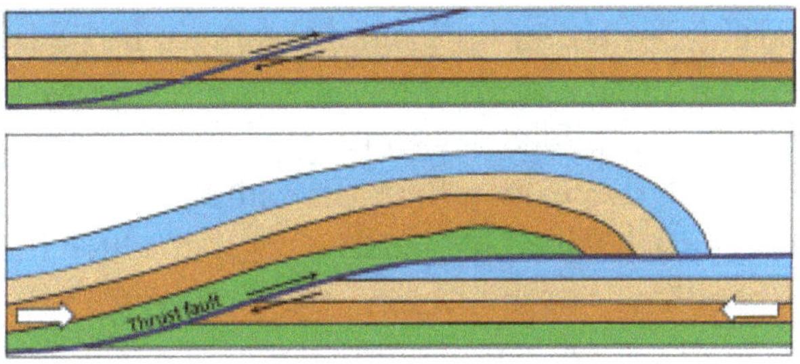

Figure 15: Depiction a thrust fault. Top: prior to faulting. Bottom: after significant fault offset. [SE]

There are numerous thrust faults in the Rocky Mountains, and a well-known example is the McConnell Thrust, along which a sequence of sedimentary rocks about 800 m thick has been pushed for about 40 km from west to east (Figure 16). The thrusted rocks range in age from Cambrian to Cretaceous, so in the area around Mt. Yamnuska Cambrian-aged rock (around 500 Ma) has been thrust over, and now lies on top of Cretaceous-aged rock (around 75 Ma) (Figure 17).

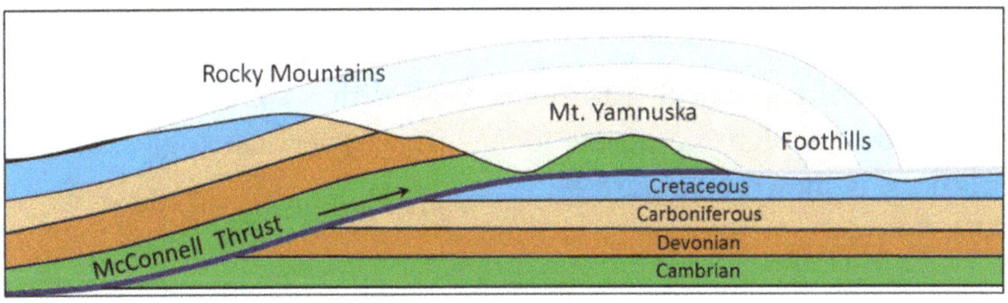

Figure 16: Depiction of the McConnell Thrust in the eastern part of the Rockies. The rock within the faded area has been eroded.

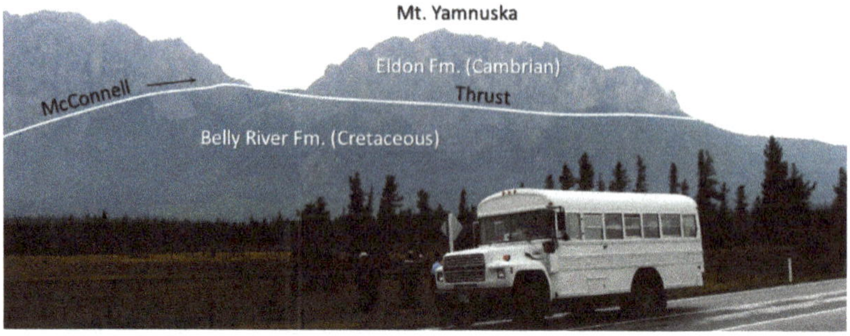

Figure 17: The McConnell Thrust at Mt. Yamnuska near Exshaw, Alberta. Carbonate rocks (limestone) of Cambrian age have been thrust over top of Cretaceous mudstone. [SE]

12.4 MEASURING GEOLOGICAL STRUCTURES

Geologists take great pains to measure and record geological structures because they are critically important to understanding the geological history of a region. One of the key features to measure is the orientation, or attitude, of bedding. We know that sedimentary beds are deposited in horizontal layers, so if the layers are no longer horizontal, then we can infer that they have been affected by tectonic forces and have become either tilted, or folded. We can express the orientation of a bed (or any other planar feature) with two values: first, the compass orientation of a horizontal line on the surface—the strike—and second, the angle at which the surface dips from the horizontal, (perpendicular to the strike)—the dip (Figure 18).

It may help to imagine a vertical surface, such as a wall in your house. The strike is the compass orientation of the wall and the dip is 90° from horizontal. If you could push the wall so it's leaning over, but still attached to the floor, the strike direction would be the same, but the dip angle would be less than 90°. If you pushed the wall over completely so it was lying on the floor, it would no longer have a strike direction and its dip would be 0°. When describing the dip it is important to include the direction. In other words. if the strike is 0° (i.e., north) and the dip is 30°, it would be necessary to say "to the west" or "to the east." Similarly if the strike is 45° (i.e., northeast) and the dip is 60°, it would be necessary to say "to the northwest" or "to the southeast."

Measurement of geological features is done with a special compass that has a built-in clinometer, which is a device for measuring vertical angles. An example of how this is done is shown on Figure 19.

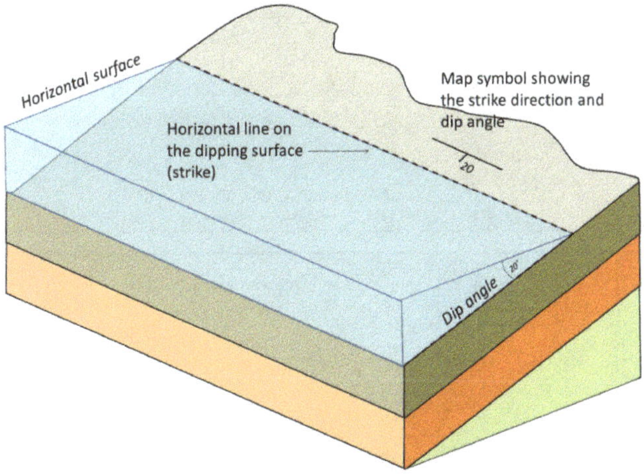

Figure 18: A depiction of the strike and dip of some tilted sedimentary beds partially covered with water. The notation for expressing strike and dip on a map is shown.

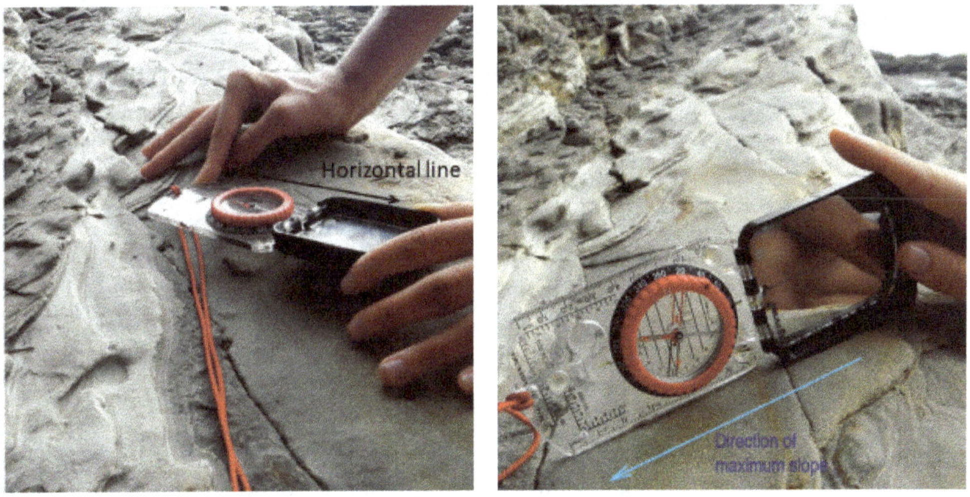

Figure 19: Measurement of strike (left) and dip (right) using a geological compass with a clinometer.

Strike and dip are also used to describe any other planar features, including joints, faults, dykes, sills, and even the foliation planes in metamorphic rocks. Figure 20 shows an example of how we would depict the beds that make up an anticline on a map.

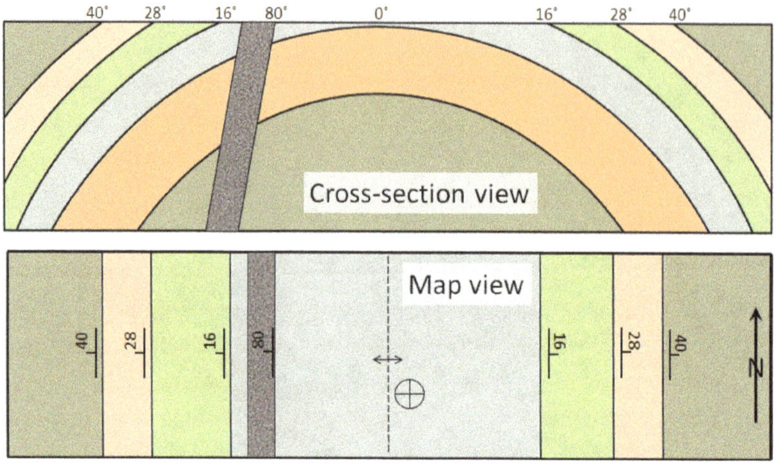

Figure 20: A depiction of an anticline and a dyke in cross-section (looking from the side) and in map view (a.k.a. plan view) with the appropriate strike-dip and anticline symbols. [SE]

The beds on the west (left) side of the map are dipping at various angles to the west. The beds on the east side are dipping to the east. The middle bed (light grey) is horizontal; this is denoted by a cross within a circle. The dyke is dipping at 80° to the west. The hinge of the fold is denoted with a dashed line with two arrows point away from it. If it were a syncline, the arrows would point towards the line.

AT A GLANCE

Stress and Strain	Stress within rocks, which includes compression, extension and shearing, typically originates from plate-boundary processes. Rock that is stressed responds with either elastic or plastic strain, and will eventually break. The way a rock responds to stress depends on its composition and structure, the rate at which strain is applied, and also to the temperature of the rock body and the presence of water.
Folding	Folding is generally a plastic response to compressive stress, although some brittle behaviour can happen during folding. An upward fold is an anticline. A downward fold is a syncline. The axis of a fold can be vertical, inclined, or even horizontal. If we know that the folded beds have not been overturned, then we can use the more specific terms: anticline and syncline.
Fracturing and Faulting	Fractures (joints) typically form during extension, but can also form during compression. Faulting, which involves the displacement of rock, can take place during compression or extension, as well as during shearing at transform boundaries. Thrust faulting is a special form of reverse faulting.
Measuring Geological Structures	It is important to be able to measure the strike and dip of planar surfaces, such as a bedding planes, fractures or faults. Special symbols are used to show the orientation of structural features on geological maps.

CHAPTER

STREAMS AND FLOODS 13

Streams are the most important agents of erosion and transportation of sediments on Earth's surface. They are responsible for the creation of much of the topography that we see around us. They are also places of great beauty and tranquility, and of course, they provide much of the water that is essential to our existence. But streams are not always peaceful and soothing. During large storms and rapid snowmelts, they can become raging torrents capable of moving cars and houses and destroying roads and bridges. When they spill over their banks, they can flood huge areas, devastating populations and infrastructure. Over the past century, many of the most damaging natural disasters in Canada have been floods, and we can expect them to become even more severe as the climate changes.

Figure 1: A small waterfall on Johnston Creek in Johnston Canyon, Banff National Park, AB

13.1 HYDROLOGICAL CYCLE

Water is constantly on the move. It is evaporated from the oceans, lakes, streams, the surface of the land, and plants (transpiration) by solar energy (Figure 2). It is moved through the atmosphere by winds and condenses to form clouds of water droplets or ice crystals. It comes back down as rain or snow and then flows through streams, into lakes, and eventually back to the oceans. Water on the surface and in streams and lakes infiltrates the ground to become groundwater. Groundwater slowly moves through the rock and surficial materials. Some groundwater returns to other streams and lakes, and some goes directly back to the oceans.

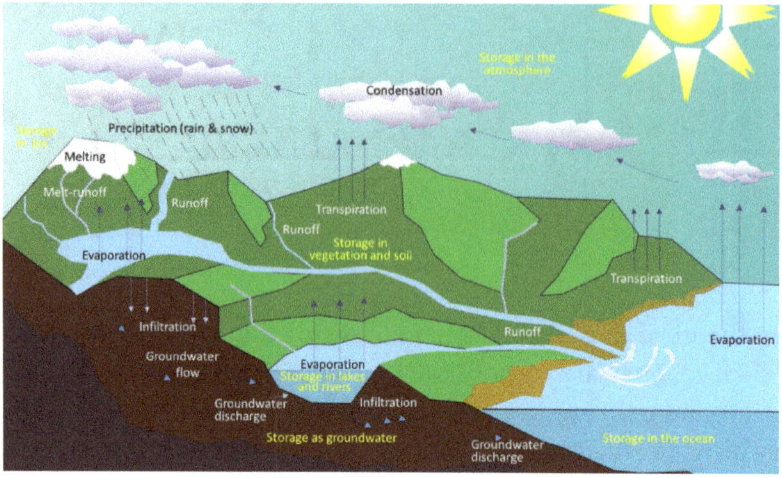

Figure 2: The various components of the water cycle. Black or white text indicates the movement or transfer of water from one reservoir to another. Yellow text indicates the storage of water.

Even while it's moving around, water is stored in various reservoirs. The largest, by far, is the oceans, accounting for 97% of the volume (Figure 3). Of course, that water is salty. The remaining 3% is fresh water. Two-thirds of our fresh water is stored in ice and one-third is stored in the ground. The remaining fresh water — about 0.03% of the total — is stored in lakes, streams, vegetation, and the atmosphere.

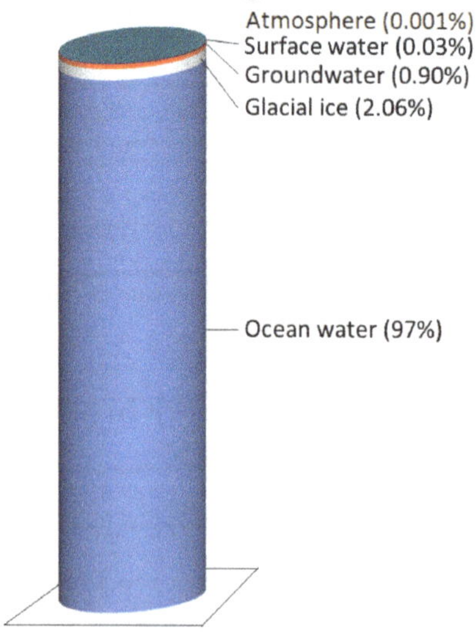

Figure 3a: The storage reservoirs for water on Earth. Glacial ice is represented by the white band, groundwater the red band, and surface water the very thin blue band at the top. The 0.001% stored in the atmosphere is not shown

To put that in perspective, let's think about putting all of Earth's water into a 1 L jug (Figure 3b). We start by almost filling the jug with 970 ml of water and 34 g of salt. Then we add one regular-sized (~20 mL) ice cube (representing glacial ice) and two teaspoons (~10 mL) of groundwater. All of the water that we see around us in lakes and streams and up in the sky can be represented by adding three more drops from an eyedropper.

Figure 3b: Representation of the Earth's water as a 1 L jug. The three drops represent all of the fresh water in lakes, streams, and wetlands, plus all of the water in the atmosphere.

Although the proportion of Earth's water that is in the atmosphere is tiny, the actual volume is huge. At any given time, there is the equivalent of approximately 13,000 km^3 of water in the air in the form of water vapour and water droplets in clouds. Water is evaporated from the oceans, vegetation, and lakes at a rate of 1,580 km^3 per day, and just about exactly the same volume falls as rain and snow every day — over both the oceans and land. The precipitation that falls on land goes back to the ocean in the form of stream flow (117 km^3/day) and groundwater flow (6 km^3/day). Most of the rest of this chapter is about that 117 km^3/day of streamflow. The average discharge of the Fraser River into the ocean is approximately 0.31 km^3/day, or 0.26% of the world's total.

Exercise 1 How Long Does Water Stay in the Atmosphere?

The residence time of a water molecule in the atmosphere (or any of the other reservoirs) can be estimated by dividing the amount that is there by the rate at which it is transferred in and out. For the atmosphere, we know that the reservoir size is 13,000 km^3, and the rate of flux is 1,580 km^3/day. If we divide 13,000 by 1,580, we get

8.22 days. This means that, *on average*, a molecule of water stays in the atmosphere for just over eight days. "Average" needs to be emphasized here because obviously some molecules stay in the air for only a few hours, while others may stay up there for weeks.

The volume of the oceans is 1,338,000,000 km^3 and the flux rate is approximately the same (1,580 km^3/day). What is the average residence time of a water molecule in the ocean?

13.2 DRAINAGE BASINS

A stream is a body of flowing surface water of any size, ranging from a tiny trickle to a mighty river. The area from which the water flows to form a stream is known as its drainage basin. All of the precipitation (rain or snow) that falls within a drainage basin eventually flows into its stream, unless some of that water is able to cross into an adjacent drainage basin via groundwater flow. An example of a drainage basin is shown in Figure 4.

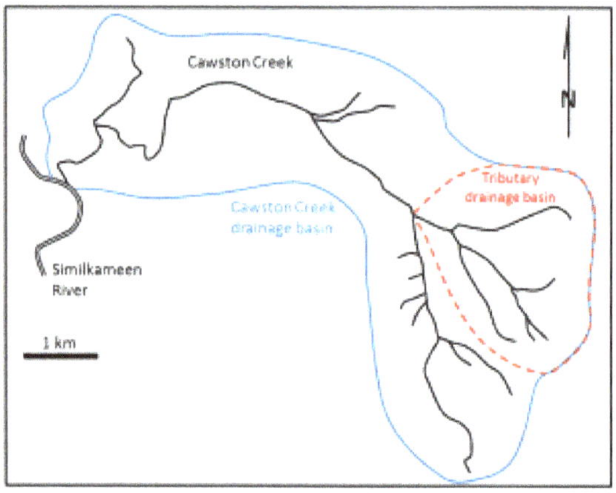

Figure 4: Cawston Creek near Keremeos, B.C. The blue line shows the extent of the drainage basin. The dashed red line is the drainage basin of one of its tributaries.

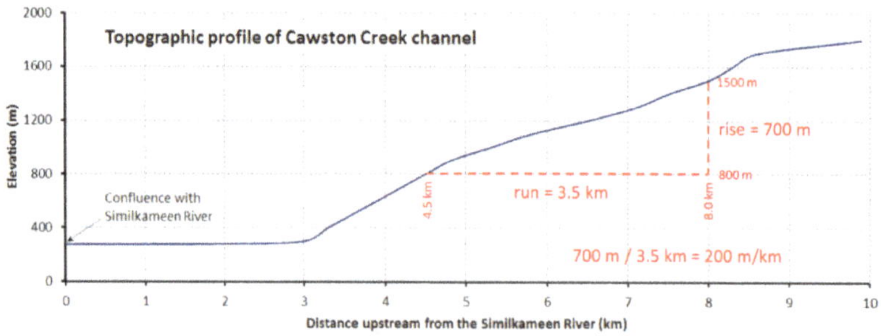

Figure 5: Profile of the main stem of Cawston Creek near Keremeos, B.C. The maximum elevation of the drainage basin is about 1,840 m, near Mount Kobau. The base level is 275 m, at the Similkameen River. As shown, the gradient of the

stream can be determined by dividing the change in elevation between any two points (rise) by the distance between those two points (run).

Cawston Creek is a typical small drainage basin (approximately 25 km2) within a very steep glaciated valley. As shown in Figure 13.5, the upper and middle parts of the creek have steep gradients (averaging about 200 m/km but ranging from 100 to 350 m/km), and the lower part, within the valley of the Similkameen River, is relatively flat (<5 m/km). The shape of the valley has been controlled first by tectonic uplift (related to plate convergence), then by pre-glacial stream erosion and mass wasting, then by several episodes of glacial erosion, and finally by post-glacial stream erosion. The lowest elevation of Cawston Creek (275 m at the Similkameen River) is its base level. Cawston Creek cannot erode below that level unless the Similkameen River erodes deeper into its flood plain (the area that is inundated during a flood).

Metro Vancouver's water supply comes from three large drainage basins on the north shore of Burrard Inlet, as shown in Figure 6. This map illustrates the concept of a drainage basin divide. The boundary between two drainage basins is the height of land between them. A drop of water falling on the boundary between the Capilano and Seymour drainage basins (a.k.a., watersheds), for example, could flow into either one of them.

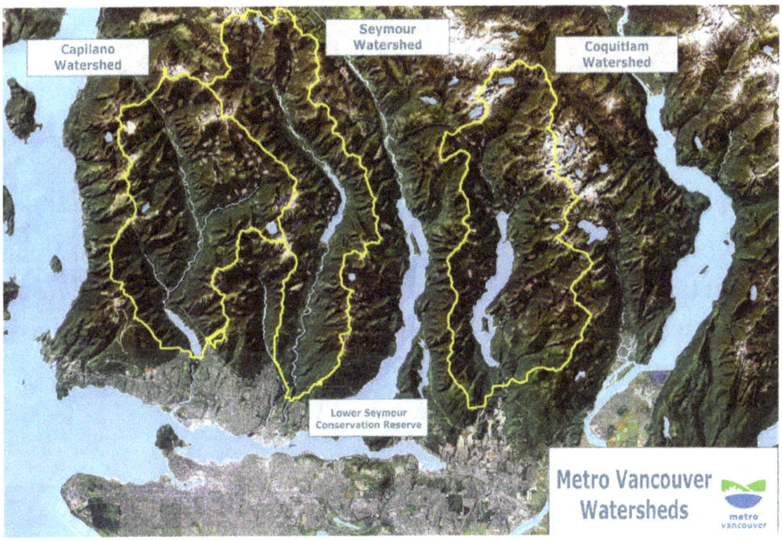

Figure 6: The three drainage basins that are used for the Metro Vancouver water supply.

The pattern of tributaries within a drainage basin depends largely on the type of rock beneath, and on structures within that rock (folds, fractures, faults, etc.). The three main types of drainage patterns are illustrated in Figure 7. Dendritic patterns, which are by far the most common, develop in areas where the rock (or unconsolidated material) beneath the stream has no particular fabric or structure and can be eroded equally easily in all directions. Examples would be granite, gneiss, volcanic rock, and sedimentary rock that has not been folded. Most areas of British Columbia have dendritic patterns, as do most areas of the prairies and the Canadian Shield. Trellis drainage patterns typically develop where sedimentary rocks have been folded or tilted and then eroded to varying degrees depending on their strength. The Rocky Mountains of B.C. and Alberta are a good example of this, and many of the drainage systems within the Rockies have trellis patterns. Rectangular patterns develop in areas that have very little topography and a system of bedding planes, fractures, or faults that form a rectangular network. Rectangular drainage patterns are rare in Canada.

In many parts of Canada, especially relatively flat areas with thick glacial sediments, and throughout much of Canadian Shield in eastern and central Canada, drainage patterns are chaotic, or what is known as deranged (Figure 8, left). Lakes and wetlands are common in this type of environment.

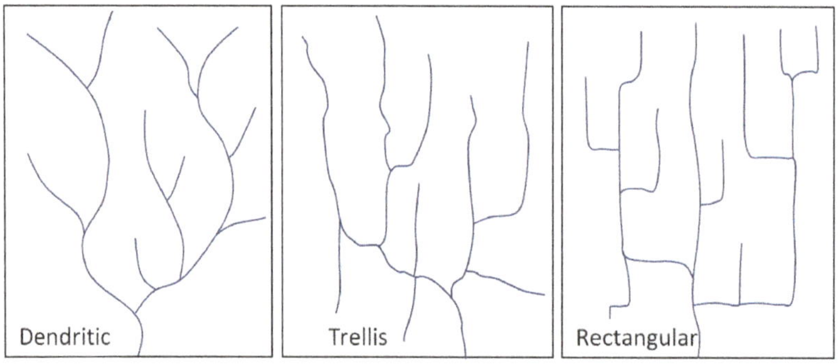

Figure 7: Typical dendritic, trellis, and rectangular stream drainage patterns.

A fourth type of drainage pattern, which is not specific to a drainage basin, is known as radial (Figure 8, right). Radial patterns form around isolated mountains (such as volcanoes) or hills, and the individual streams typically have dendritic drainage patterns.

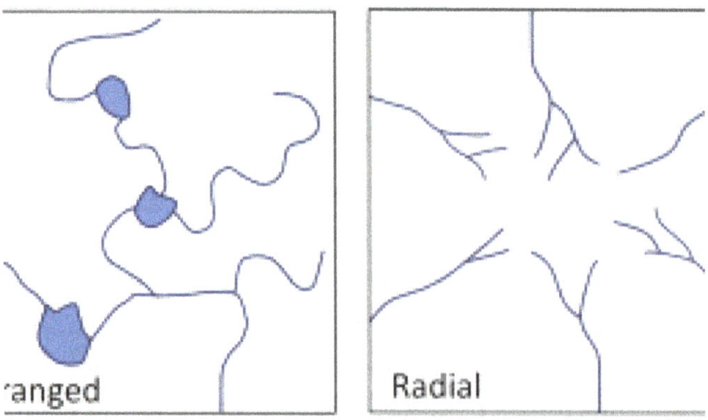

Figure 8: Left: a typical deranged pattern; right: a typical radial drainage pattern developed around a mountain or hill.

Over geological time, a stream will erode its drainage basin into a smooth profile similar to that shown in Figure 9. If we compare this with an ungraded stream like Cawston Creek (Figure 5), we can see that graded streams are steepest in their headwaters and their gradient gradually decreases toward their mouths. Ungraded streams have steep sections at various points, and typically have rapids and waterfalls at numerous locations along their lengths.

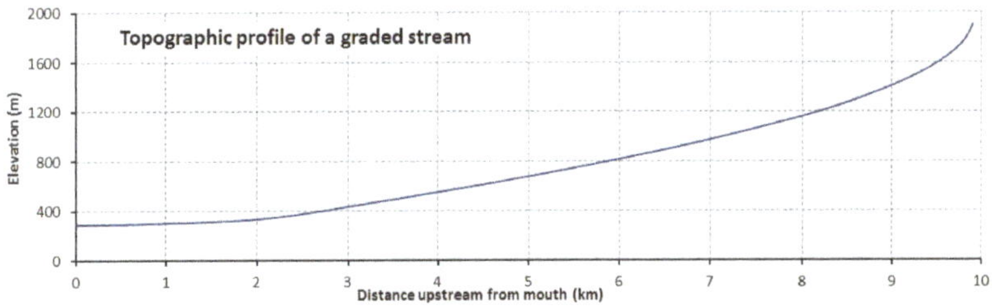

Figure 9: The topographic profile of a typical graded stream.

A graded stream can become ungraded if there is renewed tectonic uplift, or if there is a change in the base level, either because of tectonic uplift or some other reason. As stated earlier, the base level of Cawston Creek is defined by the level of the Similkameen River, but this can change, and has done so in the past. Figure 10 shows the valley of the Similkameen River in the Keremeos area. The river channel is just beyond the row of trees. The green field in the distance is underlain by material eroded from the hills behind and deposited by a small creek (not Cawston Creek) adjacent to the Similkameen River when its level was higher than it is now. Sometime in the past several centuries, the Similkameen River eroded down through these deposits (forming the steep bank on the other side of the river), and the base level of the small creek was lowered by about 10 m. Over the next few centuries, this creek will seek to become graded again by eroding down through its own alluvial fan.

Figure 10: An example of a change in the base level of a small stream that flows into the Similkameen river near Keremeos. The previous base level was near the top of the sandy bank. The current base level is the river.

Another example of a change in base level can be seen along the Juan de Fuca Trail on southwestern Vancouver Island. As shown in Figure 11, many of the small streams along this part of the coast flow into the ocean as waterfalls. It is evident that the land in this area has risen by about 5 m in the past few thousand years, probably in response to deglaciation. The streams that used to flow directly into the ocean now have a lot of down-cutting to do to become regraded.

Figure 11: Two streams with a lowered base level on the Juan de Fuca Trail, southwestern Vancouver Island.

The ocean is the ultimate base level, but lakes and other rivers act as base levels for many smaller streams. We can create an artificial base level on a stream by constructing a dam.

13.3 STREAM EROSION AND DEPOSITION

Flowing water is a very important mechanism for both erosion and deposition. Water flow in a stream is primarily related to the stream's gradient, but it is also controlled by the geometry of the stream channel. As shown in Figure 12, water flow velocity is decreased by friction along the stream bed, so it is slowest at the bottom and edges and fastest near the surface and in the middle. In fact, the velocity just below the surface is typically a little higher than right at the surface because of friction between the water and the air. On a curved section of a stream, flow is fastest on the outside and slowest on the inside.

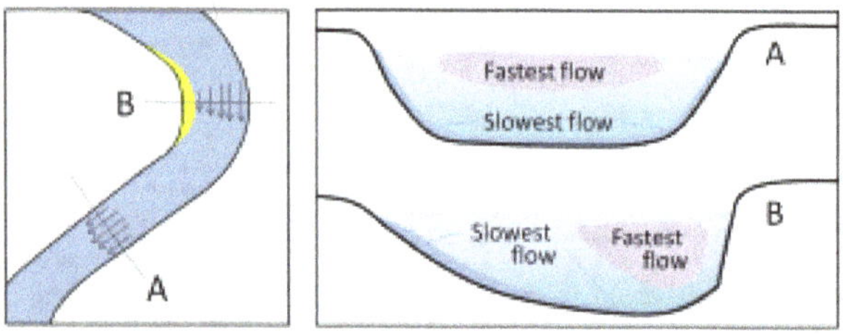

Figure 12: The relative velocity of stream flow depending on whether the stream channel is straight or curved (left), and with respect to the water depth (right).

Other factors that affect stream-water velocity are the size of sediments on the stream bed — because large particles tend to slow the flow more than small ones — and the discharge, or volume of water passing a point in a unit of time (e.g., m^3/

second). During a flood, the water level always rises, so there is more cross-sectional area for the water to flow in; however, as long as a river remains confined to its channel, the velocity of the water flow also increases.

Figure 13 shows the nature of sediment transportation in a stream. Large particles rest on the bottom —bedload — and may only be moved during rapid flows under flood conditions. They can be moved by saltation (bouncing) and by traction (being pushed along by the force of the flow).

Smaller particles may rest on the bottom some of the time, where they can be moved by saltation and traction, but they can also be held in suspension in the flowing water, especially at higher velocities. As you know from intuition and from experience, streams that flow fast tend to be turbulent (flow paths are chaotic and the water surface appears rough) and the water may be muddy, while those that flow more slowly tend to have laminar flow (straight-line flow and a smooth water surface) and clear water. Turbulent flow is more effective than laminar flow at keeping sediments in suspension.

Stream water also has a dissolved load, which represents (on average) about 15% of the mass of material transported, and includes ions such as calcium (Ca^{+2}) and chloride (Cl-) in solution. The solubility of these ions is not affected by flow velocity.

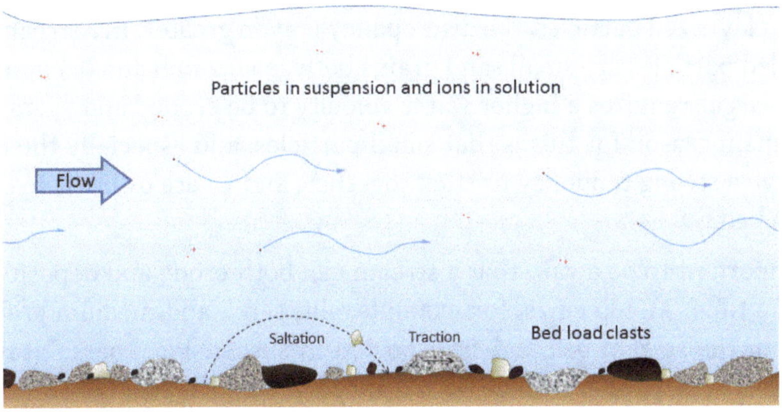

Figure 13: Modes of transportation of sediments and dissolved ions (represented by red dots with + and – signs) in a stream.

The faster the water is flowing, the larger the particles that can be kept in suspension and transported within the flowing water. However, as Swedish geographer Filip Hjulström discovered in the 1940s, the relationship between grain size and the likelihood of a grain being eroded, transported, or deposited is not as simple as one might imagine (Figure 14). Consider, for example, a 1 mm grain of sand. If it is resting on the bottom, it will remain there until the velocity is high enough to erode it, around 20 cm/s. But once it is in suspension, that same 1 mm particle will remain in suspension as long as the velocity doesn't drop below 10 cm/s. For a 10 mm gravel grain, the velocity is 105 cm/s to be eroded from the bed but only 80 cm/s to remain in suspension.

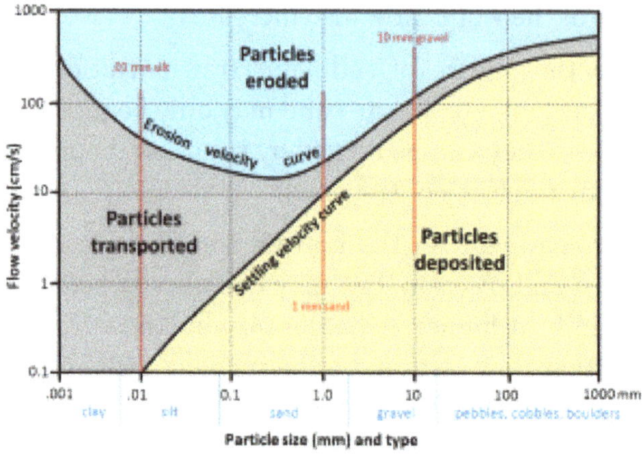

Figure 14: The Hjulström-Sundborg diagram showing the relationships between particle size and the tendency to be eroded, transported, or deposited at different current velocities

On the other hand, a 0.01 mm silt particle only needs a velocity of 0.1 cm/s to remain in suspension, but requires 60 cm/s to be eroded. In other words, a tiny silt grain requires a greater velocity to be eroded than a grain of sand that is 100 times larger! For clay-sized particles, the discrepancy is even greater. In a stream, the most easily eroded particles are small sand grains between 0.2 mm and 0.5 mm. Anything smaller or larger requires a higher water velocity to be eroded and entrained in the flow. The main reason for this is that small particles, and especially the tiny grains of clay, have a strong tendency to stick together, and so are difficult to erode from the stream bed.

It is important to be aware that a stream can both erode and deposit sediments at the same time. At 100 cm/s, for example, silt, sand, and medium gravel will be eroded from the stream bed and transported in suspension, coarse gravel will be held in suspension, pebbles will be both transported and deposited, and cobbles and boulders will remain stationary on the stream bed.

13.4 TYPES OF STREAM

Stream channels can be straight or curved, deep and slow, or rapid and choked with coarse sediments. The cycle of erosion has some influence on the nature of a stream, but there are several other factors that are important.

Youthful streams that are actively down-cutting their channels tend to be relatively straight and are typically ungraded (meaning that rapids and falls are common). As shown in Figures 1 and 15, youthful streams commonly have a step-pool morphology, meaning that the stream consists of a series of pools connected by rapids and waterfalls. They also have steep gradients and steep and narrow V-shaped valleys — in some cases steep enough to be called canyons.

Figure 15: The Cascade Falls area of the Kettle River, near Christina Lake, B.C. This stream has a step-pool morphology and a deep bedrock channel.

In mountainous terrain, such as that in western Alberta and B.C., steep youthful streams typically flow into wide and relatively low-gradient U-shaped glaciated valleys. The youthful streams have high sediment loads, and when they flow into the lower-gradient glacial valleys where the velocity isn't high enough to carry all of the sediment, braided patterns develop, characterized by a series of narrow channels separated by gravel bars (Figure 16).

Figure 16: The braided channel of the Kicking Horse River at Field, B.C.

Braided streams can develop anywhere there is more sediment than a stream is able to transport. One such environment is in volcanic regions, where explosive eruptions produce large amounts of unconsolidated material that gets washed into streams. The Coldwater River next to Mt. St. Helens in Washington State is a good example of this (Figure 17).

Figure 17: The braided Coldwater River, Mt. St. Helens, Washington.

A stream that occupies a wide, flat flood plain with a low gradient typically carries only sand-sized and finer sediments and develops a sinuous flow pattern. As you saw in Figure 14, when a stream flows around a corner, the water on the outside has farther to go and tends to flow faster. This leads to erosion of the banks on the outside of the curve, deposition on the inside, and formation of a point bar (Figure 18). Over time, the sinuosity of the stream becomes increasingly exaggerated, and the channel migrates around within its flood plain, forming a meandering pattern.

Figure 18: The meandering channel of the Bonnell Creek, Nanoose, B.C. The stream is flowing toward the viewer. The sand and gravel point bar must have formed when the creek was higher and the flow faster than it was when the photo was taken.

A well-developed meandering river is shown in Figure 19. The meander in the middle of the photo has reached the point where the thin neck of land between two parts of the channel is about to be eroded through. When this happens, another oxbow lake will form like the others in the photo.

Figure 19: The meandering channel of the Nowitna River, Alaska. Numerous oxbow lakes are present and another meander cutoff will soon take place.

13.5 FLOODING

The discharge levels of streams are highly variable depending on the time of year and on specific variations in the weather from one year to the next. In Canada, most streams show discharge variability similar to that of the Stikine River in northwestern B.C., as illustrated in Figure 20. The Stikine River has its lowest discharge levels in the depths of winter when freezing conditions persist throughout most of its drainage basin. Discharge starts to rise slowly in May, and then rises dramatically through the late spring and early summer as a winter's worth of snow melts. For the year shown, the minimum discharge on the Stikine River was 56 m3/s in March, and the maximum was 37 times higher, 2,470 m3/s, in May.

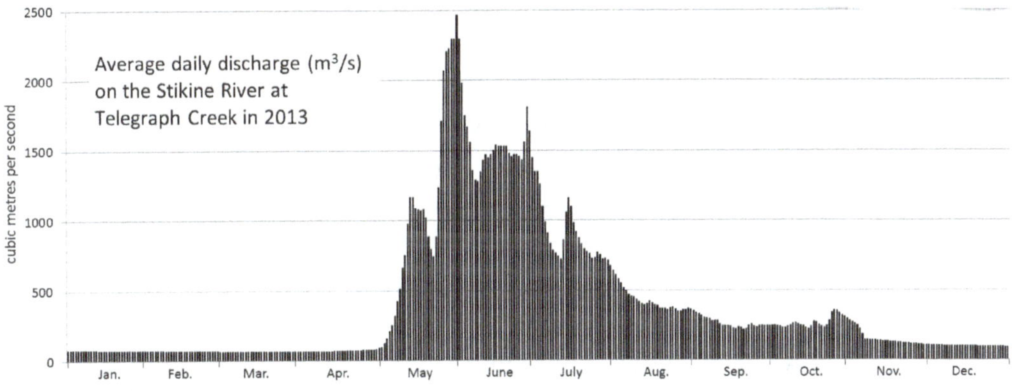

Figure 20: Variations in discharge of the Stikine River during 2013.

Streams in coastal areas of southern British Columbia show a very different pattern from those in most of the rest of the country because their drainage basins do not remain entirely frozen and because they receive a lot of rain (rather than snow) during the winter. The Qualicum River on Vancouver Island typically has its highest discharge levels in January or February and its lowest levels in late summer (Figure 21). In 2013, the minimum discharge was 1.6 m3/s, in August, and the maximum was 34 times higher, 53 m3/s, in March.

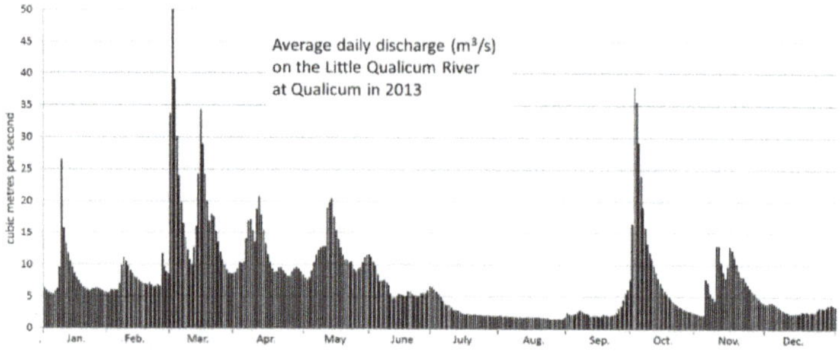

Figure 21: Variations in discharge of the Qualicum River during 2013.

When a stream's discharge increases, both the water level (stage) and the velocity increase as well. Rapidly flowing streams become muddy and large volumes of sediment are transported both in suspension and along the stream bed. In extreme situations, the water level reaches the top of the stream's banks (the bank-full stage, see Figure 17), and if it rises any more, it floods the surrounding terrain. In the case of mature or old-age streams, this could include a vast area of relatively flat ground known as a flood plain, which is the area that is typically covered with water during a major flood. Because fine river sediments are deposited on flood plains, they are ideally suited for agriculture, and thus are typically occupied by farms and residences, and in many cases, by towns or cities. Such infrastructure is highly vulnerable to damage from flooding, and the people that live and work there are at risk.

Most streams in Canada have the greatest risk of flooding in the late spring and early summer when stream discharges rise in response to melting snow. In some cases, this is exacerbated by spring storms. In years when melting is especially fast and/or spring storms are particularly intense, flooding can be very severe.

One of the worst floods in Canadian history took place in the Fraser Valley in late May and early June of 1948. The early spring of that year had been cold, and a large snow pack in the interior was slow to melt. In mid-May, temperatures rose quickly and melting was accelerated by rainfall. Fraser River discharge levels rose rapidly over several days during late May, and the dykes built to protect the valley were breached in a dozen places. Approximately one-third of the flood plain was inundated and many homes and other buildings were destroyed, but there were no deaths. The Fraser River flood of 1948, which was the highest in the past century, was followed by very high river levels in 1950 and 1972 and by relatively high levels several times since then, the most recent being 2007 (Table 1). In the years following 1948, millions of dollars were spent repairing and raising the existing dykes and building new ones; since then damage from flooding in the Fraser Valley has been relatively limited.

Table1 Ranking of the maximum stage and discharge values for the Fraser River at Hope between 1948 and 2008. Typical discharge levels are around 1,000 m3/s.

[From date in Mannerstrom, 2008Mannerström, M, 2008, Comprehensive Review of Fraser River at Hope Flood Hydrology and Flows Scoping Study, Report prepared for the B.C. Ministry of the Environment. Available at: http://www.env.gov.bc.ca/wsd/public_safety/flood/pdfs_word/review_fraser_flood_flows_hope.pdf]

Rank	Year	Month	Date	Stage (m)	Discharge (m3/s)
1	1948	May	31	11.0	15,200
2	1972	Jun	16	10.1	12,900
3	1950	Jun	20	9.9	12,500
4	1964	Jun	21	9.6	11,600
5	1997	Jun	5	9.5	11,300
6	1955	Jun	29	9.4	11,300
7	1999	Jun	22	9.4	11,000
8	2007	Jun	10	9.3	10,850
9	1974	Jun	22	9.3	10,800
10	2002	Jun	21	9.2	10,600

Table 1 Ranking of the maximum stage and discharge values for the Fraser River at Hope between 1948 and 2008. Typical discharge levels are around 1,000 m3/s. [From date in Mannerstrom, 2008Mannerström, M, 2008, Comprehensive Review of Fraser River at Hope Flood Hydrology and Flows Scoping Study,

Serious flooding happened in July in 1996 in the Saguenay-Lac St. Jean region of Quebec. In this case, the floods were caused by two weeks of heavy rainfall followed by one day of exceptional rainfall. July 19 saw 270 mm of rain, equivalent to the region's normal rainfall for the entire month of July. Ten deaths were attributed to the Saguenay floods, and the economic toll was estimated at $1.5 billion.

Just a year after the Saguenay floods, the Red River in Minnesota, North Dakota, and Manitoba reached its highest level since 1826. As is typical for the Red River, the 1997 flooding was due to rapid snowmelt. Because of the south to north flow of the river, the flooding starts in Minnesota and North Dakota, where melting starts earlier, and builds toward the north. The residents of Manitoba had plenty of warning that the 1997 flood was coming because there was severe flooding at several locations on the U.S. side of the border.

After the 1950 Red River flood, the Manitoba government built a channel around the city of Winnipeg to reduce the potential of flooding in the city (Figure 22). Known as the Red River Floodway, the channel was completed in 1964 at a cost of $63 million. Since then it has been used many times to alleviate flooding in Winnipeg, and is estimated to have saved many billions of dollars in flood damage. The massive 1997 flood was almost too much for the floodway; in fact the amount of water diverted was greater than the designed capacity. The floodway has recently been expanded so that it can be used to divert more of the Red River's flow away from Winnipeg.

Figure 22: Map of the Red River Floodway around Winnipeg, Manitoba (left), and aerial view of the southern (inlet) end of the floodway (right).

Canada's most costly flood ever was the June 2013 flood in southern Alberta. The flooding was initiated by snowmelt and worsened by heavy rains in the Rockies due to an anomalous flow of moist air from the Pacific and the Caribbean. At Canmore, rainfall amounts exceeded 200 mm in 36 hours, and at High River, 325 mm of rain fell in 48 hours.

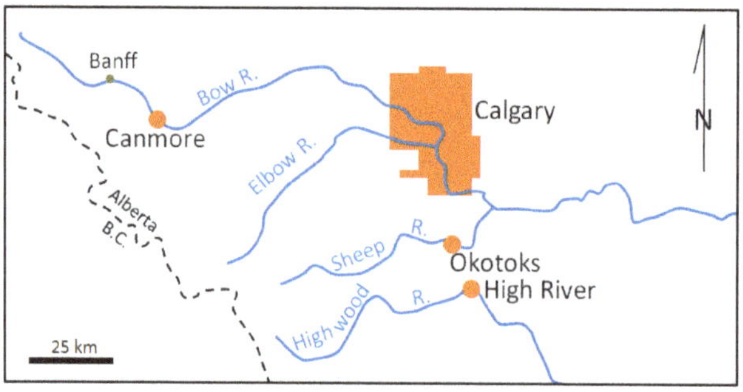

Figure 23: Map of the communities most affected by the 2013 Alberta floods (in orange)

In late June and early July, the discharges of several rivers in the area, including the Bow River in Banff, Canmore, and Exshaw, the Bow and Elbow Rivers in Calgary, the Sheep River in Okotoks, and the Highwood River in High River, reached levels that were 5 to 10 times higher than normal for the time of year (see Exercise 5). Large areas of Calgary, Okotoks, and High River were flooded and five people died (see Figures 18 and 23). The cost of the 2013 flood is estimated to be approximately $5 billion.

Figure 24: Flooding in Calgary (June 21, left) and Okotoks (June 20, right) during the 2013 southern Alberta flood

AT A GLANCE

The Hydrological Cycle	Water is stored in the oceans, glacial ice, the ground, lakes, rivers, and the atmosphere. Its movement is powered by the sun and gravity.
Drainage Basins	All of the precipitation that falls within a drainage basin flows into the stream that drains that area. Stream drainage patterns are determined by the type of rock within the basin. Over geological time, streams change the landscape that they flow within, and eventually they become graded, meaning their profile is a smooth curve. A stream can lose that gradation if there is renewed uplift or if their base level changes for some reason.
Stream Erosion and Deposition	Erosion and deposition of particles within streams is primarily determined by the velocity of the water. Erosion and deposition of different-sized particles can happen at the same time. Some particles are moved along the bottom of a river while some are suspended in the water. It takes a greater velocity of water to erode a particle from a stream bed than it does to keep it in suspension. Ions are also transported in solution. When a stream rises and then occupies its flood plain, the velocity slows and natural levées form along the edges of the channel.
Stream Types	Youthful streams in steep areas erode rapidly, and they tend to have steep, rocky, and relatively straight channels. Where sediment-rich streams empty into areas with lower gradients, braided streams can form. In areas with even lower gradients, and where silt and sand are the dominant sediments, meanders are common. Deltas form where streams flow into standing water.
Flooding	Most streams in Canada have their highest discharge rates in spring and early summer, although many of B.C.'s coastal streams are highest in the winter. Floods happen when a stream rises high enough to spill over its banks and spread across its flood plain. Some of the more significant floods in Canada include the Fraser River flood of 1948, the Saguenay River flood of 1996, the Red River flood of 1997, and the Alberta floods of 2013. We can estimate the probability of a specific flood level based on the record of past floods, and we can take steps to minimize the impacts of flooding.

CHAPTER

GROUNDWATER 14

Fresh water makes up only 3% of the water on Earth. Approximately two-thirds of that are glacial ice and most of the rest is groundwater. We can't live without water, and it's easy to see that groundwater represents a critically important component of our water supply. Groundwater is not as easily accessed as surface water, but it is also not as easily contaminated as surface water. If more than 7 billion of us want to continue living comfortably here on Earth, we have to take great care of our groundwater and learn how to use it sustainably.

Figure 1: A spring flowing from a limestone cave on Quadra Island, B.C.

14.1 GROUNDWATER AND AQUIFERS

Groundwater is stored in the open spaces within rocks and within unconsolidated sediments. Rocks and sediments near the surface are under less pressure than those at significant depth and therefore tend to have more open space. For this reason, and because it's expensive to drill deep wells, most of the groundwater that is accessed by individual users is within the first 100 m of the surface. Some municipal, agricultural, and industrial groundwater users get their water from greater depth, but deeper groundwater tends to be of lower quality than shallow groundwater, so there is a limit as to how deep we can go.

Porosity is the percentage of open space within an unconsolidated sediment or a rock. Primary porosity is represented by the spaces between grains in a sediment or sedimentary rock. Secondary porosity is porosity that has developed after the rock has formed. It can include fracture porosity — space within fractures in any kind of rock. Some volcanic rock has a special type of porosity related to vesicles, and some limestone has extra porosity related to cavities within fossils.

Porosity is expressed as a percentage calculated from the volume of open space in a rock compared with the total volume of rock. The typical ranges in porosity of a number of different geological materials are shown in Figure 2. Unconsolidated sediments tend to have higher porosity than consolidated ones because they have no cement, and most have not been strongly compressed. Finer-grained materials (e.g., silt and clay) tend to have greater porosity — some as high as 70% — than coarser materials (e.g., gravel). Primary porosity tends to be higher in well-sorted sediments compared to poorly sorted sediments, where there is a range of smaller particles to fill the spaces made by the larger particles. Glacial till, which has a wide range of grain sizes and is typically formed under compression beneath glacial ice, has relatively low porosity.

Consolidation and cementation during the process of lithification of unconsolidated sediments into sedimentary rocks reduces primary porosity. Sedimentary rocks generally have porosities in the range of 10% to 30%, some of which may be secondary (fracture) porosity. The grain size, sorting, compaction, and degree of cementation of the rocks all influence primary porosity. For example, poorly sorted and well-cemented sandstone and well-compressed mudstone can have very low porosity. Igneous or metamorphic rocks have the lowest primary porosity because they commonly form at depth and have interlocking crystals. Most of their porosity comes in the form of secondary porosity in fractures. Of the consolidated rocks, well-fractured volcanic rocks and limestone that has cavernous openings produced by dissolution have the highest potential porosity, while intrusive igneous and metamorphic rocks, which formed under great pressure, have the lowest.

Porosity is a measure of how much water can be stored in geological materials. Almost all rocks contain some porosity and therefore contain groundwater. Groundwater is found under your feet and everywhere on the planet. Considering that sedimentary rocks and unconsolidated sediments cover about 75% of the continental crust with an average thickness of a few hundred metres, and that they are likely to have around 20% porosity on average, it is easy to see that a huge volume of water can be stored in the ground.

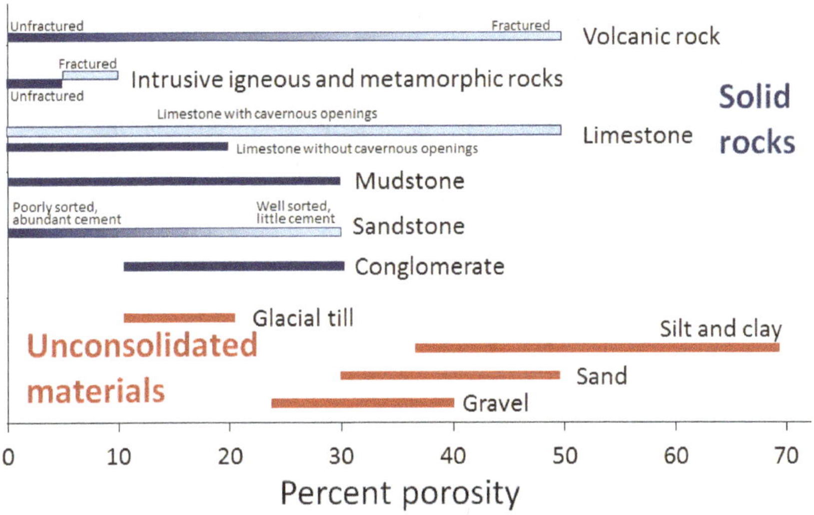

Figure 2: Variations in porosity of unconsolidated materials (in red) and rocks (in blue)

Porosity is a description of how much space there could be to hold water under the ground, and permeability describes how those pores are shaped and interconnected. This determines how easy it is for water to flow from one pore to the next. Larger pores mean there is less friction between flowing water and the sides of the pores. Smaller pores mean more friction along pore walls, but also more twists and turns for the water to have to flow-through. A permeable material has a greater number of larger, well-connected pores spaces, whereas an impermeable material has fewer, smaller pores that are poorly connected. Permeability is the most important variable in groundwater. Permeability describes how easily water can flow through the rock or unconsolidated sediment and how easy it will be to extract the water for our purposes. The characteristic of permeability of a geological material is quantified by geoscientists and engineers using a number of different units, but the most common is the hydraulic conductivity. The symbol used for hydraulic conductivity is K. Although hydraulic conductivity can be expressed in a range of different units, in this book, we will always use m/s.

The materials in Figure 3 show that there is a wide range of permeability in geological materials from 10-12 m/s (0.000000000001 m/s) to around 1 m/s. Unconsolidated materials are generally more permeable than the corresponding rocks (compare sand with sandstone, for example), and the coarser materials are much more permeable than the finer ones. The least permeable rocks are unfractured intrusive igneous and metamorphic rocks, followed by unfractured mudstone, sandstone, and limestone. The permeability of sandstone can vary widely depending on the degree of sorting and the amount of cement that is present. Fractured igneous and metamorphic rocks, and especially fractured volcanic rocks, can be highly permeable, as can limestone that has been dissolved along fractures and bedding planes to create solutional openings.

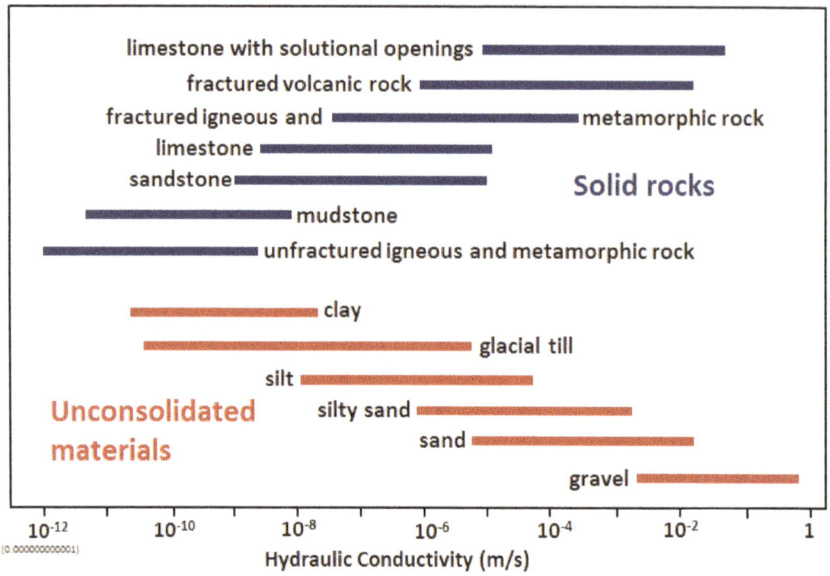

Figure 3: Variations in hydraulic conductivity (in metres/second) of unconsolidated materials (in red) and of rocks (in blue)

Why is clay porous but not permeable?

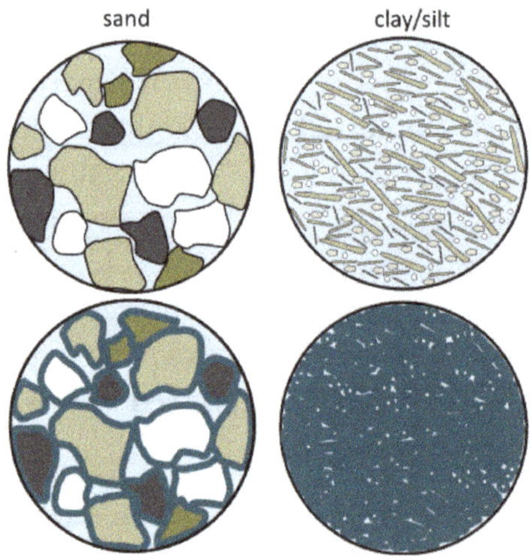

Both sand and clay deposits (and sandstone and mudstone) are quite porous (30% to 50% for sand and 40% to 70% for silt and clay), but while sand can be quite permeable, clay and mudstone are not.

The surface of most silicate mineral grains has a slight negative charge due to imperfections in the mineral structure. Water (H_2O) is a polar molecule. This means that while it has no overall electrical charge, one side of the molecule has a slight positive charge (the side with the two hydrogens), compared to a slight negative charge on the other side. Water is strongly attracted to all mineral grains and water within that bound water layer (a few microns around each grain) is not able to move and flow along with the rest of the groundwater. In the lower diagrams shown here, the bound water is represented by dark blue lines around each grain and the water that can move is light blue. In the sand, there is still a lot of water that is able to move through the sediment, but in the clay/silt almost all of the water is held tightly to the grains and this reduces the permeability. [SE]

We have now seen that there is a wide range of porosity in geological materials and an even wider range of permeability. Groundwater exists everywhere there is porosity. However, whether that groundwater is able to flow in significant quantities depends on the permeability. An aquifer is defined as a body of rock or unconsolidated sediment that has sufficient permeability to allow water to flow through it. Unconsolidated materials like gravel, sand, and even silt make relatively good aquifers, as do rocks like sandstone. Other rocks can be good aquifers if they are well fractured. An aquitard is a body that does not allow transmission of a significant amount of water, such as a clay, a till, or a poorly fractured igneous or metamorphic rock. These are relative terms, not absolute, and are usually defined based on someone's desire to pump groundwater; what is an aquifer to someone who does not need a lot of water, may be an aquitard to someone else who does. An aquifer that is exposed at the ground surface is called an unconfined aquifer. An aquifer where there is a lower permeability material between the aquifer and the ground surface is known as a confined aquifer, and the aquitard separating ground surface and the aquifer is known as the confining layer.

Figure 4 shows a cross-section of a series of rocks and unconsolidated materials, some of which might serve as aquifers and others as aquitards or confining layers. The granite is much less permeable than the other materials, and so is an aquitard in this context. The yellow layer is very permeable and would make an ideal aquifer. The overlying grey layer is a confining layer.

The upper buff-coloured layer (K = 10-2 m/s) does not have a confining layer and is an unconfined aquifer. The yellow layer (K = 10-1 m/s) is "confined" by the confining layer (K = 10-4 m/s), and is a confined aquifer. The confined aquifer gets most of its water from the upper part of the hill where it is exposed at the surface, and relatively little by seepage through the fine silt layer.

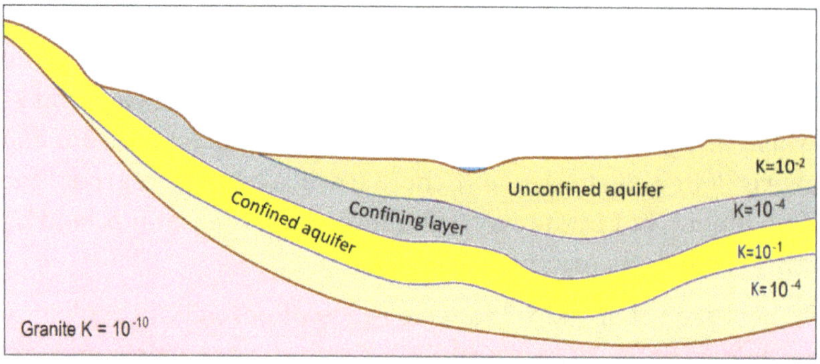

Figure 4: A cross-section showing materials that might serve as aquifers and confining layers. The relative permeabilities are denoted by hydraulic conductivity (K = m/s). The pink rock is granite; the other layers are various sedimentary layers.

14.2 GROUNDWATER FLOW

If you go out into your garden or into a forest or a park and start digging, you will find that the soil is moist (unless you're in a desert), but it's not saturated with water. This means that some of the pore space in the soil is occupied by water, and some of the pore space is occupied by air (unless you're in a swamp). This is known as the unsaturated zone. If you could dig down far enough, you would get to the point where all of the pore spaces are 100% filled with water (saturated) and the bottom of your hole would fill up with water. The level of water in the hole represents the water table, which is the surface of the saturated zone. In most parts of British Columbia, the water table is several metres below the surface.

Water falling on the ground surface as precipitation (rain, snow, hail, fog, etc.) may flow off a hill slope directly to a stream in the form of runoff, or it may infiltrate the ground, where it is stored in the unsaturated zone. The water in the unsaturated zone may be used by plants (transpiration), evaporate from the soil (evaporation), or continue past the root zone and flow downward to the water table, where it recharges the groundwater.

A cross-section of a typical hillside with an unconfined aquifer is illustrated in Figure 14.5. In areas with topographic relief, the water table generally follows the land surface, but tends to come closer to surface in valleys, and intersects the surface where there are streams or lakes. The water table can be determined from the depth of water in a well that isn't being pumped, although, as described below, that only applies if the well is within an unconfined aquifer. In this case, most of the hillside forms the recharge area, where water from precipitation flows downward through the unsaturated zone to reach the water table. The area at the stream or lake to which the groundwater is flowing is a discharge area.

What makes water flow from the recharge areas to the discharge areas? Recall that water is flowing in pores where there is friction, which means it takes work to move the water. There is also some friction between water molecules themselves, which is determined by the viscosity. Water has a low viscosity, but friction is still a factor. All flowing fluids are always losing energy to friction with their surroundings. Water will flow from areas with high energy to those with low energy. Recharge areas are at higher elevations, where the water has high gravitational energy. It was energy from the sun that evaporated the water into the atmosphere and lifted it up to the recharge area. The water loses this gravitational energy as it flows from the recharge area to the discharge area.

In Figure 5, the water table is sloping; that slope represents the change in gravitational potential energy of the water at the water table. The water table is higher under the recharge area (90 m) and lower at the discharge area (82 m). Imagine how much work it would be to lift water 8 m high in the air. That is the energy that was lost to friction as the groundwater flowed from the top of the hill to the stream.

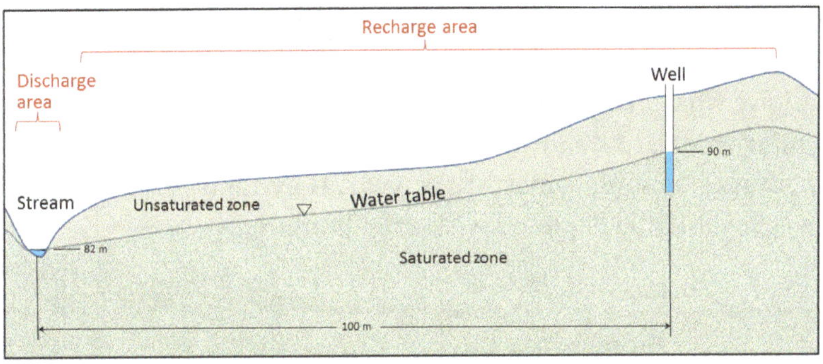

Figure 5: A depiction of the water table in cross-section, with the saturated zone below and the unsaturated zone above. The water table is denoted with a small upside-down triangle.

The situation gets a lot more complicated in the case of confined aquifers, but they are important sources of water so we need to understand how they work. As shown in Figure 6, there is always a water table, and that applies even if the geological materials at the surface have very low permeability. Where there is a confined aquifer — meaning one that is separated from the surface by a confining layer — this aquifer will have its own "water table," which is actually called a potentiometric surface, as it is a measure of the total potential energy of the water. The red dashed line in Figure 14.6 is the potentiometric surface for the confined aquifer, and it describes the total energy that water is under within the confined aquifer. If we drill a well into the unconfined aquifer, the water will rise to the level of the water table (well A in Figure 6). But if we drill a well through both the unconfined aquifer and the confining layer and into the confined aquifer, the water will rise above the top of the confined aquifer to the level of its potentiometric surface (well B in Figure 6). This is known as an artesian well, because the water rises above the top of the aquifer. In some situations, the potentiometric surface may be above the ground level. The water in a well drilled into the confined aquifer in this situation would rise above ground level, and flow out, if it's not capped (well C in Figure 6). This is known as a flowing artesian well.

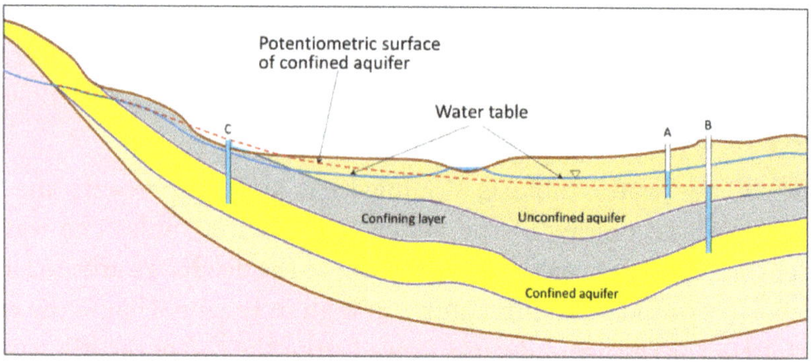

Figure 6: A depiction of the water table and the potentiometric surface of a confined aquifer.

In situations where there is an aquitard of limited extent, it is possible for a perched aquifer to exist as shown in Figure 7. Although perched aquifers may be good water sources at some times of the year, they tend to be relatively thin and small, and so can easily be depleted with over-pumping.

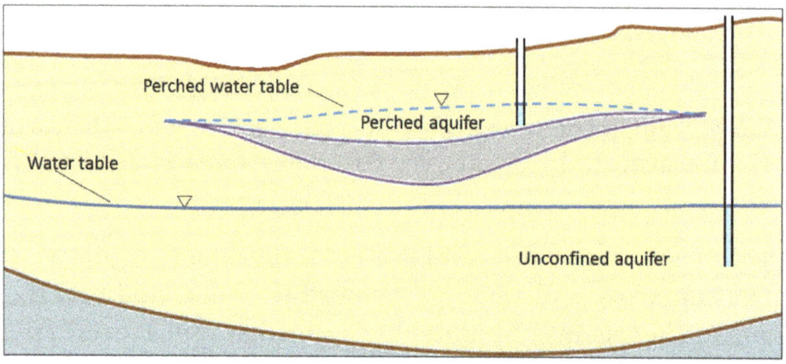

Figure 7: A perched aquifer above a regular unconfined aquifer by gravity. The water below the water table does behave like true groundwater.

In 1856, French engineer Henri Darcy carried out some experiments from which he derived a method for estimating the rate of groundwater flow based on the hydraulic gradient and the permeability of an aquifer, expressed using K, the hydraulic conductivity. Darcy's equation, which has been used widely by hydrogeologists ever since, looks like this:

$$V=K*i$$

(where V is the velocity of the groundwater flow, K is the hydraulic conductivity, and i is the hydraulic gradient).

We can apply this equation to the scenario in Figure 5. If we assume that the permeability is 0.00001 m/s we get: $V = 0.00001 * 0.08 = 0.0000008$ m/s. That is equivalent to 0.000048 m/min, 0.0029 m/hour or 0.069 m/day. That means it would take 1,450 days (nearly four years) for water to travel the 100 m from the vicinity of the well to the stream. Groundwater moves slowly, and that is a reasonable amount of time for water to move that distance. In fact it would likely take longer than that, because it doesn't travel in a straight line.

14.3 EXTRACTION OF GROUNDWATER

Except in areas where groundwater comes naturally to the surface at a spring (a place where the water table intersects the ground surface), we have to construct wells in order to extract it. If the water table is relatively close to the surface, a well can be dug by hand or with an excavator, but in most cases we need to use a drill to go down deep enough. There are many types of drills that can be used; an example is shown in Figure 8. A well has to be drilled at least as deep as the water table, but in fact must go much deeper; first, because the water table may change from season to season and from year to year, and second, because when water is being pumped, the water level will drop, at least temporarily.

Figure 8: A water-well drilling rig in operation in the Cassidy area, near Nanaimo, B.C. In the photo on the right the well is being test-pumped with air pressure. The casing (yellow arrow) is about 40 cm in diameter.

Where a well is drilled in unconsolidated sediments or relatively weak rock, it has to be lined with casing (steel pipe in most cases) in order to ensure that it doesn't cave in. A specially designed well screen is installed at the bottom of the casing. The size of the holes in the screen is carefully chosen to make sure that it allows the water to move into the well freely, but prevents aquifer particles from entering the well. A submersible pump is typically used to lift water from within the well up to the where it is needed. The well shown in Figure 8 has casing that is about 40 cm in diameter, which might be typical for a municipal water supply well, or a very large well for irrigation. Most domestic wells have 15 cm casing.

Pumping water from the well removes water from inside the well at first. That lowers the water level inside the well. This means that water will flow from the surrounding aquifer (higher groundwater head) toward the pumping well where the groundwater head is now lower. That is how a well gets water from the ground. The water table, or potentiometric surface, will slope in toward the well where the water is being withdrawn. That indicates the energy gradient that is allowing water to flow toward the well. This creates a shape known as a cone of depression surrounding the well, as illustrated in Figure 9. If pumping from a well continues for hours to days, the cone of depression may result in a loss of water in nearby wells. As shown in Figure 10, pumping of well C has contributed to well B going dry. If pumping continues in well C, it too may go dry.

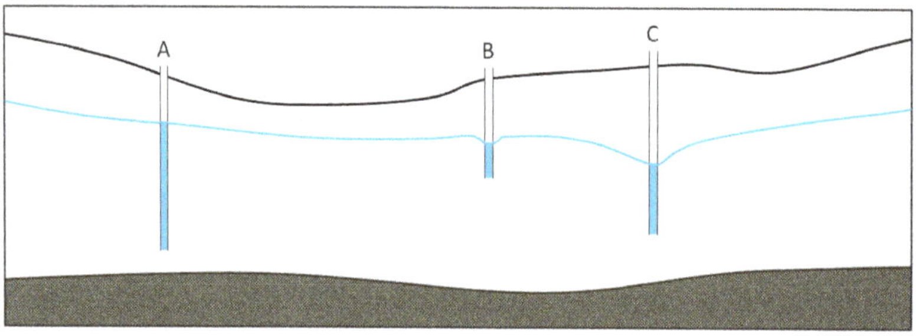

Figure 9: Three wells in an unconfined aquifer. Well A is not being pumped. Well B is being pumped at a slow rate and well C, which has a larger cone of depression, is being pumped at a faster rate.

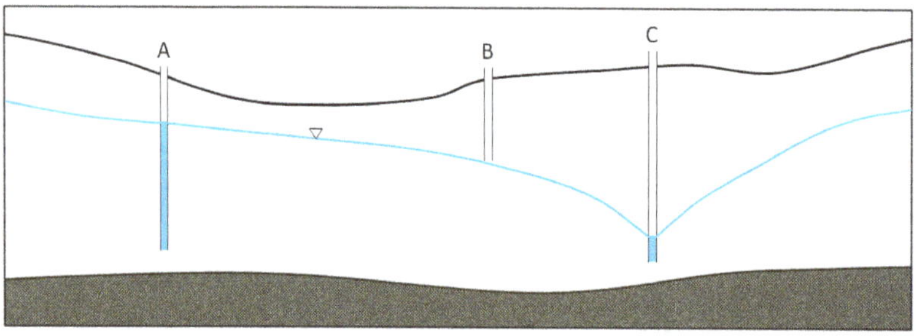

Figure 10: A similar scenario to that in Figure 11, but in this case, wells B and C have been pumped unsustainably for a long time. The cone of depression from well C has reached well B and has contributed to it going dry.

Impermeable Surfaces

Even if groundwater supplies are not being depleted by overuse, or by a changing climate, we are continuing to put stress on aquifers by covering vast areas with impermeable surfaces that don't allow rain and snowmelt to infiltrate and become groundwater. Instead, water that falls on these surfaces is channelled into drainage systems, then into storm sewers, and then directly into rivers and the ocean. In cities and their suburbs, the biggest culprits are parking lots, roads, and highways. While it would great if we didn't dedicate such huge areas to cars, that's not about to change quickly, so we need to think about ways that we can improve surface water infiltration in cities. One way is to use road and parking surfaces that will allow water to seep through, although this is not practical in many cases. Another way is to ensure that runoff from pavement is channelled into existing or constructed wetlands that serve to decontaminate the water, and then allow it to infiltrate into the ground.

14.4 GROUNDWATER QUALITY

Groundwater as a source of water is that it is not as easily contaminated as surface water is. But there are two caveats to that: one is that groundwater can become naturally contaminated because of its very close connection to the materials of its aquifer, and the second is that once contaminated by human activities, groundwater is very difficult to clean up.

Natural Groundwater Pollution

Groundwater moves slowly through an aquifer, and unlike the surface water of a stream, it has a lot of contact with the surrounding rock or sediment. In most aquifers, the geological materials that make up the aquifer are relatively inert, or are made up of minerals that dissolve very slowly into the groundwater. Over time, however, all groundwater gradually has more and more material dissolved within it as it remains in contact with the aquifer. In some areas, that rock or sediment includes some minerals that could potentially contaminate the water with elements that might make the water less than ideal for human consumption or agricultural use. Examples include copper, arsenic, mercury, fluorine, sodium, and boron. In some cases, contamination may occur because the aquifer material has particularly high levels of the element in question. In other cases, the aquifer material is just normal rock or sediment, but some particular feature of the water or the aquifer allows the contaminant to build up to significant levels.

An example of natural contamination takes place in the bedrock aquifers of the east coast of Vancouver Island and the adjacent Gulf Islands. The aquifer is the Cretaceous (90 Ma to 65 Ma) Nanaimo Group, which is made up of sandstone, mudstone, and conglomerate (Figure 11).

Figure 11: Cretaceous Nanaimo Group sandstone exposed in a Nanaimo parking lot

The rocks of the Nanaimo Group are not particularly enriched in any trace elements, but the submarine-fan sandstone that makes up much of the group is a lithic wacke, and therefore has relatively high levels of clay (for a sandstone). This clay is good at adsorbing"Adsorb" (with a "d") is not the same as "absorb" (with a "b"). Water can be absorbed by a sponge. Ions dissolved in water can be adsorbed onto — or desorbed from — the surfaces of clay minerals. some elements from the water and desorbing others, and in the process, its pH goes up (it becomes alkaline). At high pH levels (some as high as 9 in the Nanaimo Group), the element fluorine that is present naturally in the rock (as it is in almost any rock) has an increased tendency to dissolve in the water. In some areas, groundwater in the Nanaimo Group has fluorine levels that are well above recommended levels for drinking water. The World Health Organization (WHO) maximum acceptable concentration (MAC) for fluorine is 1.5 mg/L (milligrams per litre). Between 5% and 10% of the domestic wells around Nanaimo and adjacent Gabriola Island have more than that, some as much as 10 mg/L. A small amount of fluorine in the human diet is considered important for maintaining dental health, but high levels can lead to malformation and discolouration of teeth, and long-term exposure can lead to other more serious health effects such as skeletal problems.

Nanaimo Group groundwater can also have elevated levels of boron, again related to pH and adsorption from clay minerals. While boron at the levels found there is not toxic to humans, there is enough boron in some wells to be toxic to plants, and the water cannot be used for irrigation.

Rural residents in the densely populated country of Bangladesh (over 1,000 residents/km2, compared with 3.4/km2 in Canada) used to rely mostly on surface supplies for their drinking water, and many of these were subject to bacterial contamination. Infant mortality rates were among the highest in the world and other illnesses such as diarrhea, dysentery, typhoid, cholera, and hepatitis were common. In the 1970s, international agencies, including UNICEF, started a program of drilling wells to access abundant groundwater supplies at depths of 20 m to 100 m. Eventually over 8 million such wells were drilled. Infant mortality and illness rates dropped dramatically, but it was later discovered that the water from a high proportion of these wells has arsenic above safe levels (Figure 12).

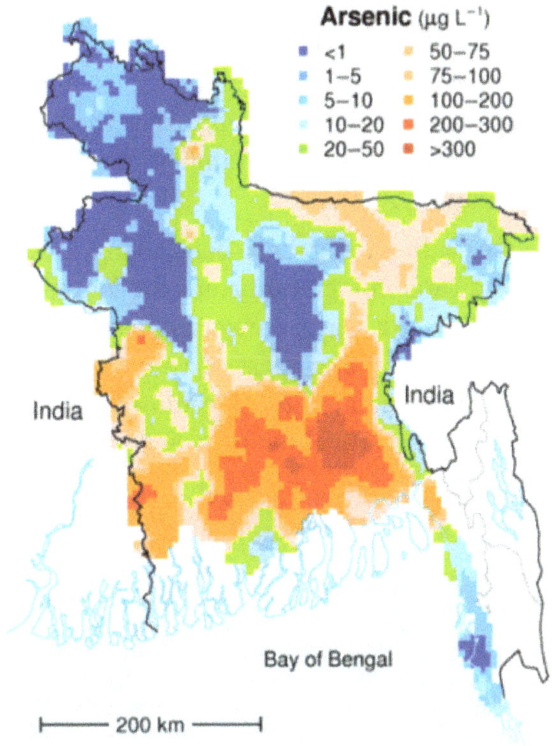

Figure 12: The distribution of arsenic in groundwater in Bangladesh. The WHO recommended safe level for arsenic is 10 µg/L. All of the green, orange, and red areas on the map exceed that limit.

Most of the wells in the affected areas are drilled into relatively recent sediments of the vast delta of the Ganges and Brahmaputra Rivers. While these sediments are not particularly enriched in arsenic, they have enough organic matter in them to use up any oxygen present. This leads to water with a naturally low oxidation potential (anoxic conditions); arsenic is highly soluble under these conditions, and so any arsenic present in the sediments easily gets dissolved into the groundwater. Arsenic poisoning leads to headaches, confusion, and diarrhea, and eventually to vomiting, stomach pain, and convulsions. If not treated, the final outcomes are heart disease, stroke, cancer, diabetes, coma, and death. There are ways to treat arsenic-rich groundwater, but it is a challenge in Bangladesh to implement the simple and effective technology that is available.

Anthropogenic Contamination of Groundwater

Groundwater can become contaminated by pollution at the surface (or at depth), and there are many different anthropogenic (human-caused) sources of contamination.

The vulnerability of aquifers to pollution depends on several factors, including the depth to the water table, the permeability of the material between the surface and the aquifer, the permeability of the aquifer, the slope of the surface, and the amount of precipitation. Confined aquifers tend to be much less vulnerable than unconfined ones, and deeper aquifers are less vulnerable than shallow ones. Steeper slopes mean that surface water tends to run off rather than infiltrate (and this can reduce the possibility of contamination). Contamination risk is also less in dry areas than in areas with heavy rainfall.

Studies of groundwater vulnerability have been completed for various regions of British Columbia. A groundwater vulnerability map for southern Vancouver Island is shown in Figure 13. The yellow to red areas are considered to have high vulnerability to pollution from surface sources, and most of these are where the aquifers are unconfined in quite permeable unconsolidated sediments of either glacial or fluvial origin, where the water table is relatively shallow and the terrain is relatively flat.

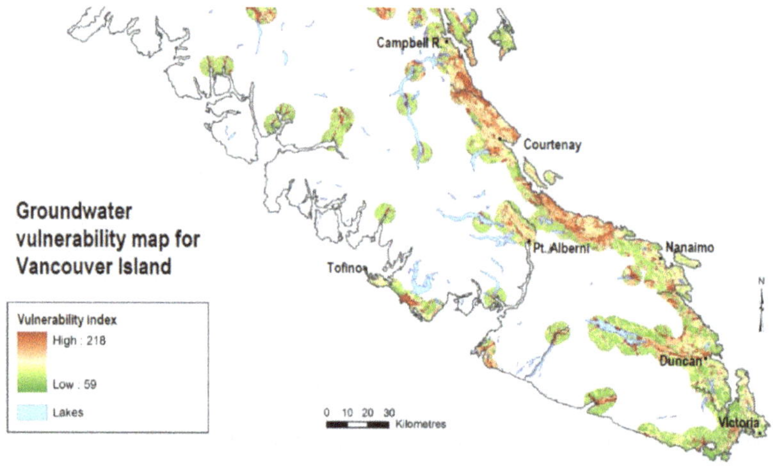

Figure 13: The vulnerability to anthropogenic contamination of aquifers on southern Vancouver Island. Much of the island is not mapped (shown as white) because of a lack of aquifer information in areas without wells. [From: Newton, P. and Gilchrist, A. 2010. Technical summary of intrinsic vulnerability mapping methods of Vancouver Island, Vancouver Island Water Resources Vulnerability Mapping Project, Vancouver Island University, 45pp. Used with permission.

The important sources of anthropogenic groundwater contamination include the following:

- Chemicals and animal waste related to agriculture, and chemicals applied to golf courses and domestic gardens
- Landfills
- Industrial operations
- Mines, quarries, and other rock excavations
- Leaking fuel storage tanks (especially those at gas stations)

- Septic systems
- Runoff from roads (e.g., winter salting) or chemical spills of materials being transported

Agriculture

Intensive agricultural operations and golf courses can have a significant impact on the environment, especially where chemicals and other materials are used to enhance growth or control pests. An example of agricultural contamination is in the Abbotsford area of the Fraser Valley, where nitrate levels above the 44 mg/L maximum acceptable level (expressed as nitrate) in the Abbotsford-Sumas aquifer have been observed since the 1950s; however, the problem became much worse as agriculture intensity increased in the 1980s. By 2004, groundwater with nitrate levels in excess of 44 mg/L was reported over an area of about 75 km2 around Abbotsford, and the problem extended across the border into the Sumas area of Washington State.

This region is intensively used for berry crops (especially raspberries and blueberries) and large poultry operations, as well as lesser amounts of grazing and forage crops. Chicken manure is typically stored in fields adjacent to chicken barns, and may release nitrogen to the environment from runoff water, and from releases of ammonia gas. Over decades, both chemical fertilizers and chicken manure and other manures have been applied to the berry crops to provide extra nitrogen to help maximize berry growth. If the fertilizer added is in excess of what the plants need, or is poorly timed compared to when it is needed, then the extra nitrogen may be leached into the groundwater below. Berry crops are irrigated over the summer to help the crops grow. Summer irrigation and winter rainfall may carry excess nitrate from the near surface to the aquifer below.

Since the 1990s, agricultural practices have been tightened up to reduce the rate of groundwater contamination, but it will take decades for nitrate levels to drop in the Abbotsford-Sumas aquifer. Agriculture and Agri-Food Canada and many others are conducting research on better irrigation and nitrate management techniques to reduce the amount of nitrogen that leaches to groundwater.

Landfills

In the past, domestic and commercial refuse was commonly trucked to a "dump" (typically a hole in the ground), and when the hole was filled, it was covered with soil and forgotten. In situations like this, rain and melting snow can easily pass through the soil used to cover the refuse. This water passes into the waste itself, and the resulting landfill leachate that flows from the bottom of the landfill can seriously contaminate the surrounding groundwater and surface water. In the past few decades, regulations around refuse disposal have been significantly strengthened, and important steps have been taken to reduce the amount of landfill waste by diverting recyclable and compostable materials to other locations.

A modern engineered landfill has an impermeable liner (typically heavy plastic, although engineered clay liners or natural clay may be adequate in some cases), a plumbing system for draining leachate (the rainwater that flows through the refuse and becomes contaminated), and a network of monitoring wells both within and around the landfill (Figure 14). Once part or all of a landfill site is full, it is sealed over with a plastic cover, and a system is put in place to extract landfill gas (typically a mixture of carbon dioxide and methane). That gas can be sent to a nearby location where it is burned to create heat or used to generate electricity. The leachate must be treated, and that can be done in a normal sewage treatment plant.

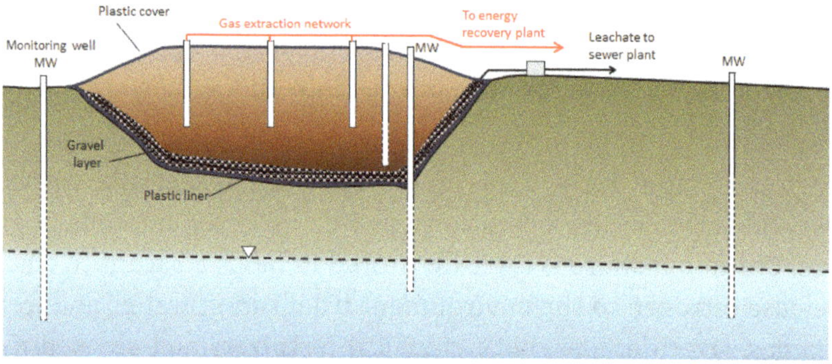

Figure 14: A cross-section of a typical modern landfill [SE]

The monitoring wells are used to assess the level of the water table around the landfill and to collect groundwater samples so that any leakage can be detected. Because some leakage is almost inevitable, the ideal placement for landfills is in areas where the depth to the water table is significant (tens of metres if possible) and where the aquifer material is relatively impermeable. Landfills should also be situated far from streams, lakes, or wetlands so that contamination of aquatic habitats can be avoided.

Today there are hundreds of abandoned dumps scattered across the country; most have been left to contaminate groundwater that we might wish to use sometime in the future. In many cases, it's unlikely that we'll be able to do so.

Industrial Operations

Although western Canada doesn't have the same extent of industrial pollution as other parts of the country, there are still seriously contaminated sites in the west, most with the potential to contaminate groundwater. One example is the lead and zinc smelter at Trail, B.C. The largest in the world, it has been operating for over 100 years and has left a residue of metal contamination around the region (Figure 15). In some parts of Trail, the contamination is serious enough that existing soil has been removed from residential properties and replaced with clean soil brought in from elsewhere. This contaminated soil has contributed to contamination of groundwater in the Trail area. Groundwater beneath the actual smelter site is contaminated, and the operator (Teck Resources) is currently working on plans to prevent that water from reaching the nearby Columbia River.

Figure 15: The Trail lead-zinc smelter in 1929

Mines, Quarries, and Rock Excavations

Mines and other operations that involve the excavation of large amounts of rock (e.g., highway construction) have the potential to create serious environmental damage. The exposure of rock that has previously not been exposed to air and water can lead to the oxidation of sulphide-bearing minerals, such a pyrite (FeS2), within the rock. The combination of pyrite, water, oxygen, and a special type of bacteria (*Acidithiobacillus ferrooxidans*) that thrives in acidic conditions leads to the generation of acidity, in some cases to pH less than 2. Water that acidic is hazardous by itself, but the low pH also has the property of increasing the solubility of certain heavy metals. The water that is generated by this process is known as acid rock drainage (ARD). ARD can occur naturally where sulphide-bearing rocks are near the surface. The issue of ARD is a major environmental concern at both operating mines and abandoned mines (see Chapter 20). In streams around the Mt. Washington Mine on Vancouver Island (Figure 17), copper levels are high enough to be toxic to fish. Groundwater adjacent to the contaminated streams in the area is very likely contaminated as well.

Figure 17: Acidic runoff at the abandoned Mt. Washington Mine near Courtenay, B.C.

Leaking Fuel Tanks

Underground storage tanks (USTs) are used to store fuel at gas stations, industrial sites, airports, and anywhere that large volumes of fuel are used. They do not last forever, and eventually they start to leak their contents into the ground. This is a particular problem at older gas stations — although it may also become a future problem at newer gas stations. You may have noticed gas stations that have been closed and then surrounded by chain-link fence (Figure 18). In virtually all such cases the closure has been triggered by the discovery of leaking USTs and the requirement to cease operations and remediate the site.

Figure 18: A closed and fenced gas station site in Nanaimo, B.C. The white pipes in the background are wells for monitoring groundwater contamination on the site.

Petroleum fuels are complex mixtures of hydrocarbon compounds and the properties of their components — such as density, viscosity, solubility in water, and volatility — tend to vary widely. As a result, a petroleum spill is like several spills for the price of one. The petroleum liquid slowly settles through the unsaturated zone and then tends to float on the surface of the groundwater (Figure 19). The more readily soluble components of the spill dissolve in the groundwater and are dispersed along with the normal groundwater flow, and the more volatile components of the spill rise toward the surface, potentially contaminating buildings.

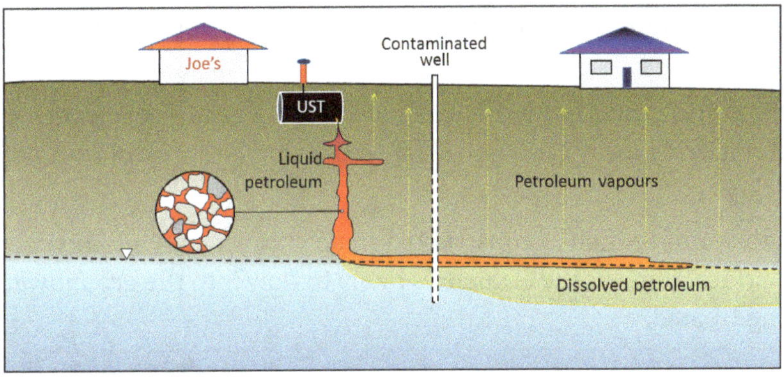

Figure 19: A depiction of the fate of different components of a petroleum spill from an underground storage tank.

Septic Systems

In areas that are not served by sewage networks leading to a central sewage treatment plant, most homeowners rely on septic systems for disposal of sewage. There are two primary components to a simple septic system, the septic tank and the drainage field (Figure 20). A typical septic tank is constructed of either concrete or plastic and has a volume of 5,000 L to 10,000 L (5 m3 to 10 m3). This forms the first treatment and is designed to be anaerobic (without oxygen). That promotes the activity of certain bacteria that help break down the waste. As the waste is degraded, some portions tend to sink to form sludge at the base of the tank, and others float to the surface, forming a scum layer. A septic tank may be divided into two parts to keep the sludge at the bottom and the scum on the top from draining out. The water then moves to the drainage field, which provides the right conditions for a different set of bacteria that operate in aerobic conditions. The drainage field includes an array of plastic pipes that are perforated to allow the effluent to drain out over a large area and seep slowly into the ground. In order to install a drainage field, it is first necessary to test the soil below, as it must be sufficiently permeable to allow the effluent to percolate away, but not so permeable that it flows too quickly and the soil is not able to filter out the pathogenic bacteria.

If they are properly installed and used, and if the sludge is periodically removed from the tank, a septic system should be effective in treating the sewage for decades. The anaerobic and aerobic bacteria should be able to break down the incoming waste and there should be little risk to the surface environment or groundwater. But many things can go wrong with a septic system, including the following:

1. If inappropriate chemicals are added to the waste stream, they may interfere with the natural breakdown of the sewage.
2. If the tank is not periodically pumped out, solids can get into the drainage field and compromise the drainage, resulting in the flow of effluent toward the surface.
3. If the soil is either not sufficiently permeable or too permeable, the effluent will not drain away (and will start to pool at the surface) or it will drain too quickly.
4. If the drainage field is constructed in an area where the water table is close to surface, some of the effluent is likely to flow into the groundwater without being treated.

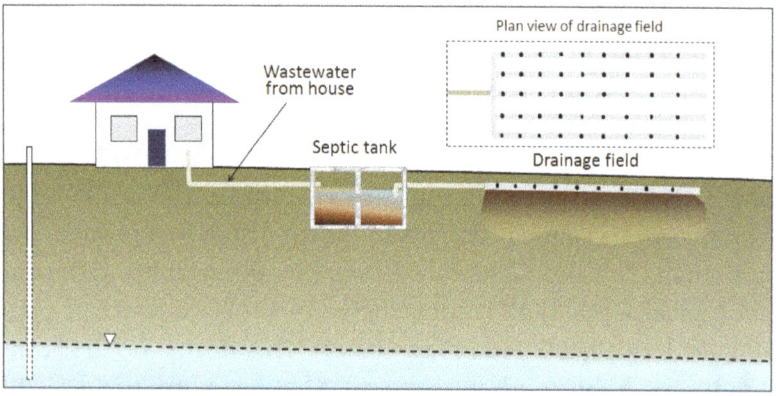

Figure 20: A typical septic system. [SE]

Prevention of Groundwater Contamination

As illustrated in the landfill example above, there are two fairly simple ways to significantly reduce the chance and degree of groundwater contamination from surface sources. One is to prevent rainwater from infiltrating down to the water table and picking up contaminants; this can be achieved by simply capping or roofing over the landfill, mine tailings, or spill site. The second is to provide an impermeable barrier beneath the contaminant. Modern landfills and mine tailings impoundments are all built using some combination of clay and engineered plastic barriers. Both of these solutions — caps and liners — are subject to failure due to leaks.

Once contaminants are in the groundwater, the main form of remediation is to pump out the contaminated water and treat it at the surface. This can be a slow process, and preventing the contaminant from travelling significantly during this process can be accomplished by manipulating local groundwater flow through the extraction or injection of water at certain locations. Consider this in the exercise below.

Exercise 6 Manipulating a Contaminant Plume

This diagram shows a groundwater contaminant plume in red. The source of the contamination has been removed but if the plume is not dealt with, it will eventually enter the stream and threaten the health of wildlife. Pumping the contaminant from well B for treatment will not be sufficient to prevent some of the contamination from making it to the stream.

What could you do at wells A and C to prevent this? Explain and use the diagram below to illustrate the expected changes to the water table and the movement of the plume.

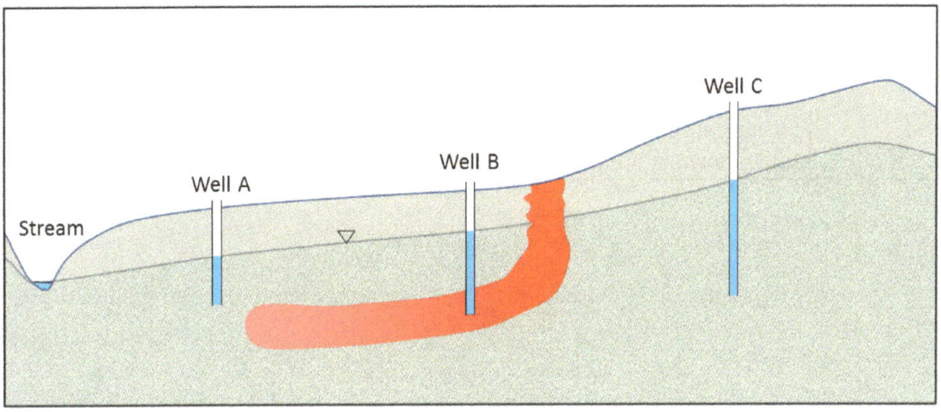

AT A GLANCE

Groundwater and Aquifers	Porosity is the percentage of open space within a rock or unconsolidated sedimentary deposit, while permeability is the facility with which water can be transmitted through that material. An aquifer is a body of rock or sediment that has sufficient permeability for water to be extracted, while an aquitard is an impermeable body. An aquifer is described as confined if it is overlain by an impermeable layer (confining layer), or unconfined if it has no such confining layer.
Groundwater Flow	The water table is the upper surface of the saturated zone in an unconfined aquifer. A confined aquifer has a potentiometric surface (instead of a water table), which is defined as the level to which water would rise if a well were drilled into the confined aquifer. Change in groundwater head over distance is the hydraulic gradient. The theoretical velocity of flow in an aquifer is defined by Darcy's law as the hydraulic conductivity (a measure of permeability) times the hydraulic gradient ($V = K * i$). It is possible to predict groundwater flow paths if we can draw equipotential lines within an aquifer. In areas where limestone has solutional openings (e.g., caves), water flow is determined by gravity above the water table and by the hydraulic gradient below the water table.

Groundwater Extraction	Groundwater can be extracted at springs, but in most cases, wells are needed to ensure a steady supply. Pumping groundwater from a well lowers groundwater head near the well, creating flow toward the well. This creates a cone of depression around the well. Excessive pumping can lead to a well running dry or to a lack of water in nearby wells. During extended periods of dry weather, or if consistent over-pumping occurs, aquifers may be depleted. Observation wells are used to monitor short-term and long-term changes in water levels that can indicate changes in aquifer health.
Groundwater Quality	The quality of groundwater can be compromised by both natural and anthropogenic contamination. Natural contamination can be caused by particularly high levels of contaminants within the aquifer itself, but is more commonly a result of enhanced solubility of contaminants due to the aquifer chemistry. Some common sources of anthropogenic contamination include agriculture, industry, mining, landfills, and leaking underground storage tanks. We can assess the vulnerability of aquifers to contamination by mapping regional variations in parameters such as depth to the water table, permeability, slope, and precipitation.

CHAPTER

MASS WASTING 15

Early in the morning on January 9, 1965, 47 million cubic metres of rock broke away from the steep upper slopes of Johnson Peak (16 km southeast of Hope) and roared 2,000 m down the mountain, gouging out the contents of a small lake at the bottom, and continuing a few hundred metres up the other side (Figure 15.1). Four people, who had been stopped on the highway by a snow avalanche, were killed. Many more might have become victims, except that a Greyhound bus driver, en route to Vancouver, turned his bus around on seeing the avalanche. The rock failed along weakened foliation planes of the metamorphic rock on Johnson Peak, in an area that had been eroded into a steep slope by glacial ice. There is no evidence that it was triggered by any specific event, and there was no warning that it was about to happen. Even if there had been warning, nothing could have been done to prevent it. There are hundreds of similar situations throughout British Columbia.

What can we learn from the Hope Slide? In general, we cannot prevent most mass wasting, and significant effort is required if an event is to be predicted with any level of certainty. Understanding the geology is critical to understanding mass wasting. Although failures are inevitable in a region with steep slopes, larger ones happen less frequently than smaller ones, and the consequences vary depending on the downslope conditions, such as the presence of people, buildings, roads, or fish-bearing streams.

An important reason for learning about mass wasting is to understand the nature of the materials that fail, and how and why they fail so that we can minimize risks from similar events in the future. For this reason, we need to be able to classify mass-wasting events, and we need to know the terms that geologists, engineers, and others use to communicate about them.

Mass wasting, which is synonymous with "slope failure," is the failure and downslope movement of rock or unconsolidated materials in response to gravity. The term "landslide" is almost synonymous with mass wasting, but not quite because some people reserve "landslide" for relatively rapid slope failures, while others do not. Because of that ambiguity, we will avoid the use of "landslide" in this textbook.

Figure 1: The site of the 1965 Hope Slide as seen in 2014. The initial failure is thought to have taken place along the foliation planes and sill within the area shown in the inset.

15.1 FACTORS THAT CONTROL SLOPE STABILITY

Mass wasting happens because tectonic processes have created uplift. Erosion, driven by gravity, is the inevitable response to that uplift, and various types of erosion, including mass wasting, have created slopes in the uplifted regions. Slope stability is ultimately determined by two factors: the angle of the slope and the strength of the materials on it.

In Figure 2 a block of rock situated on a rock slope is being pulled toward Earth's centre (vertically down) by gravity. We can split the vertical gravitational force into two components relative to the slope: one pushing the block down the slope (the shear force), and the other pushing into the slope (the normal force). The shear force, which wants to push the block down the slope, has to overcome the strength of the connection between the block and the slope, which may be quite weak if the block has split away from the main body of rock, or may be very strong if the block is still a part of the rock. This is the shear strength, and in Figure 15.2a, it greater than the shear force, so the block should not move. In Figure 15.2b the slope is steeper and the shear force is approximately equal to the shear strength. The block may or may not move under these circumstances. In Figure 15.2c, the slope is steeper still, so the shear force is considerably greater than the shear strength, and the block will very likely move.

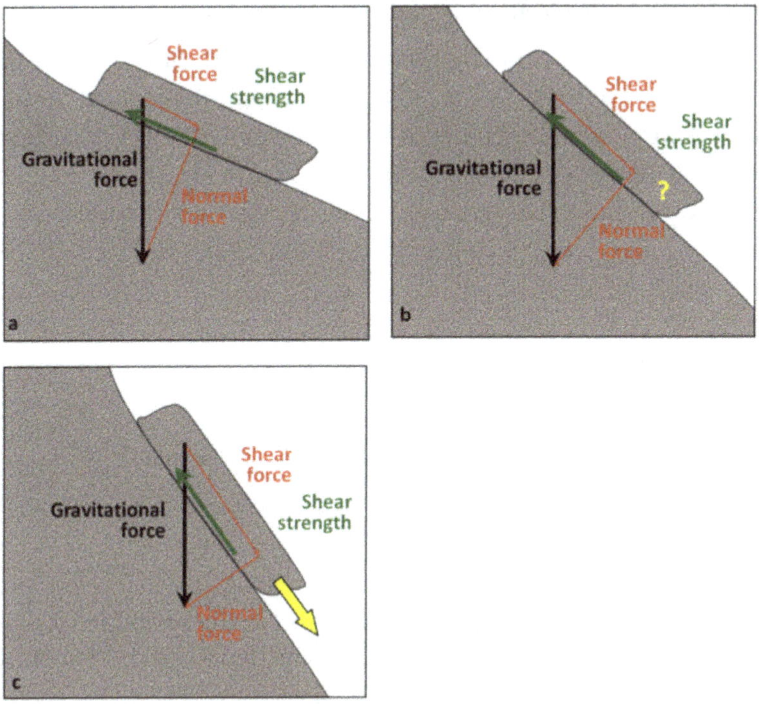

Figure 2: Differences in the shear and normal components of the gravitational force on slopes with differing steepness. The gravitational force is the same in all three cases. In (a) the shear force is substantially less than the shear strength, so the block should be stable. In (b) the shear force and shear strength are about equal, so the block may or may not move. In (c) the shear force is substantially greater than the shear strength, so the block is very likely to move.

As already noted, slopes are created by uplift followed by erosion. In areas with relatively recent uplift (such as most of British Columbia and the western part of Alberta), slopes tend to be quite steep. This is especially true where glaciation has taken place because glaciers in mountainous terrain create steep-sided valleys. In areas without recent uplift (such as central Canada), slopes are less steep because hundreds of millions of years of erosion (including mass wasting) has made them that way. However, as we'll see, some mass wasting can happen even on relatively gentle slopes.

The strength of the materials on slopes can vary widely. Solid rocks tend to be strong, but there is a very wide range of rock strength. If we consider just the strength of the rocks, and ignore issues like fracturing and layering, then most crystalline rocks — like granite, basalt, or gneiss — are very strong, while some metamorphic rocks — like schist — are moderately strong. Sedimentary rocks have variable strength. Dolostone and some limestone are strong, most sandstone and conglomerate are moderately strong, and some sandstone and all mudstones are quite weak.

Fractures, metamorphic foliation, or bedding can significantly reduce the strength of a body of rock, and in the context of mass wasting, this is most critical if the planes of weakness are parallel to the slope and least critical if they are perpendicular to the slope. This is illustrated in Figure 3. At locations A and B the bedding is nearly perpendicular to the slope and the situation is relatively stable. At location D the bedding is nearly parallel to the slope and the situation is quite unstable. At location C the bedding is nearly horizontal and the stability is intermediate between the other two extremes.

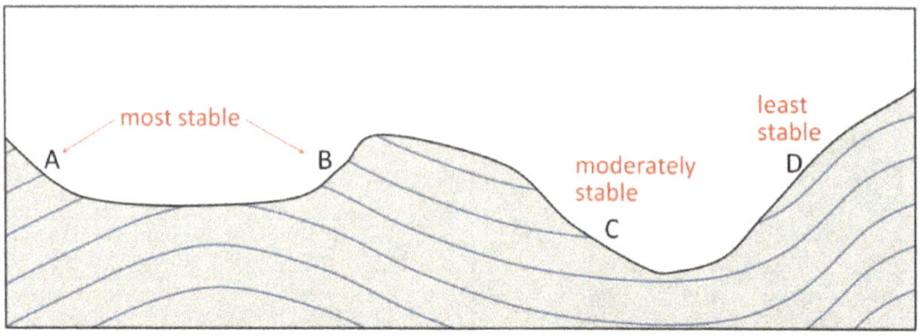

Figure 3: Relative stability of slopes as a function of the orientation of weaknesses (in this case bedding planes) relative to the slope orientations.

Internal variations in the composition and structure of rocks can significantly affect their strength. Schist, for example, may have layers that are rich in sheet silicates (mica or chlorite) and these will tend to be weaker than other layers. Some minerals tend to be more susceptible to weathering than others, and the weathered products are commonly quite weak (e.g., the clay formed from feldspar). The side of Johnson Peak that failed in 1965 (Hope Slide) is made up of chlorite schist (metamorphosed sea-floor basalt) that has feldspar-bearing sills within it (they are evident within the inset area of Figure 15.1). The foliation and the sills are parallel to the steep slope. The schist is relatively weak to begin with, and the feldspar in the sills, which has been altered to clay, makes it even weaker.

Unconsolidated sediments are generally weaker than sedimentary rocks because they are not cemented and, in most cases, have not been significantly compressed by overlying materials. This binding property of sediment is sometimes referred to as *cohesion.* Sand and silt tend to be particularly weak, clay is generally a little stronger, and sand mixed with clay can be stronger still. The deposits that make up the cliffs at Point Grey in Vancouver include sand, silt, and clay overlain by sand. As shown in Figure 4 (left) the finer deposits are relatively strong (they maintain a steep slope), while the overlying sand is relatively weak, and has a shallower slope that has recently failed. Glacial till — typically a mixture of clay, silt, sand, gravel, and larger clasts — forms and is compressed beneath tens to thousands of metres of glacial ice so it can be as strong as some sedimentary rock (Figure 4, right).

Figure 4: Left: Glacial outwash deposits at Point Grey, in Vancouver. The dark lower layer is made up of sand, silt, and clay. The light upper layer is well-sorted sand. Right: Glacial till on Quadra Island, B.C. The till is strong enough to have formed a near-vertical slope. [SE]

Apart from the type of material on a slope, the amount of water that the material contains is the most important factor controlling its strength. This is especially true for unconsolidated materials, like those shown in Figure 4, but it also applies to bodies of rock. Granular sediments, like the sand at Point Grey, have lots of spaces between the grains. Those spaces may be completely dry (filled only with air); or moist (often meaning that some spaces are water filled, some grains have a film of water around them, and small amounts of water are present where grains are touching each other); or completely saturated (Figure 15.5). Unconsolidated sediments tend to be strongest when they are moist because the small amounts of water at the grain boundaries hold the grains together with surface tension. Dry sediments are held together only by the friction between grains, and if they are well sorted or well rounded, or both, that cohesion is weak. Saturated sediments tend to be the weakest of all because the large amount of water actually pushes the grains apart, reducing the mount friction between grains. This is especially true if the water is under pressure.

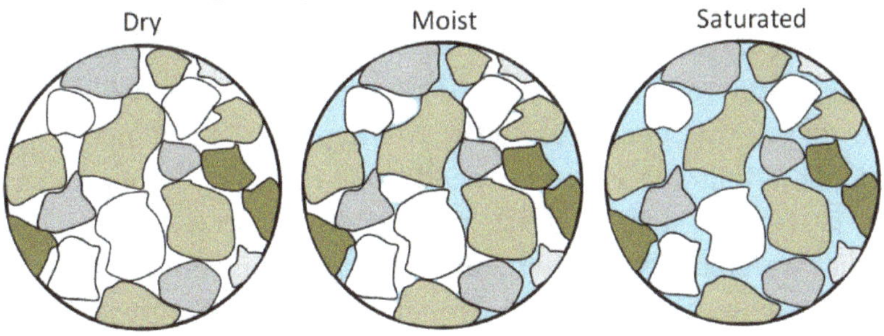

Figure 5: Depiction of dry, moist, and saturated sand

Exercise 1 Sand and Water

If you've ever been to the beach, you'll already know that sand behaves differently when it's dry than it does when it's wet, but it's worth taking a systematic look at the differences in its behaviour. Find about half a cup of clean, dry sand (or get some wet sand and dry it out), and then pour it from your hand onto a piece of paper. You should be able to make a cone-shaped pile that has a slope of around 30°. If you pour more sand on the pile, it will get bigger, but the slope should remain the same. Now add some water to the sand so that it is moist. An easy way to do this is to make it completely wet and then let the water drain away for a minute. You should be able to form this moist sand into a steep pile (with slopes of around 80°). Finally, put the same sand into a cup and fill the cup with water so the sand is just covered. Swirl it around so that the sand remains in suspension, and then quickly tip it out onto a flat surface (best to do this outside). It should spread out over a wide area, forming a pile with a slope of only a few degrees. [SE]

Water will also reduce the strength of solid rock, especially if it has fractures, bedding planes, or clay-bearing zones. This effect is even more significant when the water is under pressure, which is why you'll often see holes drilled into rocks on road cuts to relieve this pressure. One of the hypotheses advanced to explain the 1965 Hope Slide is that the very cold conditions that winter caused small springs in the lower part of the slope to freeze over, preventing water from flowing out. It is possible that water pressure gradually built up within the slope, weakening the rock mass to the extent that the shear strength was no longer greater than the shear force.

Water also has a particular effect on clay-bearing materials. All clay minerals will absorb a little bit of water, and this reduces their strength. The smectite clays (such as the bentonite used in cat litter) can absorb a lot of water, and that water pushes the sheets apart at a molecular level and makes the mineral swell. Smectite that has expanded in this way has almost no strength; it is extremely slippery.

And finally, water can significantly increase the mass of the material on a slope, which increases the gravitational force pushing it down. A body of sediment that has 25% porosity and is saturated with water weighs approximately 13% more than it does when it is completely dry, so the gravitational shear force is also 13% higher. In the situation shown in Figure 15.2b, a 13% increase in the shear force could easily be enough to tip the balance between shear force and shear strength.

Mass-Wasting Triggers

Shear force is primarily related to slope angle, and this does not change quickly. But shear strength can change quickly for a variety of reasons, and events that lead to a rapid reduction in shear strength are considered to be triggers for mass wasting.

An increase in water content is the most common mass-wasting trigger. This can result from rapid melting of snow or ice, heavy rain, or some type of event that changes the pattern of water flow on the surface. Rapid melting can be caused by a dramatic increase in temperature (e.g., in spring or early summer) or by a volcanic eruption. Heavy rains are typically related to storms. Changes in water flow patterns can be caused by earthquakes, previous slope failures that dam up streams, or human structures that interfere with runoff (e.g., buildings, roads, or parking lots). An example of this is the deadly 2005 debris flow in North Vancouver (Figure 6). The 2005 failure took place in an area that had failed previously, and a report written in 1980 recommended that the municipal authorities and residents take steps to address surface and slope drainage issues. Little was done to improve the situation.

Figure 6: The debris flow in the Riverside Drive area of North Vancouver in January, 2005 happened during a rainy period, but was likely triggered by excess runoff related to the roads at the top of this slope and by landscape features, including a pool, in the area surrounding the house visible here.

In some cases, a *decrease* in water content can lead to failure. This is most common with clean sand deposits (e.g., the upper layer in Figure 15.4 (left)), which lose strength when there is no more water around the grains.

Freezing and thawing can also trigger some forms of mass wasting. More specifically, the thawing can release a block of rock that was attached to a slope by a film of ice.

One other process that can weaken a body of rock or sediment is shaking. The most obvious source of shaking is an earthquake, but shaking from highway traffic, construction, or mining will also do the job. Several deadly mass-wasting events (including snow avalanches) were trigged by the M7.8 earthquake in Nepal in April 2015.

15.2 MASS WASTING CLASSIFICATION

It's important to classify slope failures so that we can understand what causes them and learn how to mitigate their effects. The three criteria used to describe slope failures are:

- The type of material that failed (typically either bedrock or unconsolidated sediment)
- The mechanism of the failure (how the material moved)
- The rate at which it moved

The type of motion is the most important characteristic of a slope failure, and there are three different types of motion:

- If the material drops through the air, vertically or nearly vertically, it's known as a fall.
- If the material moves as a mass along a sloping surface (without internal motion within the mass), it's a slide.
- If the material has internal motion, like a fluid, it's a flow.

Unfortunately it's not normally that simple. Many slope failures involve two of these types of motion, some involve all three, and in many cases, it's not easy to tell how the material moved. The types of slope failure that we'll cover here are summarized in Table 1.

Table 1: Classification of slope failures based on type of material and type of motion [SE]

Failure Type	Type of Material	Type of Motion	Rate of Motion
Rock fall	Rock fragments	Vertical or near-vertical fall (plus bouncing in many cases)	Very fast (>10s m/s)

Failure Type	Type of Material	Type of Motion	Rate of Motion
Rock slide	A large rock body	Motion as a unit along a planar surface (translational sliding)	Typically very slow (mm/y to cm/y), but some can be faster
Rock avalanche	A large rock body that slides and then breaks into small fragments	Flow (at high speeds, the mass of rock fragments is suspended on a cushion of air)	Very fast (>10s m/s)
Creep or solifluction	Soil or other overburden; in some cases, mixed with ice	Flow (although sliding motion may also occur)	Very slow (mm/y to cm/y)
Slump	Thick deposits (m to 10s of m) of unconsolidated sediment	Motion as a unit along a curved surface (rotational sliding)	Slow (cm/y to m/y)
Mudflow	Loose sediment with a significant component of silt and clay	Flow (a mixture of sediment and water moves down a channel)	Moderate to fast (cm/s to m/s)
Debris flow	Sand, gravel, and larger fragments	Flow (similar to a mudflow, but typically faster)	Fast (m/s)

Rock Fall

Rock fragments can break off relatively easily from steep bedrock slopes, most commonly due to frost-wedging in areas where there are many freeze-thaw cycles per year. If you've ever hiked along a steep mountain trail on a cool morning, you might have heard the occasional fall of rock fragments onto a talus slope. This happens because the water between cracks freezes and expands overnight, and then when that same water thaws in the morning sun, the fragments that had been pushed beyond their limit by the ice fall to the slope below (Figure 7).

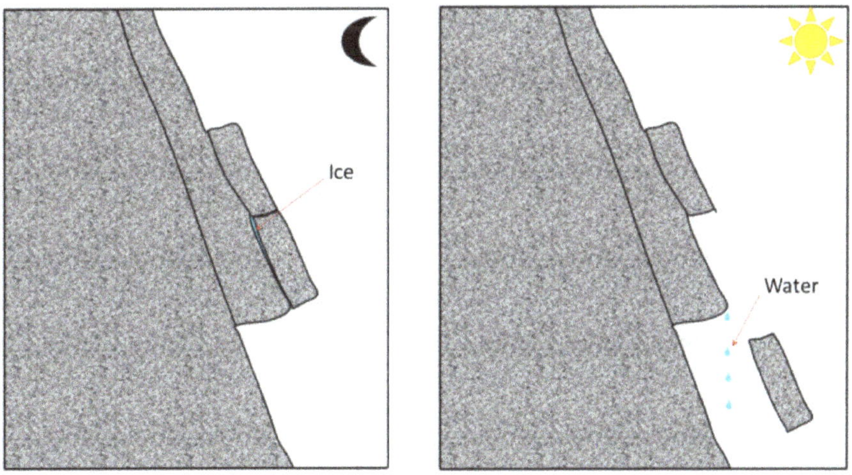

Figure 7: The contribution of freeze-thaw to rock fall

A typical talus slope, near Keremeos in southern B.C., is shown in Figure 8. In December 2014, a large block of rock split away from a cliff in this same area. It broke into smaller pieces that tumbled down the slope and crashed into the road, smashing the concrete barriers and gouging out large parts of the pavement. Luckily no one was hurt.

Figure 8 Left: A talus slope near Keremeos, B.C., formed by rock fall from the cliffs above. Right: The results of a rock fall onto a highway west of Keremeos in December 2014.

Rock Slide

A rock slide is the sliding motion of rock along a sloping surface. In most cases, the movement is parallel to a fracture, bedding, or metamorphic foliation plane, and it can range from very slow to moderately fast. The word sackung describes the very slow motion of a block of rock (mm/y to cm/y) on a slope. A good example is the Downie Slide north of Revelstoke, B.C., which is shown in Figure 15.9. In this case, a massive body of rock is very slowly sliding down a steep slope along a plane of weakness that is approximately parallel to the slope. The Downie Slide, which was recognized prior to the construction of the Revelstoke Dam, was moving very slowly at the time (a few cm/year). Geological engineers were concerned that the presence of water in the reservoir (visible in Figure 9) could further weaken the plane of failure, leading to an acceleration of the motion. The result would have been a catastrophic failure into the reservoir that would have sent a wall of water over the dam and into the community of Revelstoke. During the construction of the dam, they tunnelled into the rock at the base of the slide and drilled hundreds of drainage holes upward into the plane of failure. This allowed water to drain out so that the pressure was reduced, which reduced the rate of movement of the sliding block. BC Hydro monitors this site continuously; the slide block is currently moving more slowly than it was prior to the construction of the dam.

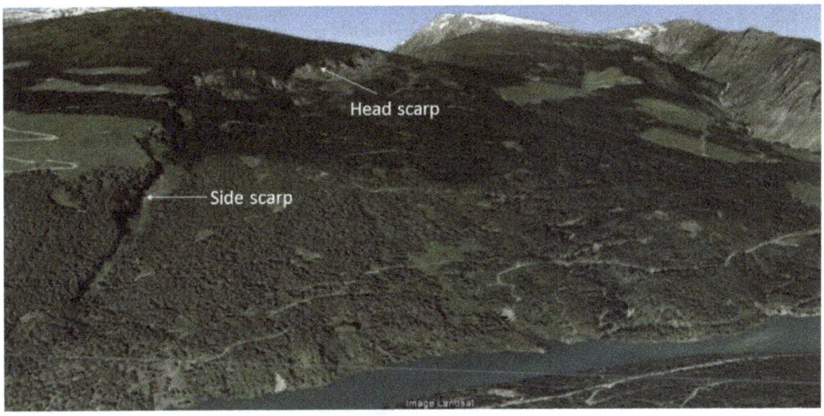

Figure 9: The Downie Slide, a sackung, on the shore of the Revelstoke Reservoir (above the Revelstoke Dam). The head scarp is visible at the top and a side-scarp along the left side.

In the summer of 2008, a large block of rock slid rapidly from a steep slope above Highway 99 near Porteau Cove (between Horseshoe Bay and Squamish). The block slammed into the highway and adjacent railway and broke into many pieces. The highway was closed for several days, and the slope was subsequently stabilized with rock bolts and drainage holes. As shown in Figure 10, the rock is fractured parallel to the slope, and this almost certainly contributed to the failure. However, it is not actually known what triggered this event as the weather was dry and warm during the preceding weeks, and there was no significant earthquake in the region.

Figure 10: Site of the 2008 rock slide at Porteau Cove. Notice the prominent fracture set parallel to the surface of the slope. The slope has been stabilized with rock bolts (top) and holes have been drilled into the rock to improve drainage

(one is visible in the lower right). Risk to passing vehicles from rock fall has been reduced by hanging mesh curtains (background).

Rock Avalanche

If a rock slides and then starts moving quickly (m/s), the rock is likely to break into many small pieces, and at that point it turns into a rock avalanche, in which the large and small fragments of rock move in a fluid manner supported by a cushion of air within and beneath the moving mass. The 1965 Hope Slide (Figure 1) was a rock avalanche, as was the famous 1903 Frank Slide in southwestern Alberta. The 2010 slide at Mt. Meager (west of Lillooet) was also a rock avalanche, and rivals the Hope Slide as the largest slope failure in Canada during historical times (Figure 11).

Figure 11: The 2010 Mt. Meager rock avalanche, showing where the slide originated (arrow, 4 km upstream). It then raced down a steep narrow valley and out into the wider valley in the foreground.

Creep (Solifluction)

The very slow — mm/y to cm/y — movement of soil or other unconsolidated material on a slope is known as creep. Creep, which normally only affects the upper several centimetres of loose material, is typically a type of very slow flow, but in some cases, sliding may take place. Creep can be facilitated by freezing and thawing because, as shown in Figure 12, particles are lifted perpendicular to the surface by the growth of ice crystals within the soil, and then let down vertically by gravity when the ice melts. The same effect can be produced by frequent wetting and drying of the soil. In cold environments, solifluction is a more intense form of freeze-thaw-triggered creep.

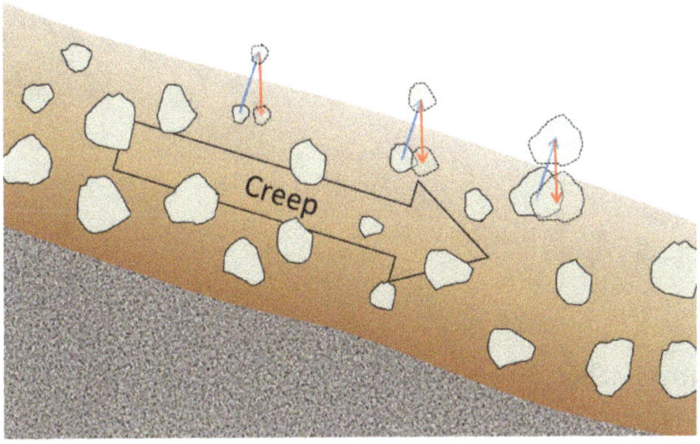

Figure 12: A depiction of the contribution of freeze-thaw to creep. The blue arrows represent uplift caused by freezing in the wet soil underneath, while the red arrows represent depression by gravity during thawing. The uplift is perpendicular to the slope, while the drop is vertical.

Creep is most noticeable on moderate-to-steep slopes where trees, fence posts, or grave markers are consistently leaning in a downhill direction (Figure 13). In the case of trees, they try to correct their lean by growing upright, and this leads to a curved lower trunk known as a "pistol butt."

Figure 13: Evidence of creep (tilted grave markers) at a cemetery in Nanaimo, B.C. [SE]

Slump

Slump is a type of slide (movement as a mass) that takes place within thick unconsolidated deposits (typically thicker than 10 m). Slumps involve movement along one or more curved failure surfaces, with downward motion near the top and outward motion toward the bottom (Figure 14). They are typically caused by an excess of water within these materials on a steep slope.

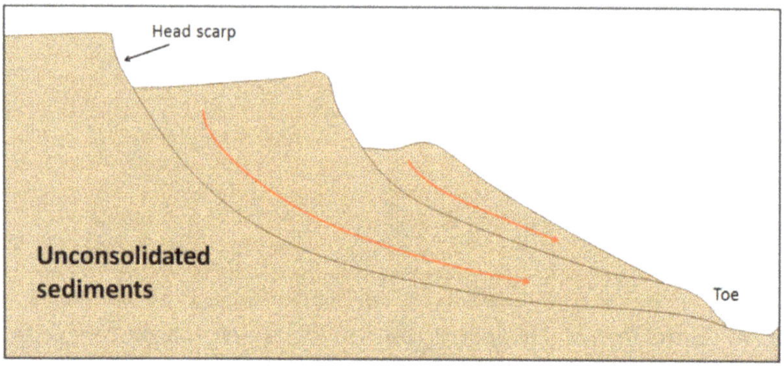

Figure 14: A depiction of the motion of unconsolidated sediments in an area of slumping [SE]

An example of a slump in the Lethbridge area of Alberta is shown in Figure 15. This feature has likely been active for many decades, and moves a little more whenever there are heavy spring rains and significant snowmelt runoff. The toe of the slump is failing because it has been eroded by the small stream at the bottom.

Figure 15: A slump along the banks of a small coulee near Lethbridge, Alberta. The main head-scarp is clearly visible at the top, and a second smaller one is visible about one-quarter of the way down. The toe of the slump is being eroded by the seasonal stream that created the coulee. [SE 2005]

Mudflows and Debris Flows

As you saw in Exercise 1, when a mass of sediment becomes completely saturated with water, the mass loses strength, to the extent that the grains are pushed apart, and it will flow, even on a gentle slope. This can happen during rapid spring snowmelt or heavy rains, and is also relatively common during volcanic eruptions because of the rapid melting of snow and ice. (A mudflow or debris flow on a volcano or during a volcanic eruption is a lahar.) If the material involved is primarily sand-sized or smaller, it is known as a mudflow, such as the one shown in Figure 16.

Figure 16: A slump (left) and an associated mudflow (centre) at the same location as Figure 15, near Lethbridge, Alberta. [SE]

If the material involved is gravel sized or larger, it is known as a debris flow. Because it takes more gravitational energy to move larger particles, a debris flow typically forms in an area with steeper slopes and more water than does a mudflow. In many cases, a debris flow takes place within a steep stream channel, and is triggered by the collapse of bank material into the stream. This creates a temporary dam, and then a major flow of water and debris when the dam breaks. This is the situation that led to the fatal debris flow at Johnsons Landing, B.C., in 2012. A typical west-coast debris flow is shown in Figure 17. This event took place in November 2006 in response to very heavy rainfall. There was enough energy to move large boulders and to knock over large trees.

Figure 17: The lower part of debris flow within a steep stream channel near Buttle Lake, B.C., in November 2006.

15.3 MASS WASTING: PREVENTING, DELAYING, MONITORING AND MITIGATING

As already noted, we cannot prevent mass wasting in the long term as it is a natural and ongoing process; however, in many situations there are actions that we can take to reduce or mitigate its damaging effects on people and infrastructure. Where we can neither delay nor mitigate mass wasting, we should consider moving out of the way.

Mass Wasting: Preventing and Delaying

It is comforting to think that we can prevent some effects of mass wasting by mechanical means, such as the rock bolts in the road cut at Porteau Cove (Figure 15.10), or the drill holes used to drain water out of a slope, as was done at the Downie Slide (Figure 9), or the building of physical barriers, such as retaining walls. What we have to remember is that the works of humans are limited compared to the works of nature. The rock bolts in the road cut at Porteau Cove will slowly start to corrode after a few years, and within a few decades, many of them will begin to lose their strength. Unless they are replaced, they will no longer support that slope. Likewise, drainage holes at the Downie Slide will eventually become plugged with sediment and chemical precipitates, and unless they are periodically unplugged, their effectiveness will decrease. Eventually, unless new holes are drilled, the drainage will be so compromised that the slide will start to move again. This is why careful slope monitoring by geological and geotechnical engineers is important at these sites. The point here is that our efforts to "prevent" mass wasting are only as good as our resolve to maintain those preventive measures.

Delaying mass wasting is a worthy endeavour, of course, because during the time that the measures are still effective they can save lives and reduce damage to property and infrastructure. The other side of the coin is that we must be careful to avoid activities that could make mass wasting more likely. One of the most common anthropogenic causes of mass wasting is road construction, and this applies both to remote gravel roads built for forestry and mining and large urban and regional highways. Road construction is a potential problem for two reasons. First, creating a flat road surface on a slope inevitably involves creating a cut bank that is steeper than the original slope. This might also involve creating a filled bank that is both steeper and weaker than the original slope (Figure 18). Second, roadways typically cut across natural drainage features, and unless great care is taken to reroute the runoff water and prevent it from forming concentrated flows, oversaturating fill of materials can result. A specific example of the contribution of construction-related impeded drainage to slope instability was shown earlier in Figure 6.

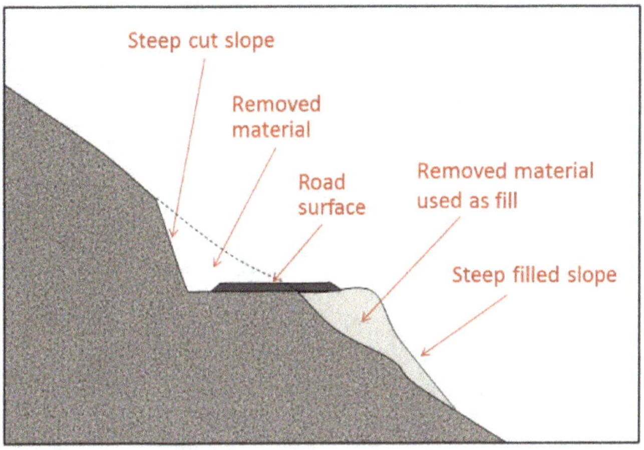

Figure 18: An example of a road constructed by cutting into a steep slope and the use of the cut material as fill. SE]

Apart from water issues, engineers building roads and other infrastructure on bedrock slopes have to be acutely aware of the geology, and especially of any weaknesses or discontinuities in the rock related to bedding, fracturing, or foliation. If possible, situations like that at Porteau Cove (Figure 10) should be avoided — by building somewhere else — rather than trying to stitch the slope back together with rock bolts.

It is widely believed that construction of buildings on the tops of steep slopes can contribute to the instability of the slope. This is probably true, but not because of the weight of the building. As you'll see by completing Exercise 3, a typical house isn't usually heavier than the fill that was removed from the hole in the ground made to build it. A more likely contributor to instability of the slope around a building is the effect that it and the changes made to the surrounding area have on drainage.

Monitoring Mass Wasting

In some areas, it is necessary to establish warning systems so that we know if conditions have changed at a known slide area, or if a rapid failure, such as a debris flow, is actually on its way downslope. The Downie Slide above the Revelstoke Resevoir is monitored 24/7 with a range of devices, such as inclinometers (slope-change detectors), bore-hole motion sensors, and GPS survey instruments. A simple mechanical device for monitoring the nearby Checkerboard Slide (which is also above the Revelstoke Reservoir) is shown in Figure 19. Both of these are very slow-moving rock slides, but it's very important to be able to detect changes in their rates of motion because at both of these locations a rapid failure would result in large bodies of rock plunging into the reservoir and sending a wall of water over the Revelstoke Dam, potentially destroying the nearby town of Revelstoke.

Figure 19: Part of a motion-monitoring device at the Checkerboard Slide near Revelstoke, B.C. The lower end of the cable is attached to a block of rock that is unstable. Any incremental motion of that block will move the cable, which will be detectable on this device.

Mt. Rainier, a glacier-covered volcano in Washington State, has the potential to produce massive mudflows or debris flows (lahars) with or without a volcanic eruption. Over 100,000 people in the Tacoma, Puyallup, and Sumner areas are in harm's way because they currently reside on deposits from past lahars (Figure 20). In 1998, a network of acoustic monitors was established around Mt. Rainier. The monitors are embedded in the ground adjacent to expected lahar paths. They are intended to provide warnings to emergency officials, and when a lahar is detected, the residents of the area will have anywhere from 40 minutes to three hours to get to safe ground.

Figure 20: Mt. Rainier, Washington, from Tacoma.

Mitigating the Impacts of Mass Wasting

In situations where we can't predict, prevent, or delay mass-wasting hazards, some effective measures can be taken to minimize the associated risk. For example, many highways in B.C. and western Alberta have avalanche shelters like that shown in Figure 21. In some parts of the world, similar features have been built to protect infrastructure from other types of mass wasting.

Figure 21: A snow avalanche shelter on the Coquihalla Highway. The expected path of the avalanche is the steep un-treed slope above.

Debris flows are inevitable, unpreventable, and unpredictable in many parts of B.C., but nowhere more so than along the Sea-to-Sky Highway between Horseshoe Bay and Squamish. The results have been deadly and expensive many times in the past. It would be very expensive to develop a new route in this region, so provincial authorities have taken steps to protect residents and traffic on the highway and the railway. Debris-flow defensive structures have been constructed in several drainage basins, as shown in Figure 22. One strategy is to allow the debris to flow quickly through to the ocean along a smooth channel. Another is to capture the debris within a constructed basin that allows the excess water to continue through, but catches the debris materials.

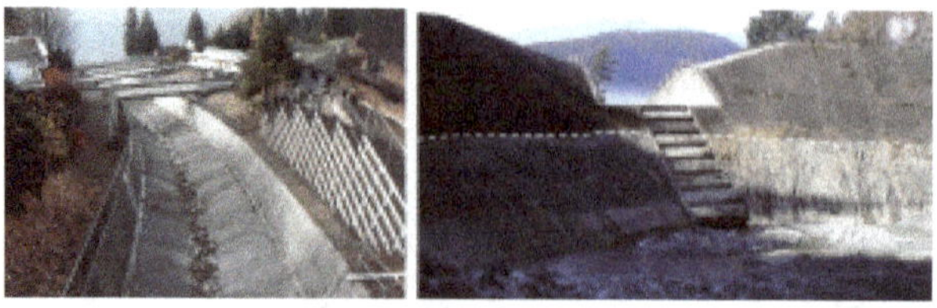

Figure 22: Two strategies for mitigating debris flows on the Sea-to-Sky Highway. Left: A concrete –lined channel on Alberta Creek allows debris to flow quickly through to the ocean. Right: A debris-flow catchment basin on Charles Creek. In 2010, a debris flow filled the basin to the level of the dotted white line.

Finally, in situations where we can't do anything to delay, predict, contain, or mitigate slope failures, we simply have to have the sense to stay away. There is a famous example of this in B.C. at a site known as Garibaldi, 25 km south of Whistler. In the early 1980s the village of Garibaldi had a population of about 100, with construction underway on some new homes, and plans for many more. In the months that followed the deadly 1980 eruption of Mt. St. Helens in Washington State, the B.C. Ministry of Transportation commissioned a geological study that revealed that a steep cliff known as The Barrier (Figure 23) had collapsed in 1855, leading to a large rock avalanche, and that it was likely to collapse again unpredictably, putting the village of Garibaldi at extreme risk. In an ensuing court case, it was ruled that the Garibaldi site was not a safe place for people to live. Those who already had homes there were compensated, and everyone else was ordered to leave.

Figure 23: The Barrier, south of Whistler, B.C., was the site of a huge rock avalanche in 1855, which extended from the cliff visible here 4 km down the valley and across the current location of the Sea-to-Sky Highway and the Cheakamus River.

AT A GLANCE

Factors That Control Stability on Slopes	Slope stability is controlled by the slope angle and the strength of the materials on the slope. Slopes are a product of tectonic uplift, and their strength is determined by the type of material on the slope and its water content. Rock strength varies widely and is determined by internal planes of weakness and their orientation with respect to the slope. In general, the more water, the greater the likelihood of failure. This is especially true for unconsolidated sediments, where excess water pushes the grains apart. Addition of water is the most common trigger of mass wasting, and can come from storms, rapid melting, or flooding.
Classification of Mass Wasting	The key criterion for classifying mass wasting is the nature of the movement that takes place. This may be a precipitous fall through the air, sliding as a solid mass along either a plane or a curved surface, or internal flow as a viscous fluid. The type of material that moves is also important – specifically whether it is solid rock or unconsolidated sediments. The important types of mass wasting are creep, slump, translational slide, rotational slide, fall, and debris flow or mudflow.
Preventing, Delaying, and Mitigating Mass Wasting	We cannot prevent mass wasting, but we can delay it through efforts to strengthen the materials on slopes. Strategies include adding mechanical devices such as rock bolts or ensuring that water can drain away. Such measures are never permanent, but may be effective for decades or even centuries. We can also avoid practices that make matters worse, such as cutting into steep slopes or impeding proper drainage. In some situations, the best approach is to mitigate the risks associated with mass wasting by constructing shelters or diversionary channels. And in other cases, where slope failure is inevitable, we should simply avoid building anything there.

CHAPTER

GLACIERS AND GLACIATION 16

A glacier is a long-lasting body of ice (decades or more) that is large enough (at least tens of metres thick and at least hundreds of metres in extent) to move under its own weight. About 10% of Earth's land surface is currently covered with glacial ice, and although the vast majority of that is in Antarctica and Greenland, there are many glaciers in Canada, especially in the mountainous parts of B.C., Alberta, and Yukon and in the far north (Figure 1). At various times during the past million years, glacial ice has been much more extensive, covering at least 30% of the land surface at times.

Glaciers represent the largest repository of fresh water on Earth (~69% of all fresh water), and they are highly sensitive to changes in climate. In the current warming climate, glaciers are melting rapidly worldwide, and although some of the larger glacial masses will last for centuries more, many smaller glaciers, including many in western Canada, will be gone within decades, and in some cases, within years. That is more than just a troubling thought for western Canadians because we rely on glacial ice for our water supplies — if not for water to drink, then for water to grow food. Irrigation systems in B.C. and across Alberta and Saskatchewan are replenished by meltwater originating from glaciers in the Coast Range and the Rocky Mountains.

Figure 1: Glaciers in the Alberta Rockies: Athabasca Glacier (centre left), Dome Glacier (right), and the Columbia Icefield (visible above both glaciers). The Athabasca Glacier has prominent lateral moraines on both sides. [SE]

16.1 GLACIAL PERIODS IN EARTH'S

We are currently in the middle of a glacial period (although it's less intense now than it was 20,000 years ago) but this is not the only period of glaciation in Earth's history; there have been many in the distant past, as illustrated in Figure 2. In general, however, Earth has been warm enough to be ice-free for much more of the time than it has been cold enough to be glaciated.

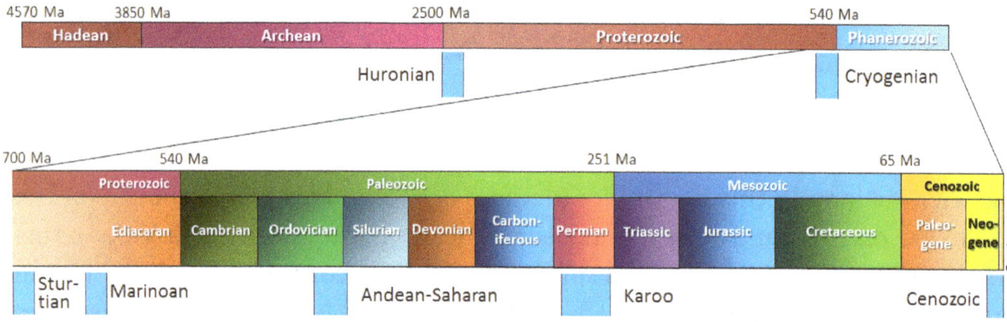

Figure 2: The record of major past glaciations during Earth's history. [SE]

The oldest known glacial period is the Huronian. Based on evidence of glacial deposits from the area around Lake Huron in Ontario and elsewhere, it is evident that the Huronian Glaciation lasted from approximately 2,400 to 2,100 Ma. Because rocks of that age are rare, we don't know much about the intensity or global extent of this glaciation.

Late in the Proterozoic, for reasons that are not fully understood, the climate cooled dramatically and Earth was seized by what appears to be its most intense glaciation. The glaciations of the Cryogenian Period (*cryo* is Latin for icy cold) are also known as the "Snowball Earth" glaciations, because it is hypothesized that the entire planet was frozen — even in equatorial regions — with ice on the oceans up to 1 km thick. A visitor to our planet at that time might not have held out much hope for its inhabitability, although life still survived in the oceans. There were two main glacial periods within the Cryogenian, each lasting for about 20 million years: the Sturtian at around 700 Ma and the Marinoan at 650 Ma. There is also evidence of some shorter glaciations both before and after these. The end of the Cryogenian glaciations coincides with the evolution of relatively large and complex life forms on Earth. This started during the Ediacaran Period, and then continued with the so-called explosion of life forms in the Cambrian. Some geologists think that the changing environmental conditions of the Cryogenian are what actually triggered the evolution of large and complex life.

There have been three major glaciations during the Phanerozoic (the past 540 million years), including the Andean/Saharan (recorded in rocks of South America and Africa), the Karoo (named for rocks in southern Africa), and the Cenozoic glaciations. The Karoo was the longest of the Phanerozoic glaciations, persisting for much of the time that the supercontinent Gondwana was situated over the South Pole (~360 to 260 Ma). It covered large parts of Africa, South America, Australia, and Antarctica (see Figure 4). As you might recall from Chapter 10, this widespread glaciation, across continents that are now far apart, was an important component of Alfred Wegener's evidence for continental drift. Unlike the Cryogenian glaciations, the Andean/Saharan, Karoo, and Cenozoic glaciations only affected parts of Earth. During Karoo times, for example, what is now North America was near the equator and remained unglaciated.

Earth was warm and essentially unglaciated throughout the Mesozoic. Although there may have been some alpine glaciation at this time, there is no longer any record of it. The dinosaurs, which dominated terrestrial habitats during the Mesozoic, did not have to endure icy conditions.

A warm climate persisted into the Cenozoic; in fact there is evidence that the Paleocene (~50 to 60 Ma) was the warmest part of the Phanerozoic since the Cambrian (Figure 3). A number of tectonic events during the Cenozoic contributed to persistent and significant planetary cooling since 50 Ma. For example, the collision of India with Asia and the formation of the Himalayan range and the Tibetan Plateau resulted in a dramatic increase in the rate of weathering and erosion. Higher than normal rates of weathering of rocks with silicate minerals, especially feldspar, consumes carbon dioxide from the atmosphere and therefore reduces the greenhouse effect, resulting in long-term cooling.

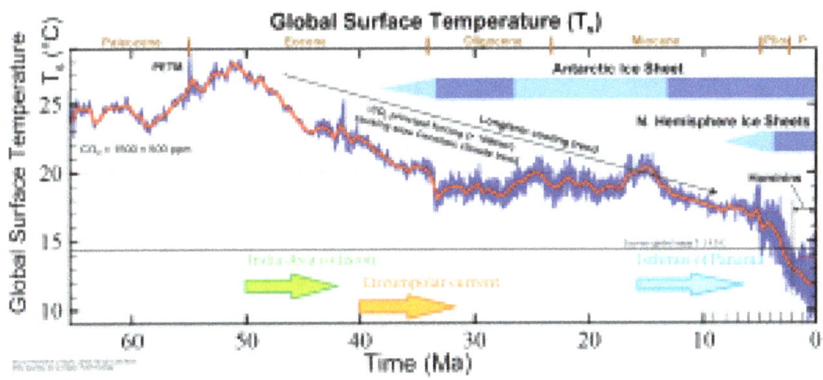

Figure 3: The global temperature trend over the past 65 Ma (the Cenozoic). From the end of the Paleocene to the height of the Pleistocene Glaciation, global average temperature dropped by about 14°C. (PETM is the Palecene-Eocene thermal maximum

At 40 Ma, ongoing plate motion widened the narrow gap between South America and Antarctica, resulting in the opening of the Drake Passage. This allowed for the unrestricted west-to-east flow of water around Antarctica, the Antarctic Circumpolar Current (Figure 4), which effectively isolated the southern ocean from the warmer waters of the Pacific, Atlantic, and Indian Oceans. The region cooled significantly, and by 35 Ma (Oligocene) glaciers had started to form on Antarctica.

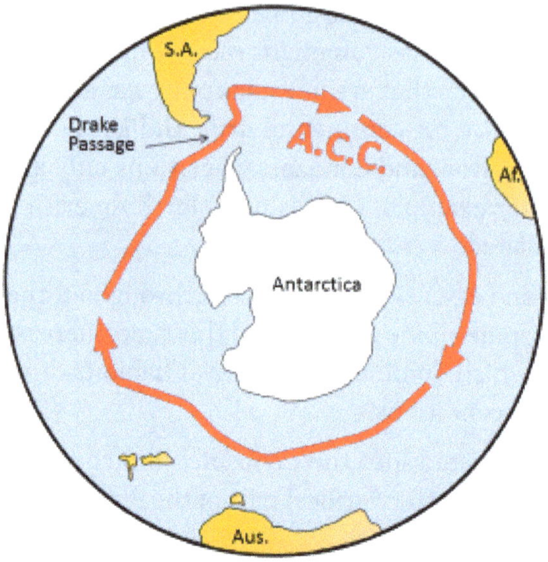

Figure 4: The Antarctic Circumpolar Current (red arrows) prevents warm water from the rest of Earth's oceans from getting close to Antarctica.

Global temperatures remained relatively steady during the Oligocene and early Miocene, and the Antarctic glaciation waned during that time. At around 15 Ma, subduction-related volcanism between central and South America created the connection between North and South America, preventing water from flowing between the Pacific and Atlantic Oceans. This further restricted the transfer of heat from the tropics to the poles, leading to a rejuvenation of the Antarctic glaciation. The expansion of that ice sheet increased Earth's reflectivity enough to promote a positive feedback loop of further cooling: more reflective glacial ice, more cooling, more ice, etc. By the Pliocene (~5 Ma) ice sheets had started to grow in North America and northern Europe (Figure 5). The most intense part of the current glaciation — and the coldest climate — has been during the past million years (the last one-third of the Pleistocene), but if we count Antarctic glaciation, it really extends from the Oligocene to the Holocene, and will likely continue into the future.

The Pleistocene has been characterized by significant temperature variations (through a range of almost 10°C) on time scales of 40,000 to 100,000 years, and corresponding expansion and contraction of ice sheets. These variations are attributed to subtle changes in Earth's orbital parameters (Milankovitch cycles), which are explained in more detail in Chapter 21. Over the past million years, the glaciation cycles have been approximately 100,000 years; this variability is visible in Figure 5.

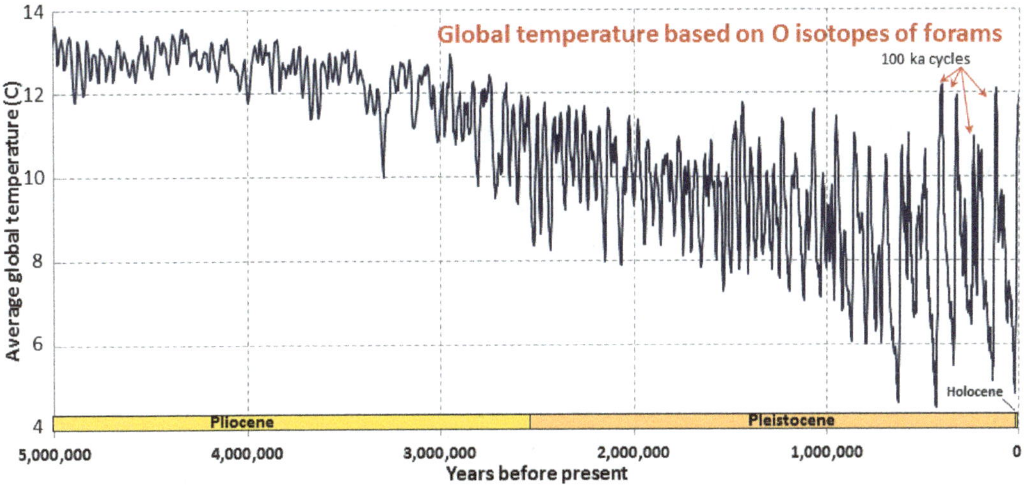

Figure 5: Foram oxygen isotope record for the past 5 million years based on O isotope data from sea-floor sediments

Exercise 1 Pleistocene Glacials and Interglacials

This diagram shows the past 500,000 years of the same data set as that shown in Figure 5. The last five glacial periods are marked with snowflakes. The most recent one, which peaked at around 20 ka, is known as the Wisconsin Glaciation. Describe the nature of temperature change that followed each of these glacial periods.

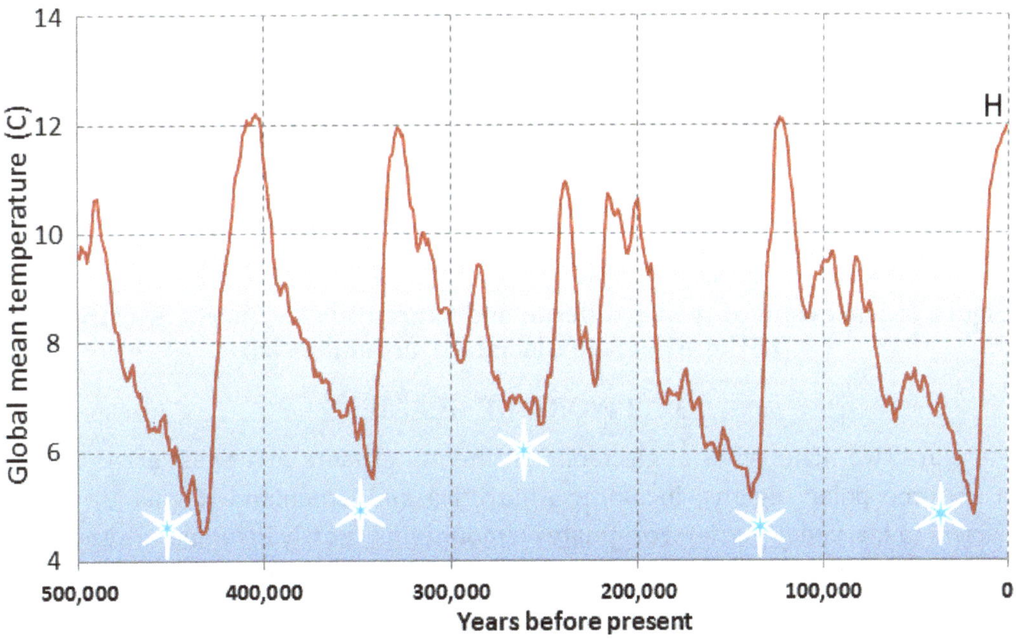

The current interglacial (Holocene) is marked with an H. Point out the previous five interglacial periods.

At the height of the last glaciation (Wisconsin Glaciation), massive ice sheets covered almost all of Canada and much of the northern United States (Figure 6). The massive Laurentide Ice Sheet covered most of eastern Canada, as far west as the Rockies, and the smaller Cordilleran Ice Sheet covered most of the western region. At various other glacial peaks during the Pleistocene and Pliocene, the ice extent was similar to this, and in some cases, even more extensive. The combined Laurentide and Cordilleran Ice Sheets were comparable in volume to the current Antarctic Ice Sheet.

Figure 6: The extent of the Cordilleran and Laurentide Ice Sheets near the peak of the Wisconsin Glaciation, around 15 ka.

16.2 WORK OF GLACIERS

There are two main types of glaciers. Continental glaciers cover vast areas of land in extreme polar regions, including Antarctica and Greenland (Figure 7). Alpine glaciers (a.k.a. valley glaciers) originate on mountains, mostly in temperate and polar regions (Figure 1), but even in tropical regions if the mountains are high enough.

Figure 7: Part of the continental ice sheet in Greenland, with some outflow alpine glaciers in the foreground.

Earth's two great continental glaciers, on Antarctica and Greenland, comprise about 99% of all of the world's glacial ice, and approximately 68% of all of Earth's fresh water. As is evident from Figure 8, the Antarctic Ice Sheet is vastly bigger than the Greenland Ice Sheet; it contains about 17 times as much ice. If the entire Antarctic Ice Sheet were to melt, sea level would rise by about 80 m and most of Earth's major cities would be submerged.

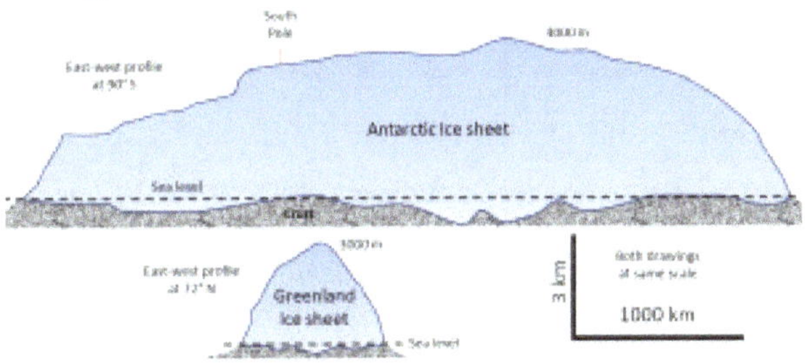

Figure 8: Simplified cross-sectional profiles the continental ice sheets in Greenland and Antarctica – both drawn to the same scale.

Continental glaciers do not flow "downhill" because the large areas that they cover are generally flat. Instead, ice flows from the region where it is thickest toward the edges where it is thinner, as shown in Figure 9. This means that in the central thickest parts, the ice flows almost vertically down toward the base, while in the peripheral parts, it flows out toward the margins. In continental glaciers like Antarctica and Greenland, the thickest parts (4,000 m and 3,000 m respectively) are the areas where the rate of snowfall and therefore of ice accumulation are highest.

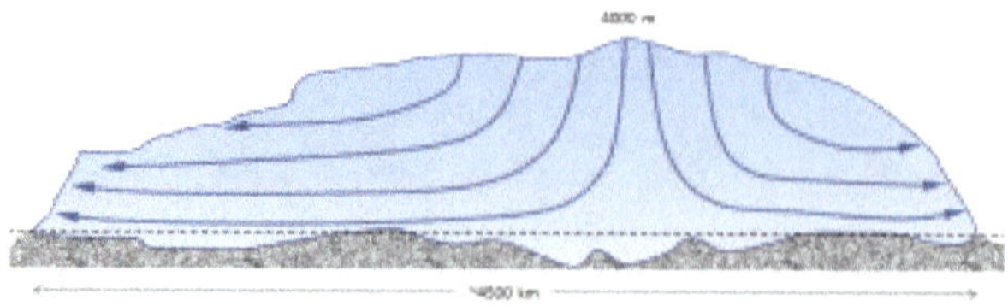

Figure 9: Schematic ice-flow diagram for the Antarctic Ice Sheet.

The flow of alpine glaciers is primarily controlled by the slope of the land beneath the ice (Figure 10). In the zone of accumulation, the rate of snowfall is greater than the rate of melting. In other words, not all of the snow that falls each winter melts during the following summer, and the ice surface is always covered with snow. In the zone of ablation, more ice melts than accumulates as snow. The equilibrium line marks the boundary between the zones of accumulation (above) and ablation (below).

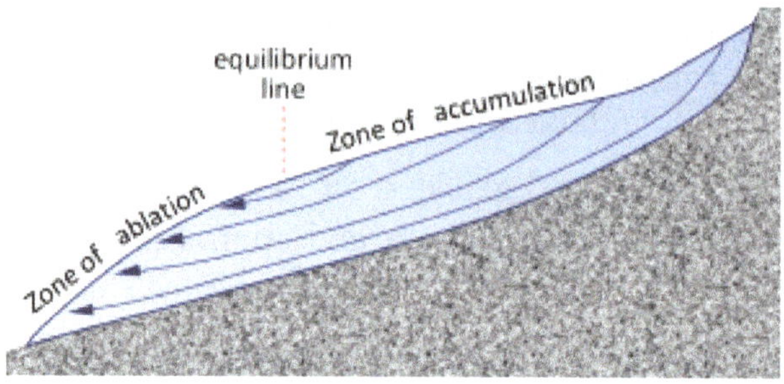

Figure 10: Schematic ice-flow diagram for an alpine glacier.

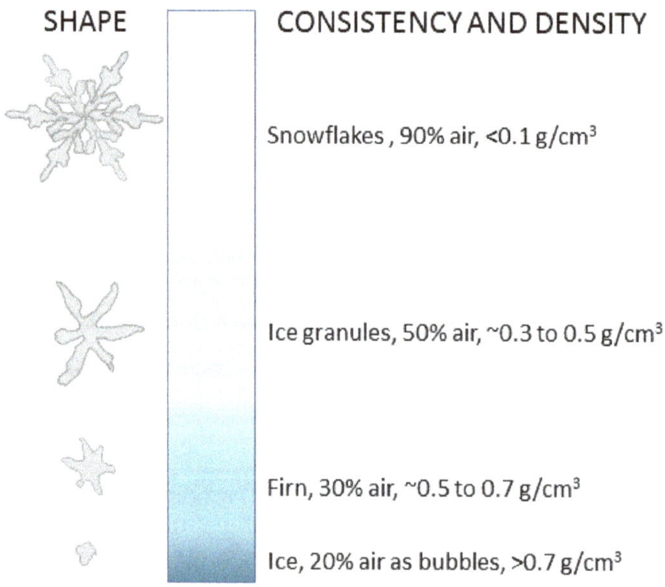

Figure 11: Steps in the process of formation of glacial ice from snow, granules, and firn.

Above the equilibrium line of a glacier, not all of the winter snow melts in the following summer, so snow gradually accumulates. The snow layer from each year is covered and compacted by subsequent snow, and it is gradually compressed and turned into firn within which the snowflakes lose their delicate shapes and become granules. With more compression, the granules are pushed together and air is squeezed out. Eventually the granules are "welded" together to create glacial ice (Figure 11). Downward percolation of water from melting taking place at the surface contributes to the process of ice formation.

The equilibrium line of a glacier near Whistler, B.C., is shown in Figure 12. Below that line, in the zone of ablation, bare ice is exposed because last winter's snow has all melted; above that line, the ice is still mostly covered with snow from last winter. The position of the equilibrium line changes from year to year as a function of the balance between snow accumulation in the winter and snowmelt during the summer. More winter snow and less summer melting obviously favours the advance of the equilibrium line (and of the glacier's leading edge), but of these two variables, it is the summer melt that matters most to a glacier's budget. Cool summers promote glacial advance and warm summers promote glacial retreat.

Figure 12: The approximate location of the equilibrium line (red) in September 2013 on the Overlord Glacier, near Whistler, B.C. [SE, after Isaac Earle, used with permission]

Glaciers move because the surface of the ice is sloped. This generates a stress on the ice, which is proportional to the slope and to the depth below the surface. As shown in Figure 12, the stresses are quite small near the ice surface but much larger at depth, and also greater in areas where the ice surface is relatively steep. Ice will deform, meaning that it will behave in a plastic manner, at stress levels of around 100 kilopascals; therefore, in the upper 50 m to 100 m of the ice (above the dashed red line), flow is not plastic (the ice is rigid), while below that depth, ice is plastic and will flow.

When the lower ice of a glacier flows, it moves the upper ice along with it, so although it might seem from the stress patterns (red numbers and red arrows) shown in Figure 13 that the lower part moves the most, in fact while the lower part deforms (and flows) and the upper part doesn't deform at all, the upper part moves the fastest because it is pushed along by the lower ice.

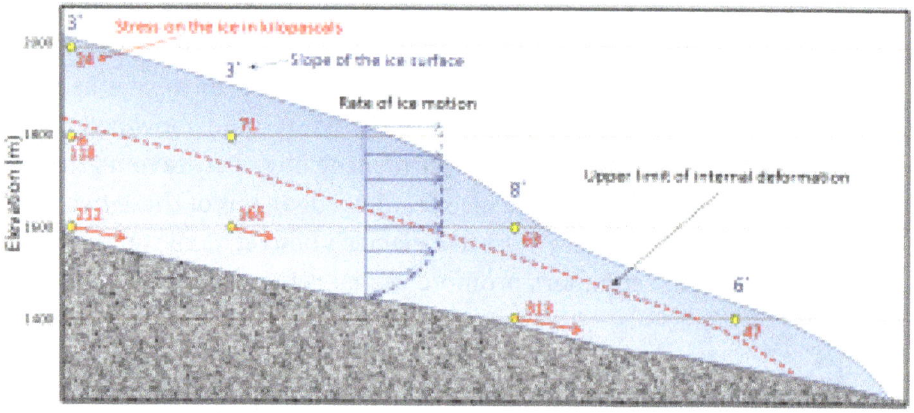

Figure 13: Stress within a valley glacier (red numbers) as determined from the slope of the ice surface and the depth within the ice. The ice will deform and flow where the stress is greater than 100 kilopascals, and the relative extent of that deformation is depicted by the red arrows. Any deformation motion in the lower ice will be transmitted to the ice above it, so although the red arrows get shorter toward the top, the ice velocity increases upward (blue arrows). The upper ice (above the red dashed line) does not flow, but it is pushed along with the lower ice. [SE]

The plastic lower ice of a glacier can flow like a very viscous fluid, and can therefore flow over irregularities in the base of the ice and around corners. However, the upper rigid ice cannot flow in this way, and because it is being carried along by the lower ice, it tends to crack where the lower ice has to flex. This leads to the development of crevasses in areas where the rate of flow of the plastic ice is changing. In the area shown in Figure 14, for example, the glacier is speeding up over the steep terrain, and the rigid surface ice has to crack to account for the change in velocity.

Figure 14: Crevasses on Overlord Glacier in the Whistler area, B.C. [Isaac Earle, used with permission]

The base of a glacier can be cold (below the freezing point of water) or warm (above the freezing point). If it is warm, there will likely be a film of water between the ice and the material underneath, and the ice will be able to slide over that surface. This is known as basal sliding (Figure 15, left). If the base is cold, the ice will be frozen to the material underneath and it will be stuck — unable to slide along its base. In this case, all of the movement of the ice will be by internal flow.

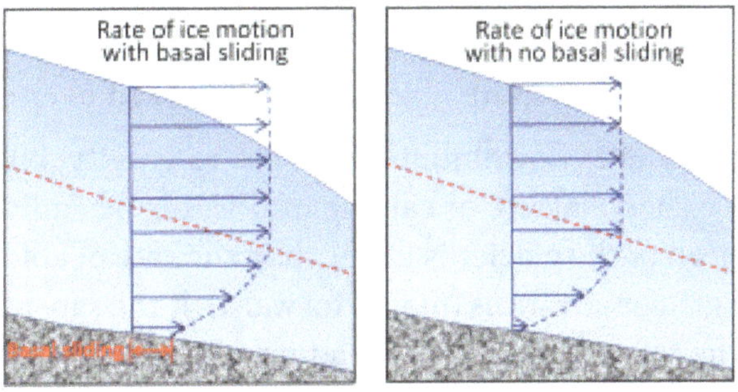

Figure 15: Differences in glacial ice motion with basal sliding (left) and without basal sliding (right). The dashed red line indicates the upper limit of plastic internal flow.

One of the factors that affects the temperature at the base of a glacier is the thickness of the ice. Ice is a good insulator. The slow transfer of heat from Earth's interior provides enough heat to warm up the base if the ice is thick, but not enough if it is thin and that heat can escape. It is typical for the leading edge of an alpine glacier to be relatively thin (see Figure 13), so it is common for that part to be frozen to its base while the rest of the glacier is still sliding. This is illustrated in Figure 16.16 for the Athabasca Glacier. Because the leading edge of the glacier is stuck to its frozen base, while the rest continues to slide, the ice coming from behind has pushed (or thrust) itself over top of the part that is stuck fast.

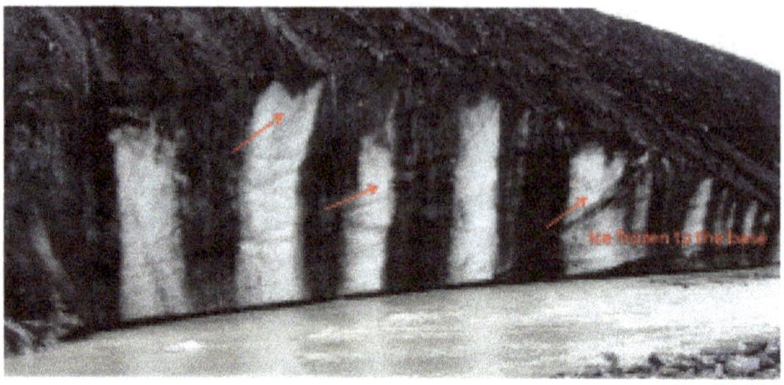

Figure 16: Thrust faults at the leading edge of the Athabasca Glacier, Alberta. The arrows show how the trailing ice has been thrust over the leading ice. (The dark vertical stripes are mud from sediments that have been washed off of the lateral moraine lying on the surface of the ice.) [SE]

Just as the base of a glacier moves more slowly than the surface, the edges, which are more affected by friction along the sides, move more slowly than the middle. If we were to place a series of markers across an alpine glacier and come back a year later, we would see that the ones in the middle had moved farther forward than the ones near the edges (Figure 17).

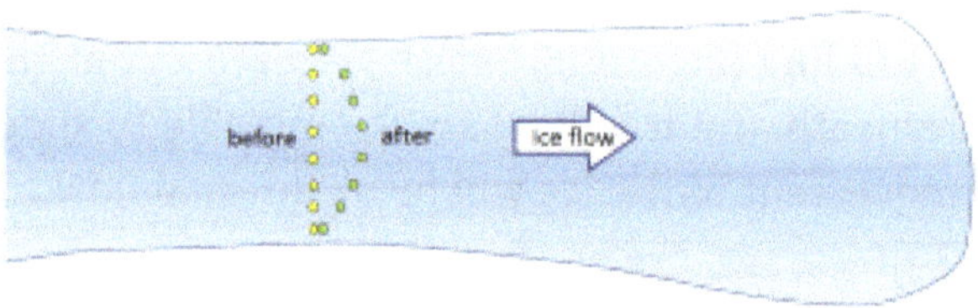

Figure 17: Markers on an alpine glacier move forward over a period of time.

Glacial ice always moves downhill, in response to gravity, but the front edge of a glacier is always either melting or calving into water (shedding icebergs). If the rate of forward motion of the glacier is faster than the rate of ablation (melting), the leading edge of the glacier advances (moves forward). If the rate of forward motion is about the same as the rate of ablation, the leading edge remains stationary, and if the rate of forward motion is slower than the rate of ablation, the leading edge retreats (moves backward).

Calving of icebergs is an important process for glaciers that terminate in lakes or the ocean. An example of such a glacier is the Berg Glacier on Mt. Robson (Figure 18), which sheds small icebergs into Berg Lake. The Berg Glacier also loses mass by melting, especially at lower elevations.

Figure 18: Mt. Robson, the tallest peak in the Canadian Rockies, Berg Glacier (centre), and Berg Lake. Although there were no icebergs visible when this photo was taken, the Berg Glacier loses mass by shedding icebergs into Berg Lake. [SE]

Exercise 2 Ice Advance and Retreat

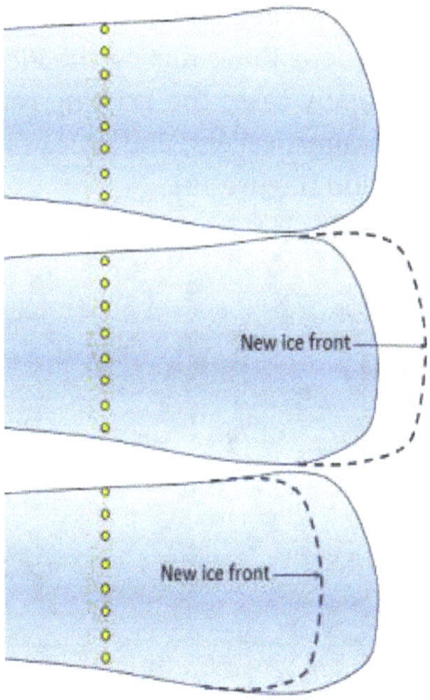

These diagrams represent a glacier with markers placed on its surface to determine the rate of ice motion over a one-year period. The ice is flowing from left to right.

1. In the middle diagram, the leading edge of the glacier has advanced. Draw in the current position of the markers.

2. In the lower diagram, the leading edge of the glacier has retreated. Draw in the current position of the markers.

16.3 GLACIAL EROSION

Glaciers are effective agents of erosion, especially in situations where the ice is not frozen to its base and can therefore slide over the bedrock or other sediment. The ice itself is not particularly effective at erosion because it is relatively soft (Mohs hardness 1.5 at 0°C); instead, it is the rock fragments embedded in the ice and pushed down onto the underlying surfaces that do most of the erosion. A useful analogy would be to compare the effect of a piece of paper being rubbed against a wooden surface, as opposed to a piece of sandpaper that has embedded angular fragments of garnet.

The results of glacial erosion are different in areas with continental glaciation versus alpine glaciation. Continental glaciation tends to produce relatively flat bedrock surfaces, especially where the rock beneath is uniform in strength. In areas where there are differences in the strength of rocks, a glacier obviously tends to erode the softer and weaker rock more effectively than the harder and stronger rock. Much of central and eastern Canada, which was completely covered by the huge Laurentide Ice Sheet at various times during the Pleistocene, has been eroded to a relatively flat surface. In many cases the existing relief is due the presence of glacial deposits — such as drumlins, eskers, and moraines (all discussed below) — rather than to differential erosion (Figure 19).

Figure 19: Drumlins — streamlined hills formed beneath a glacier, here made up of sediment — in the Amundsun Gulf region of Nunavut. The drumlins are tens of metres high, a few hundred metres across, and a few kilometres long. One of them is highlighted with a dashed white line.

Alpine glaciers produce very different topography than continental glaciers, and much of the topographic variability of western Canada can be attributed to glacial erosion. In general, glaciers are much wider than rivers of similar length, and since they tend to erode more at their bases than their sides, they produce wide valleys with relatively flat bottoms and steep sides — known as U-shaped valleys (Figure 20). Howe Sound, north of Vancouver, was occupied by a large glacier that originated in the Squamish, Whistler, and Pemberton areas, and then joined the much larger glacier in the Strait of Georgia. Howe Sound and most of its tributary valleys have pronounced U-shaped profiles (Figure 21).

Figure 20: A depiction of a U-shaped valley occupied by a large glacier.

Figure 21: The view down the U-shaped valley of Mill Creek valley toward the U-shaped valley of Howe Sound, with the village of Britannia on the opposite side.

U-shaped valleys and their tributaries provide the basis for a wide range of alpine glacial topographic features, examples of which are visible on the International Space Station view of the Swiss Alps shown in Figure 22. This area was much more intensely glaciated during the past glacial maximum. At that time, the large U-shaped valley in the lower right was occupied by glacial ice, and all of the other glaciers shown here were longer and much thicker than they are now. But even at the peak of the Pleistocene Glaciation, some of the higher peaks and ridges would have been exposed and not directly affected by glacial erosion. A peak that extends above the surrounding glacier is called a nunatuk. In these areas, and in the areas above the glaciers today, most of the erosion is related to freeze-thaw effects.

Some of the important features visible in Figure 22 are arêtes: sharp ridges between U-shaped glacial valleys; cols: low points along arêtes that constitute passes between glacial valleys; horns: steep peaks that have been glacially and freeze-thaw eroded on three or more sides; cirques: bowl-shaped basins that form at the head of a glacial valley; hanging valleys: U-shaped valleys of tributary glaciers that hang above the main valley because the larger main-valley glacier eroded more deeply into the terrain; and truncated spurs (a.k.a. "spurs"): the ends of arêtes that have been eroded into steep triangle-shaped cliffs by the glacier in the corresponding main valley.

Figure 22: A view from the International Space Station of the Swiss Alps in the area of the Aletsch Glacier. The prominent peaks labelled "Horn" are the Eiger (left) and Wetterhorn (right). A variety of alpine glacial erosion features are labelled.

Some of these alpine-glaciation erosional features are also shown in Figure 16.23 in diagram form.

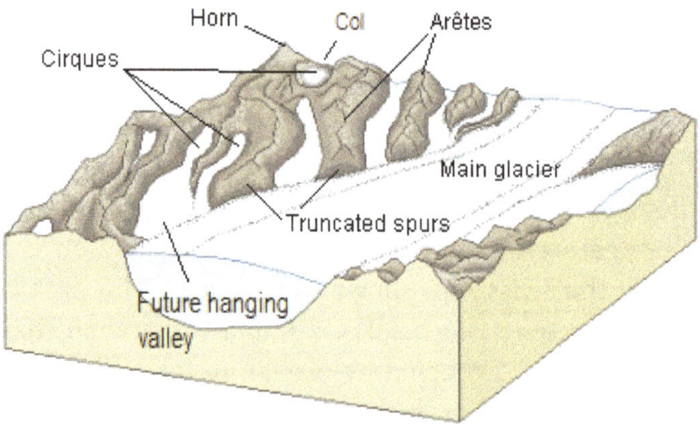

Figure 23: A diagram of some of the important alpine-glaciation erosion features.

Exercise 3 Identify Glacial Erosion Features

This is a photo of Mt. Assiniboine in the B.C. Rockies. What are the features at locations a through e? Look for one of each of the following: a horn, an arête, a truncated spur, a cirque, and a col. Try to identify some of the numerous other arêtes in this view, as well as another horn.

A number of other glacial erosion features exist at smaller scales. For example, a drumlin is an elongated feature that is streamlined at the down-ice end. The one shown in Figure 24 is larger than most, and is made up almost entirely of rock. Drumlins made up of glacial sediments are very common in some areas of continental glaciation (Figure 19).

Figure 24: Bowyer Island, a drumlin in Howe Sound, B.C. Ice flow was from right to left.

A roche moutonée is another type of elongated erosional feature that has a steep and sometimes jagged down-ice end (Figure 25, left). On a smaller scale still, glacial grooves (tens of centimetres to metres wide) and glacial striae (millimetres to centimetres wide) are created by fragments of rock embedded in the ice at the base of a glacier (Figure 25, left and right). Glacial striae are very common on rock surfaces eroded by both alpine and continental glaciers.

Figure 25: Left: Roches moutonées with glacial striae near Squamish, B.C. Right: Glacial striae at the same location near Squamish. Ice flow was from right to left in both cases.

Lakes are common features in glacial environments. A lake that is confined to a glacial cirque is known as a tarn (Figure 26). Tarns are common in areas of alpine glaciation because the ice that forms a cirque typically carves out a depression in bedrock that then fills with water. In some cases, a series of such basins will form, and the resulting lakes are called rock basin lakes or paternoster lakes.

Figure 26: Lower Thornton Lake, a tarn, in the Northern Cascades National Park, Washington.

A lake that occupies a glacial valley, but is not confined to a cirque, is known as a finger lake. In some cases, a finger lake is confined by a dam formed by an end moraine, in which case it may be called a moraine lake (Figure 16.27).

Figure 27: Peyto Lake in the Alberta Rockies, is both a finger lake and a moraine lake as it is dammed by an end moraine, on the right.

In areas of continental glaciation, the crust is depressed by the weight of glacial ice that is up to 4,000 m thick. Basins are formed along the edges of continental glaciers (except for those that cover entire continents like Antarctica and Greenland), and these basins fill with glacial meltwater. Many such lakes, some of them huge, existed at various times along the southern edge of the Laurentide Ice Sheet. One example is Glacial Lake Missoula, which formed within Idaho and Montana, just south of the B.C. border with the United States. During the latter part of the last glaciation (30 ka to 15 ka), the ice holding back Lake Missoula retreated enough to allow some of the lake water to start flowing out, which escalated into a massive and rapid outflow (over days to weeks) during which much of the volume of the lake drained along the valley of the Columbia River to the Pacific Ocean. It is estimated that this type of flooding happened at least 25 times over that period, and in many cases, the rate of outflow was equivalent to the discharge of all of Earth's current rivers combined. The record of these massive floods is preserved in the Channelled Scablands of Idaho, Washington, and Oregon (Figure 28).

Figure 28: Potholes Coulee near Wenatchee, Washington, one of many basins that received Lake Missoula floodwaters during the late Pleistocene. Here the water flowed from right to left, over the cliff and into this basin.

Another type of glacial lake is a kettle lake. These are discussed in section 16.4 in the context of glacial deposits.

16.4 GLACIAL DEPOSITION

Sediments transported and deposited during the Pleistocene glaciations are abundant throughout Canada. They are important sources of construction materials and are valuable as reservoirs for groundwater. Because they are almost all unconsolidated, they have significant implications for mass wasting.

Figure 29 illustrates some of the ways that sediments are transported and deposited. The Bering Glacier is the largest in North America, and although most of it is in Alaska, it flows from an icefield that extends into southwestern Yukon. The surface of the ice is partially, or in some cases completely, covered with rocky debris that has fallen from surrounding steep rock faces. There are muddy rivers issuing from the glacier in several locations, depositing sediment on land, into Vitus

Lake, and directly into the ocean. There are dirty icebergs shedding their sediment into the lake. And, not visible in this view, there are sediments being moved along beneath the ice.

Figure 29: Part of the Bering Glacier in southeast Alaska, the largest glacier in North America. It is about 14 km across in the centre of this view.

The formation and movement of sediments in glacial environments is shown diagrammatically in Figure 30. There are many types of glacial sediment generally classified by whether they are transported on, within, or beneath the glacial ice. The main types of sediment in a glacial environment are described below.

Supraglacial (on top of the ice) and englacial (within the ice) sediments that slide off the melting front of a stationary glacier can form a ridge of unsorted sediments called an end moraine. The end moraine that represents the farthest advance of the glacier is a terminal moraine. Sediments transported and deposited by glacial ice are known as till.

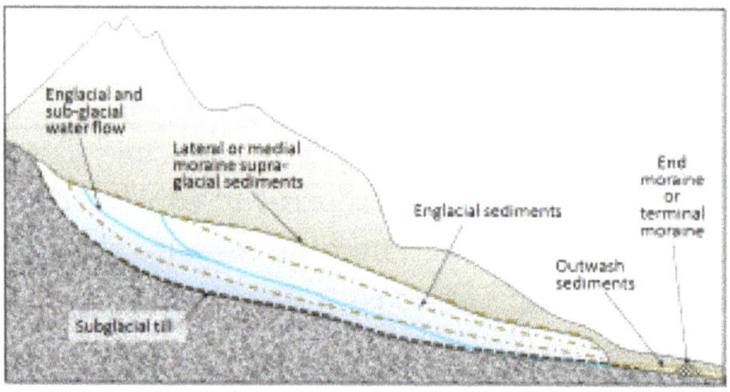

Figure 30: A depiction of the various types of sediments associated with glaciation. The glacier is shown in cross-section.

Subglacial sediment (e.g., lodgement till) is material that has been eroded from the underlying rock by the ice, and is moved by the ice. It has a wide range of grain sizes, including a relatively high proportion of silt and clay. The larger clasts (pebbles to boulders in size) tend to become partly rounded by abrasion. When a glacier eventually melts, the lodgement till is exposed as a sheet of well-compacted sediment ranging from several centimetres to many metres in thickness. Lodgement till is normally unbedded. An example is shown in Figure 31a.

Supraglacial sediments are primarily derived from freeze-thaw eroded material that has fallen onto the ice from rocky slopes above. These sediments form lateral moraines (Figure 1) and, where two glaciers meet, medial moraines. (Medial moraines are visible on the Aletsch Glacier in Figure 22.) Most of this material is deposited on the ground when the ice melts, and is therefore called ablation till, a mixture of fine and coarse angular rock fragments, with much less sand, silt, and clay than lodgement till. An example is shown in Figure 31b. When supraglacial sediments become incorporated into the body of the glacier, they are known as englacial sediments (Figure 30).

Figure 31: Examples of glacial till: a: lodgement till from the front of the Athabasca Glacier, Alberta; b: ablation till at the Horstman Glacier, Blackcomb Mountain, B.C.

Massive amounts of water flow on the surface, within, and at the base of a glacier, even in cold areas and even when the glacier is advancing. Depending on its velocity, this water is able to move sediments of various sizes and most of that material is washed out of the lower end of the glacier and deposited as outwash sediments. These sediments accumulate in a wide range of environments in the proglacial region (the area in front of a glacier), most in fluvial environments, but some in lakes and the ocean. Glaciofluvial sediments are similar to sediments deposited in normal fluvial environments, and are dominated by silt, sand, and gravel. The grains tend to be moderately well rounded, and the sediments have similar sedimentary structures (e.g., bedding, cross-bedding, clast imbrication) to those formed by non-glacial streams (Figure 32a and 32b).

Figure 32 Examples of glaciofluvial sediments: a: glaciofluvial sand of the Quadra Sand Formation at Comox, B.C.; b: glaciofluvial gravel and sand, Nanaimo, B.C.

A large proglacial plain of sediment is called a sandur (a.k.a. an outwash plain), and within that area, glaciofluvial deposits can be tens of metres thick. In situations where a glacier is receding, a block of ice might become separated from the main ice sheet and become buried in glaciofluvial sediments. When the ice block eventually melts, a depression forms, known as a kettle, and if this fills with water, it is known as a kettle lake (Figure 33).

Figure 33: A kettle lake amid vineyards and orchards in the Osoyoos area of B.C. [SE]

A subglacial stream will create its own channel within the ice, and sediments that are being transported and deposited by the stream will build up within that channel. When the ice recedes, the sediment will remain to form a long sinuous ridge known as an **esker**. Eskers are most common in areas of continental glaciation. They can be several metres high, tens of metres wide, and tens of kilometres long (Figure 34).

Figure 34: Part of an esker that formed beneath the Laurentide Ice Sheet in northern Canada.

Outwash streams commonly flow into proglacial lakes where glaciolacustrine sediments are deposited. These are dominated by silt- and clay-sized particles and are typically laminated on the millimetre scale. In some cases, varves develop; varves are series of beds with distinctive summer and winter layers: relatively coarse in the summer when melt discharge is high, and finer in the winter, when discharge is very low. Icebergs are common on proglacial lakes, and most of them contain englacial sediments of various sizes. As the bergs melt, the released clasts sink to the bottom and are incorporated into the glaciolacustrine layers as drop stones (Figure 35a).

The processes that occur in proglacial lakes can also take place where a glacier terminates in the ocean. The sediments deposited there are called glaciomarine sediments (Figure 35b).

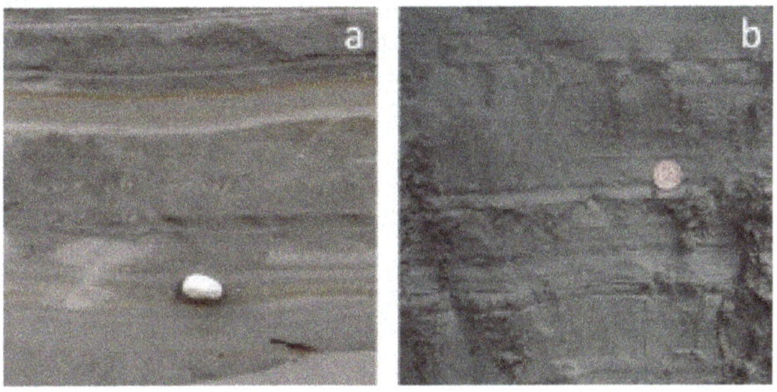

Figure 35: Examples of glacial sediments formed in quiet water: a: glaciolacustrine sediment with a drop stone, Nanaimo, B.C.; and b: a laminated glaciomarine sediment, Englishman River, B.C. Although not visible in this photo, the glaciomarine sediment has marine shell fossils.

AT A GLANCE

Glacial Periods in Earth's History	There have been many glaciations in Earth's distant past, the oldest known starting around 2,400 Ma. The late Proterozoic "Snowball Earth" glaciations were thought to be sufficiently intense to affect the entire planet. The current glacial period is known as the Pleistocene Glaciation, and while it was much more intense 20,000 years ago than it is now, we are still in the middle of it. The periodicity of the Pleistocene glaciations is related to subtle changes in Earth's orbital characteristics, which are exaggerated by a variety of positive feedback processes.
How Glaciers Work	The two main types of glaciers are continental glaciers, which cover large parts of continents, and alpine glaciers, which occupy mountainous regions. Ice accumulates at higher elevations — above the equilibrium line — where the snow that falls in winter does not all melt in summer. In continental glaciers, ice flows outward from where it is thickest. In alpine glaciers, ice flows downslope. At depth in the glacier ice, flow is by internal deformation, but glaciers that have liquid water at their base can also flow by basal sliding. Crevasses form in the rigid surface ice in places where the lower plastic ice is changing shape.

Glacial Erosion	Glaciers are important agents of erosion. Continental glaciers tend to erode the land surface into flat plains, while alpine glaciers create a wide variety of different forms. The key feature of alpine glacial erosion is the U-shaped valley. Arêtes are sharp ridges that form between two valleys, and horns form where a mountain is glacially eroded on at least three sides. Because tributary glaciers do not erode as deeply as main-valley glaciers, hanging valleys exist where the two meet. On a smaller scale, both types of glaciers form drumlins, roches moutonées, and glacial grooves or striae.
Glacial Deposition	Glacial deposits are quite varied, as materials are transported and deposited in a variety of different ways in a glacial environment. Sediments that are moved and deposited directly by ice are known as till. Glaciofluvial sediments are deposited by glacial streams, either forming eskers or large proglacial plains known as sandurs. Glaciolacustrine and glaciomarine sediments originate within glaciers and are deposited in lakes and the ocean respectively.

CHAPTER

SHORELINES 17

Most people love shorelines. They love panoramic ocean views, they love sandy beaches on crystal-clear lakes, they love to swim and surf and go out in boats, and they love watching giant waves crash onto rocky shores. While an understanding of coastal processes isn't necessary for our enjoyment of coastal regions, it can make our time there much more interesting. But an understanding of coastal processes is critical to people who live near a coast, or those who like to spend a lot of time there, because in order to be safe and avoid damage to infrastructure, we need to know how coastal processes work. We also need to understand the processes in order to avoid some of the possible consequences of changes that we might like to make in coastal areas.

Figure 1: Chesterman Beach near Tofino on the west coast Vancouver Island. The strip of sediment connecting the main beach to the rocky island is a tombolo.

17.1 WAVES

Waves form on the ocean and on lakes because energy from the wind is transferred to the water. The stronger the wind, the longer it blows, and the larger the area of water over which it blows (the fetch), the larger the waves are likely to be.

The important parameters of a wave are its wavelength (the horizontal distance between two crests or two troughs), its amplitude (the vertical distance between a trough and a crest), and its velocity (the speed at which wave crests move across the water) (Figure 2).

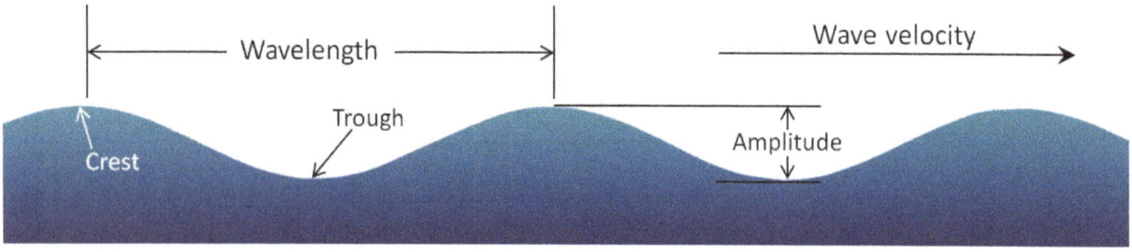

Figure 2: The parameters of water waves

The typical sizes and speeds of waves in situations where they have had long enough to develop fully are summarized in Table 1. In a situation where the fetch is short (say 19 km on a lake) and the wind is only moderate (19 km/h), the waves will develop fully within 2 hours, but they will remain quite small (average amplitude about 27 cm, wavelength 8.5 m). On a large body of water (the ocean or a very large lake) with a fetch of 139 km and winds of 37 km/h, the waves will develop fully in 10 hours; the average amplitude will be around 1.5 m and average wavelength around 34 m. In the open ocean, with strong winds (92 km/h) that blow for at least 69 hours, the waves will average nearly 15 m high and their wavelengths will be over 200 m. Small waves (amplitudes under a metre) tend to have relatively shallow slopes (amplitude is 3% to 4% of wavelength), while larger waves (amplitudes over 10 m) have much steeper slopes (amplitude is 6% to 7% of wavelength). In other words, not only are large waves bigger than small ones, they are also generally more than twice as steep, and therefore many times more impressive. It is important to recognize, however, that amplitudes decrease with distance from the area where the waves were generated. Waves on our coast that are generated by a storm near Japan will have similar wavelengths but lower amplitudes than those generated by a comparable storm offshore from Vancouver Island.

Table 1: The parameters of wind waves in situations where the wind blows in roughly the same direction for long enough for the waves to develop fully. The duration times listed are the minimum required for the waves to develop fully.

Wind Speed	Fetch	Duration	Amplitude	Wavelength	Wave Period	Wave Velocity	
km/h	km	h	m	m	s	m/s	km/h
19	19	2	0.27	8.5	3.0	2.8	10.2
37	139	10	1.5	33.8	5.7	5.9	19.5

Wind Speed	Fetch	Duration	Amplitude	Wavelength	Wave Period	Wave Velocity	
56	518	23	4.1	76.5	8.6	8.9	32.0
74	1,313	42	8.5	136	11.4	11.9	42.9
92	2,627	69	14.8	212	14.3	14.8	53.4

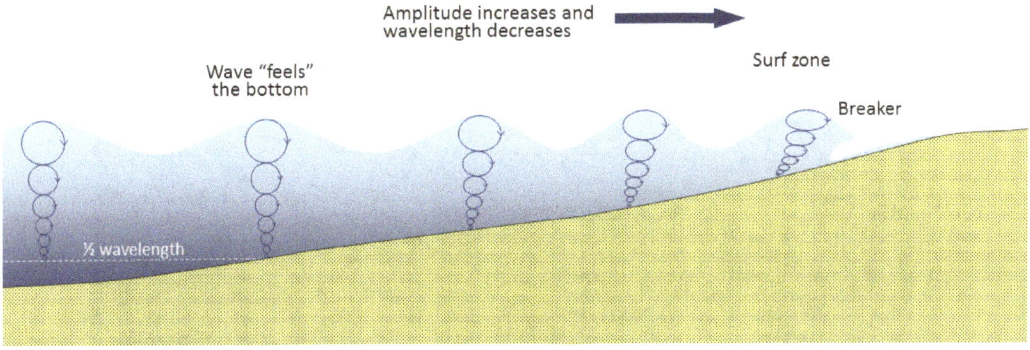

Figure 3: The effect of waves approaching a sandy shore

Figure 4: Waves breaking on the shore at Greensand Beach, Hawaii (the sand is green because it is made up mostly of the mineral olivine eroded from the nearby volcanic rocks)

Waves normally approach the shore at an angle, and this means that one part of the wave feels the bottom sooner than the rest of it, so the part that feels the bottom first slows down first. This process is illustrated in Figure 5, which is based on an aerial photograph showing actual waves approaching Long Beach on Vancouver Island. When the photo was taken, the waves (with crests shown as white lines in the diagram) were approaching at an angle of about 20° to the beach. The waves first reached shore at the southern end (right side of the image). As they moved into shallow water, they were slowed more at the southern end, and thus gradually became more parallel to the beach.

In open water, these waves had wavelengths close to 100 m. In the shallow water closer to shore, the wavelengths decreased to around 50 m, and in some cases, even less.

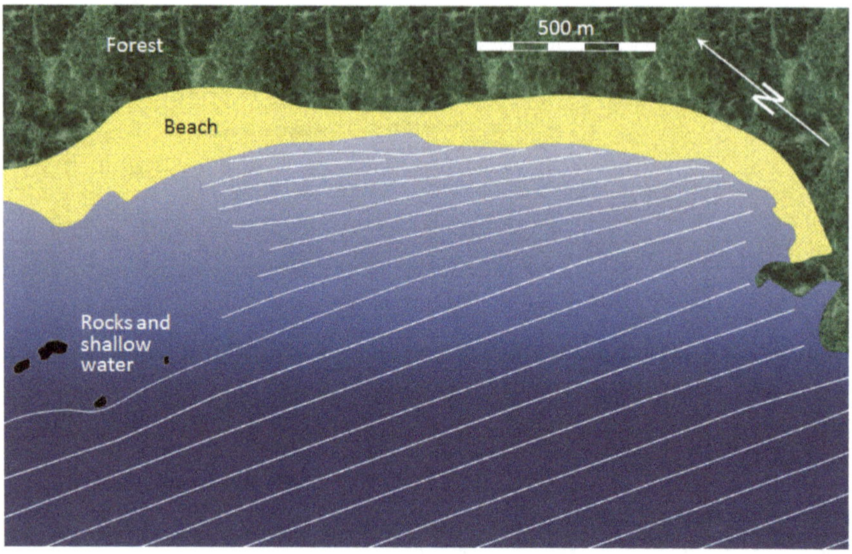

Figure 5: Waves approaching the shore of Long Beach in Pacific Rim National Park. As the waves (depicted by white lines) approach shore, they are refracted to become more parallel to the beach, and their wavelength decreases.

Even though they bend and become nearly parallel to shore, most waves still reach the shore at a small angle, and as each one arrives, it pushes water along the shore, creating what is known as a longshore current within the surf zone (the areas where waves are breaking) (Figure 6).

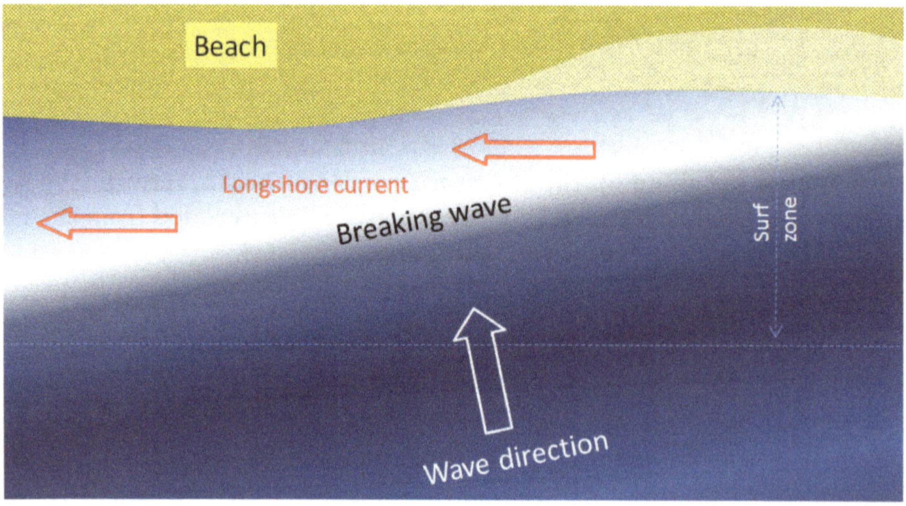

Figure 6: The generation of a longshore current by waves approaching the shore at an angle. [SE]

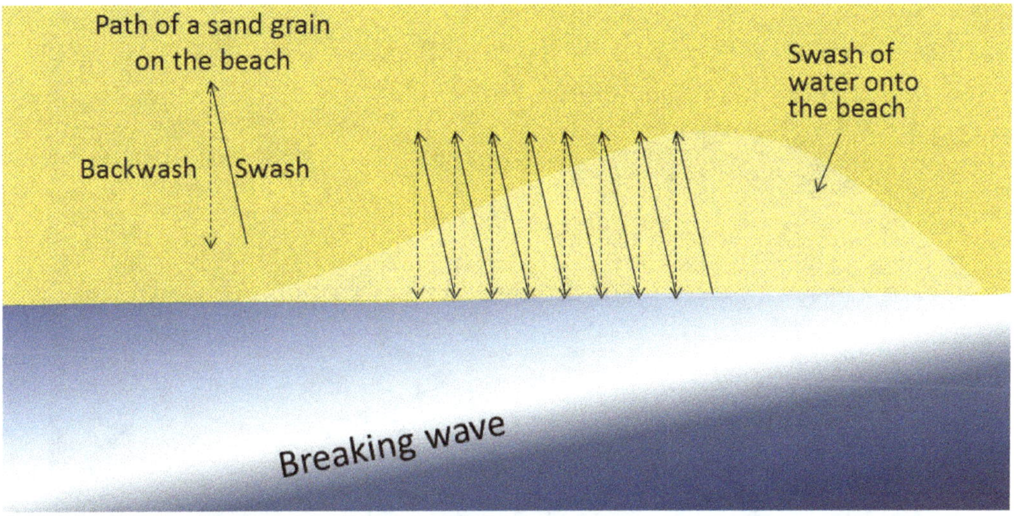

Figure 7: The movement of particles on a beach as a result of swash and backwash

A rip current is another type of current that develops in the nearshore area, and has the effect of returning water that has been pushed up to the shore by incoming waves. As shown in Figure 8, rip currents flow straight out from the shore and are fed by the longshore currents. They die out quickly just outside the surf zone, but can be dangerous to swimmers who get caught in them. If part of a beach does not have a strong unidirectional longshore current, the rip currents may be fed by longshore currents going in both directions.

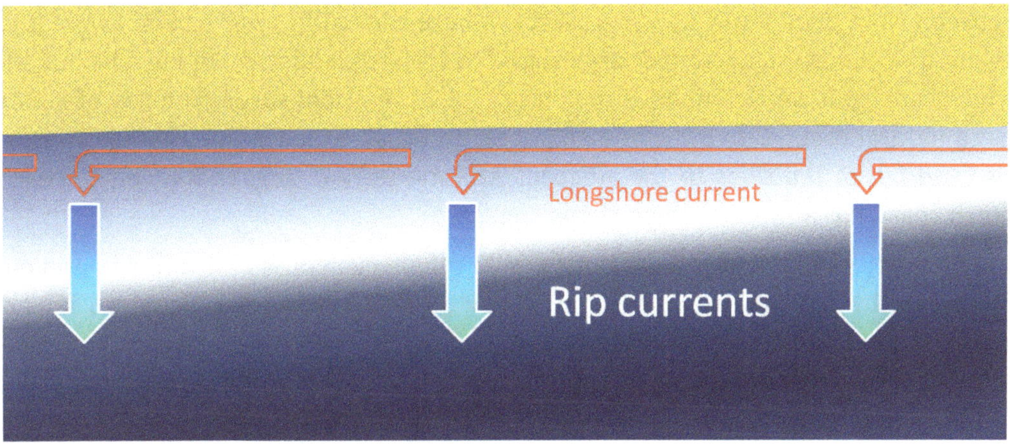

Figure 8: The formation of rip currents on a beach with strong surf

Rip currents are visible in Figure 9, a beach at Tunquen in Chile near Valparaiso. As is evident from the photo, the rips correspond with embayments in the beach profile. Three of them are indicated with arrows, but it appears that there may be several others farther along the beach.

Figure 9: Rip currents on Tunquen Beach in central Chile

Tides are related to very long-wavelength but low-amplitude waves on the ocean surface (and to a much lesser extent on very large lakes) that are caused by variations in the gravitational effects of the Sun and Moon. Tide amplitudes in shoreline areas vary quite dramatically from place to place. On the west coast of Canada, the tidal range is relatively high, in some areas as much as 6 m, while on most of the east coast the range is lower, typically around 2 m. A major exception is the Bay of Fundy between Nova Scotia and New Brunswick, where the daily range can be as great as 16 m. Anomalous tides like that are related to the shape and size of bays and inlets, which can significantly enhance the amplitude of the tidal surge. The Bay of Fundy has a natural oscillation cycle of 12.5 hours, and that matches the frequency of the rise and fall of the tides in the adjacent Atlantic Ocean. Ungava Bay, on Quebec's north coast, has a similarly high tidal range.

As the tides rise and fall they push and pull a large volume of water in and out of bays and inlets and around islands. They do not have as significant an impact on coastal erosion and deposition as wind waves do, but they have an important influence on the formation of features within the intertidal zone, as we'll see in the following sections.

17.2 LANDFORMS OF COASTAL EROSION

Large waves crashing onto a shore bring a tremendous amount of energy that has a significant eroding effect, and several unique erosion features commonly form on rocky shores with strong waves.

When waves approach an irregular shore, they are slowed down to varying degrees, depending on differences in the water depth, and as they slow, they are bent or refracted. In Figure 10, wave energy is represented by the red arrows. That energy is evenly spaced out in the deep water, but because of refraction, the energy of the waves — which moves perpendicular to the wave crests — is being focused on the headlands (Frank Island and Cox Point in this case). On irregular coasts, the headlands receive much more wave energy than the intervening bays, and thus they are more strongly eroded. The result of this is coastal straightening. An irregular coast, like the west coast of Vancouver Island, will eventually become straightened, although that process will take millions of years.

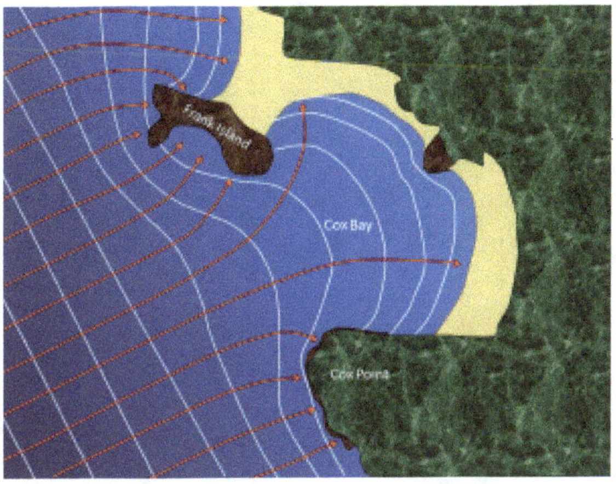

Figure 10: The approach of waves (white lines) in the Cox Bay area of Long Beach, Vancouver Island. The red arrows represent wave energy; most of that energy is focused on the headlands of Frank Island and Cox Point.

Wave erosion is greatest in the surf zone, where the wave base is impinging strongly on the sea floor and where the waves are breaking. The result is that the substrate in the surf zone is typically eroded to a flat surface known as a wave-cut platform (or wave-cut terrace) (Figure 11). A wave-cut platform extends across the intertidal zone.

Figure 11: A wave-cut platform in bedded sedimentary rock on Gabriola Island, B.C. The wave-eroded surface is submerged at high tide.

Relatively resistant rock that does not get completely eroded during the formation of a wave-cut platform will remain behind to form a stack. An example from the Juan de Fuca Trail of southwestern Vancouver Island is shown in Figure 12. Here the different layers of the sedimentary rock have different resistance to erosion. The upper part of this stack is made up of rock that resisted erosion, and that rock has protected a small pedestal of underlying softer rock. The softer rock will eventually be eroded and the big rock will become just another boulder on the beach.

Figure 12: A stack on the Juan de Fuca Trail section of the southwestern shore of Vancouver Island. The rock surrounding the stack is part of a wave-cut platform.

Arches and sea caves are related to stacks because they all form as a result of the erosion of relatively non-resistant rock. An arch in the Barachois River area of western Newfoundland is shown in Figure 13. This feature started out as a sea cave, and then, after being eroded from both sides, became an arch. During the winter of 2012/2013, the arch collapsed, leaving a small stack at the end of the point. If you look carefully at the upper photograph you can see that the hole that makes the arch developed within a layer of relatively soft and weak rock.

Figure 13: Top: An arch in tilted sedimentary rock at the mouth of the Barachois River, Newfoundland, July 2012. Bottom: The same location in June 2013. The arch has collapsed and a small stack remains.

Figure 14 summarizes the process of transformation of an irregular coast, initially produced by tectonic uplift, into a straightened coast with sea cliffs (wave-eroded escarpments) and the remnants of stacks, arches, and wave-cut platforms. The next stages of this process would be the continued landward erosion of the sea cliffs and the complete erosion of the stacks and wave-cut platforms in favour of a continuous and nearly straight sandy beach.

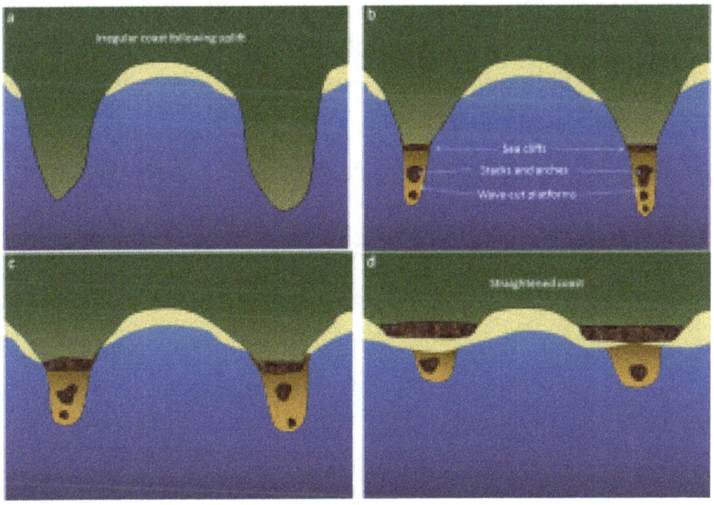

Figure 14: Evolution of a straightened coast through the erosion to stacks and arches, sea cliffs, and wave-cut platforms

17.3 LANDFORMS OF COASTAL DEPOSITION

Some coastal areas are dominated by erosion, an example being the Pacific coast of Canada and the United States, while others are dominated by deposition, examples being the Atlantic and Caribbean coasts of the United States. But on almost all coasts, both deposition and erosion are happening to varying degrees most of the time, although in different places. This is clearly evident in the Tofino area of Vancouver Island (Figure 1), where erosion is the predominant process on the rocky headlands, while depositional processes predominate within the bays. On deposition-dominant coasts, the coastal sediments are still being eroded from some areas and deposited in others.

The main factor in determining if a coast is dominated by erosion or deposition is its history of tectonic activity. A coast like that of British Columbia is tectonically active, and compression and uplift have been going on for tens of millions of years. This coast has also been uplifted during the past 15,000 years by isostatic rebound due to deglaciation. The coasts of the United States along the Atlantic and the Gulf of Mexico have not seen significant tectonic activity in a few hundred million years, and except in the northeast, have not experienced post-glacial uplift. These areas have relatively little topographic relief, and there is now minimal erosion of coastal bedrock.

On coasts that are dominated by depositional processes, most of the sediment being deposited typically comes from large rivers. An obvious example is where the Mississippi River flows into the Gulf of Mexico at New Orleans; another is the Fraser River at Vancouver. There are no large rivers bringing sandy sediments to the west coast of Vancouver Island, but there are still long and wide sandy beaches there. In this area, most of the sand comes from glaciofluvial sand deposits situated along the shore behind the beach, and some comes from the erosion of the rocks on the headlands.

The components of a typical beach are shown in Figure 15. On a sandy marine beach, the beach face is the area between the low and high tide levels. A berm is a flatter region beyond the reach of high tides; this area stays dry except during large storms.

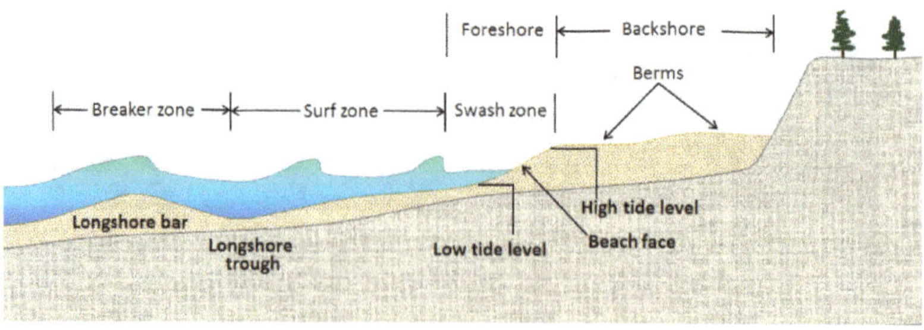

Figure 15: The components of a sandy marine beach [SE]

Most beaches go through a seasonal cycle because conditions change from summer to winter. In summer, sea conditions are relatively calm with long-wavelength, low-amplitude waves generated by distant winds. Winter conditions are rougher, with shorter-wavelength, higher-amplitude waves caused by strong local winds. As shown in Figure 16, the heavy seas of winter gradually erode sand from beaches, moving it to an underwater sandbar offshore from the beach. The gentler waves of summer gradually push this sand back toward the shore, creating a wider and flatter beach.

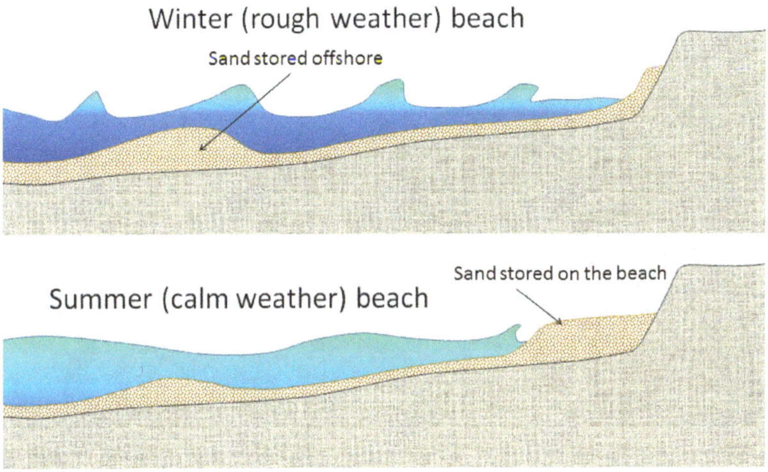

Figure 16: The differences between summer and winter on beaches in areas where the winter conditions are rougher and waves have a shorter wavelength but higher energy. In winter, sand from the beach is stored offshore.

The evolution of sandy depositional features on sea coasts is primarily influenced by waves and currents, especially longshore currents. As sediment is transported along a shore, either it is deposited on beaches, or it creates other depositional features. A spit, for example is an elongated sandy deposit that extends out into open water in the direction of a longshore current. A good example is Goose Spit at Comox on Vancouver Island (Figure 17). At this location, the longshore current typically flows toward the southwest, and the sand eroded from a 60 m high cliff of Pleistocene glaciofluvial Quadra Sand is pushed in that direction and then out into Comox Harbour.

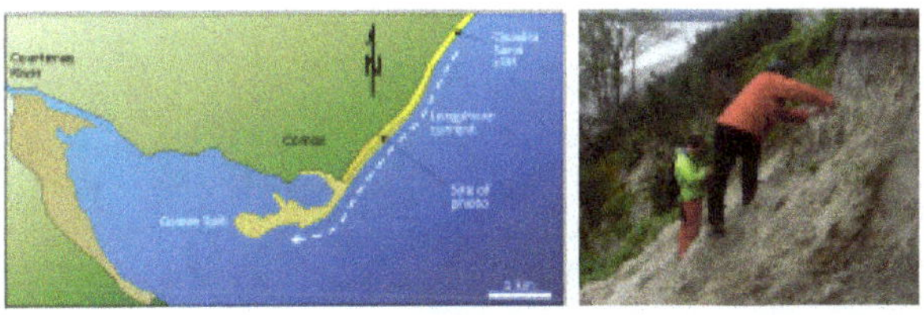

Figure 17: The formation of Goose Spit at Comox on Vancouver Island. The sand that makes up Goose Spit is derived from the erosion of Pleistocene Quadra Sand (a thick glaciofluvial sand deposit, as illustrated in the photo on the right).

The Quadra Sand at Comox is visible in Figure 19. There are numerous homes built at the top of the cliff, and the property owners have gone to considerable expense to reinforce the base of the cliff with large angular rocks (rip-rap) and concrete barriers so as to limit further erosion of their properties. One result of this will be to starve Goose Spit of sediments and eventually contribute to its erosion. Of course the rocks and concrete barriers are only temporary; they will be eroded by strong winter storms over the next few decades and the Quadra Sand will once again contribute to the maintenance of Goose Spit.

Figure 18: The Quadra Sand cliff at Comox, and the extensive concrete and rip-rap barrier that has been constructed to reduce erosion. Note that the waves (dashed lines) are approaching the shore at an angle, contributing to the longshore current. [SE]

A spit that extends across a bay to the extent of closing, or almost closing it off, is known as a baymouth bar. Most bays have streams flowing into them, and since this water has to get out, it is rare that a baymouth bar will completely close the entrance to a bay. In areas where there is sufficient sediment being transported, and there are near-shore islands, a tombolo may form (Figure 19).

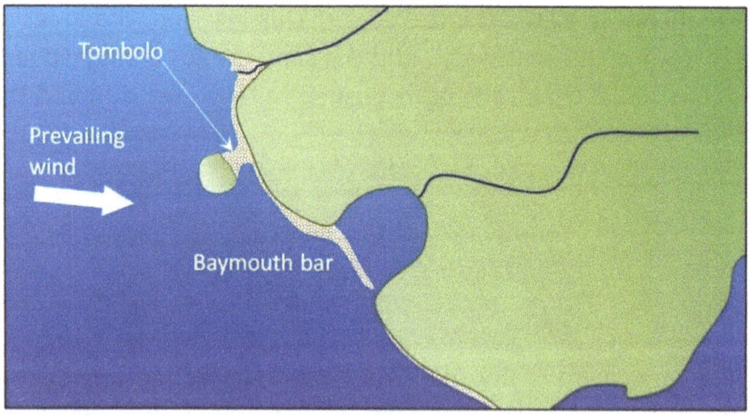

Figure 19 A depiction of a baymouth bar and a tombolo

Tombolos are common around the southern part of the coast of British Columbia, where islands are abundant, and they typically form where there is a wave shadow behind a nearshore island (Figure 20). This becomes an area with reduced energy, and so the longshore current slows and sediments accumulate. Eventually enough sediments accumulate to connect the island to the mainland with a tombolo. There is a good example of a tombolo in Figure 1, and another in Figure 21.

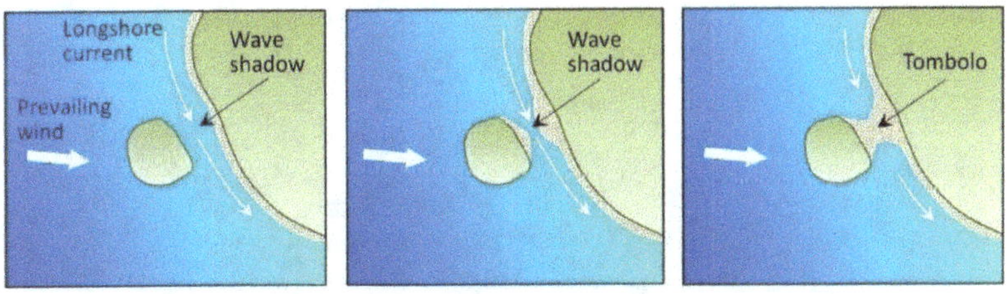

Figure 20: The process of formation of a tombolo in a wave shadow behind a nearshore island

Figure 21: A stack (with a wave-cut platform) connected to the mainland by a tombolo, Leboeuf Bay, Gabriola Island, B.C.

In areas where coastal sediments are abundant and coastal relief is low (because there has been little or no recent coastal uplift), it is common for barrier islands to form (Figure 22). Barrier islands are elongated islands composed of sand that form a few kilometres away from the mainland. They are common along the U.S. Gulf Coast from Texas to Florida, and along the U.S. Atlantic Coast from Florida to Massachusetts. North of Boston, the coast becomes rocky, partly because that area has been affected by post-glacial crustal rebound.

Figure 22: Assateague Island on the Maryland coast, U.S. This barrier island is about 60 km long and only 1 km to 2 km wide. The open Atlantic Ocean is to the right and the lagoon is to the left. This part of Assateague Island has recently been eroded by a tropical storm, which pushed massive amounts of sand into the lagoon.

Exercise 3 Beach Forms

On the map, sketch where you would expect the following to form:

- A spit
- A baymouth bar
- A tombolo

What conditions might lead to the formation of barrier islands in this area?

Some coasts in tropical regions (between 30° S and 30° N) are characterized by carbonate reefs. Reefs form in relatively shallow marine water within a few hundred to a few thousand metres of shore in areas where there is little or no input of clastic sediments from streams, and marine organisms such as corals, algae, and shelled organisms can thrive. The associated biological processes are enhanced where upwelling currents bring chemical nutrients from deeper water (but not so deep that the water is cooler than about 25°C) (Figure 23). Sediments that form in the back reef (shore side) and fore reef (ocean side) are typically dominated by carbonate fragments eroded from the reef and from organisms that thrive in the back-reef area that is protected from wave energy by the reef.

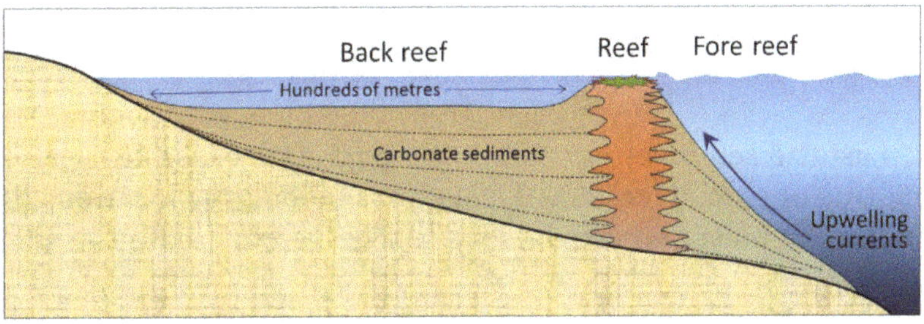

Figure 23: Cross-section through a typical barrier or fringing reef [SE]

17.4 SEA-LEVEL CHANGE

Sea-level change has been a feature on Earth for billions of years, and it has important implications for coastal processes and both erosional and depositional features. There are three main mechanisms of sea-level change, as described below.

Eustatic sea-level changes are global sea-level changes related either to changes in the volume of glacial ice on land or to changes in the shape of the sea floor caused by plate tectonic processes. For example, changes in the rate of mid-ocean spreading will change the shape of the sea floor near the ridges, and this affects sea level.

Over the past 20,000 years, there has been approximately 125 m of eustatic sea-level rise due to glacial melting. Most of that took place between 15,000 and 7,500 years ago during the major melting phase of the North American and Eurasian Ice Sheets (Figure 24). At around 7,500 years ago, the rate of glacial melting and sea-level rise decreased dramatically, and since that time, the average rate has been in the order of 0.7 mm/year. Anthropogenic climate change led to accelerating sea-level rise starting around 1870. Since that time, the average rate has been 1.1 mm/year, but it has been gradually increasing. Since 1992, the average rate has been 3.2 mm/year.

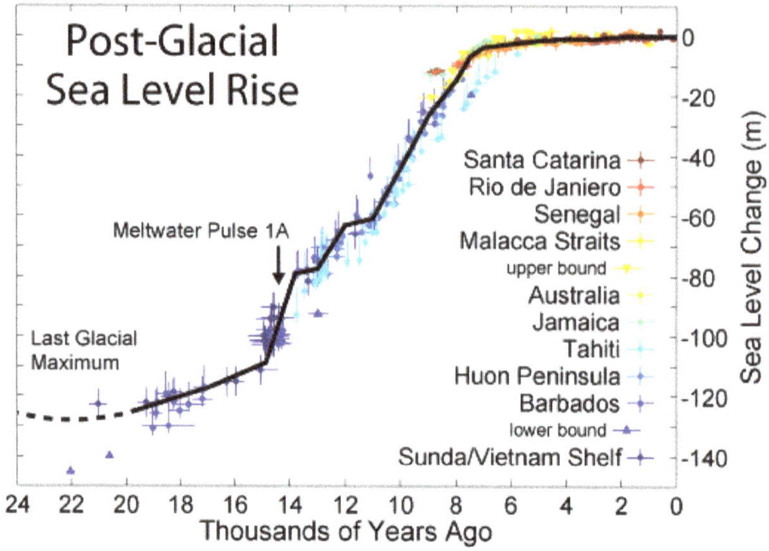

Figure 24: Eustatic sea-level curve for the past 24 ka (sea-level rise resulting from the melting of glacial ice). Sea-level rise is global; the locations listed in the caption are the places where data were acquired to create this diagram.

Isostatic sea-level changes are local changes caused by subsidence or uplift of the crust related either to changes in the amount of ice on the land, or to growth or erosion of mountains.

Almost all of Canada and parts of the northern United States were covered in thick ice sheets at the peak of the last glaciation. Following the melting of this ice, there has been an isostatic rebound of continental crust in many areas. This ranges from several hundred metres of rebound in the central part of the Laurentide Ice Sheet (around Hudson Bay) to 100 m to 200 m in the peripheral parts of the Laurentide and Cordilleran Ice Sheets — in places such as Vancouver Island and the mainland coast of B.C. In other words, although global sea level was about 130 m lower during the last glaciation, the glaciated regions were depressed at least that much in most places, and more than that in places where the ice was thickest.

There is evidence of isostatic rebound along the southwest coast of Vancouver Island, where a number of streams enter the ocean as 5 m high waterfalls, as shown in Figure 25.

Tectonic sea-level changes are local changes caused by tectonic processes. The subduction of the Juan de Fuca Plate beneath British Columbia is creating tectonic uplift (about 1 mm/year) along the western edge of Vancouver Island, although much of this uplift is likely to be reversed when the next large subduction-zone earthquake strikes.

Figure 25: This stream is on the southwest coast of Vancouver Island near Sooke. Like many other streams along this coast, it used to flow directly into the ocean, but the land has been uplifted by post-glacial isostatic rebound.

Coastlines in areas where there has been net sea-level rise in the geologically recent past are commonly characterized by estuaries and fiords. Howe Sound, north of Vancouver, is an example of a fiord (Figure 26). This valley was filled with ice during the last glaciation, and there has been a net rise in sea level here since that time. Coastlines in areas where there has been net sea-level drop in the geologically recent past are characterized by uplifted wave-cut platforms (or stream valleys as shown in Figure 26). Uplifted beach lines are another product of relative sea-level drop, although these are difficult to recognize in areas with vigorous vegetation. They are relatively common in Canada's far north.

Figure 26: Howe Sound, north of Vancouver, is a fiord with well-defined glacial erosion features. [SE]

Exercise 4 A Holocene Uplifted Shore

The blue-grey sediments in the photo contain marine fossils of early Holocene age (~12,500 years ago). Explain the role that melting ice might have played in raising this material 60 m above sea level. Is this change a result of isostatic or eustatic relative sea-level change?

Early Holocene marine sediments at 60 m elevation on Gabriola Island, B.C.

17.5 HUMAN INTERFERENCE WITH SHORELINES

There are various modifications that we make in an attempt to influence beach processes for our own purposes. Sometimes these changes are effective, and may appear to be beneficial, although in most cases there are unintended negative consequences that we don't recognize until much later.

An example is at the beach near Comox (described above), which has been armoured with rip-rap and concrete blocks in an attempt to limit the natural erosion that is threatening the properties at the top of the cliff (Figure 19). As already noted, the unintended effect of this installation will be to starve Goose Spit of sediment. As long as the armour remains in place, which might be several decades, there is a risk that the spit will start to erode, which will affect many of the organisms that use that area as their habitat, and many of the people who go there for recreation.

Seawalls, like the one around Vancouver's Stanley Park (Figure 16.27), also help to limit erosion and can be very pleasant amenities for the public, but they have geological and ecological costs. When a shoreline is "hardened" in this way, important marine habitat is lost and sediment production is reduced, and that can affect beaches elsewhere. Seawalls also affect the behaviour of waves and longshore currents, sometimes with negative results.

Figure 27: The seawall at Stanley Park, Vancouver [https://commons.wikimedia.org/wiki/File:Seawall2.jpg]

Another example is at Sunset Beach in Vancouver. As shown in Figure 29, a series of rip-rap breakwaters (structures parallel to the shore) were built in the 1990s and sand has accumulated behind them to form the beach. The breakwaters have acted as islands and the sand has been deposited in the low-energy water behind them, in the same way that a tombolo forms. This can be seen from a photograph taken from the Burrard Bridge in 2015 (Figure 29). The two benefits of this project are that a pleasant beach has been created, and some of the sediment that previously would have been moved into False Creek, and could have blocked its entrance, has been trapped in English Bay. The negative impacts are probably not well understood, but have likely involved loss of marine animal habitat.

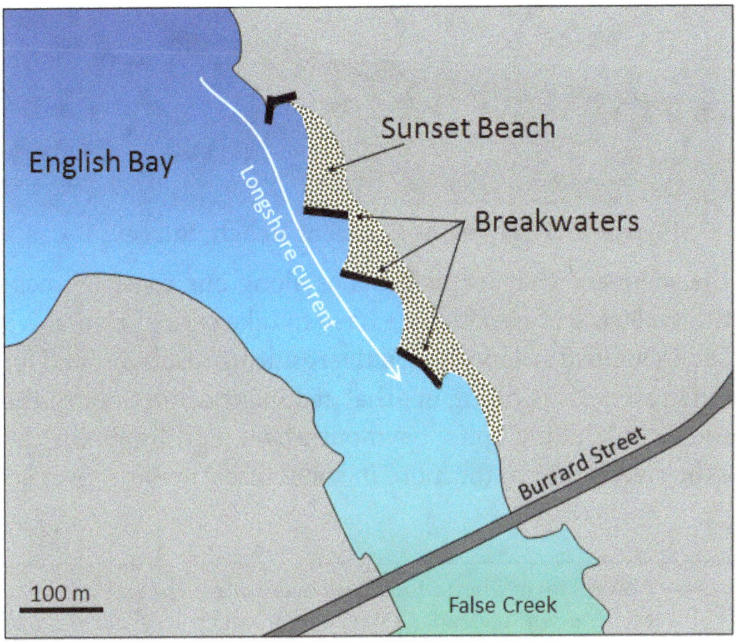

Figure 28: Map of the impact of breakwaters (or groynes) on beach formation at Sunset Beach, Vancouver

Figure 29: Photograph of the impact of breakwaters on beach development at Sunset Beach, Vancouver

Groynes (or groins in the U.S.) have an effect that is similar to that of breakwaters, although groynes are constructed perpendicular to the beach (Figure 31), and they trap sediment by slowing the longshore current.

Figure 3: A groyne at Crescent Beach, Surrey, B.C.

Most of the sediment that forms beaches along our coasts comes from rivers, so if we want to take care of beaches, we have to take care of rivers. When a river is dammed, its sediment load is deposited in the resulting reservoir, and for the century or two while the reservoir is filling up, that sediment cannot get to the sea. During that time, beaches (including spits, baymouth bars, and tombolos) within tens of kilometres of the river's mouth (or more in some cases) are at risk of erosion.

AT A GLANCE

Waves	Waves form when wind blows over water. The size of the waves depends on the wind speed, the area over which it is blowing, and time. The important parameters of a wave are its amplitude, wavelength, and speed. The water beneath a wave is disturbed to a depth of one-half the wavelength, and a wave is slowed when it approaches shallow water. A longshore current develops where waves approach the shore at an angle, and swash and backwash on a beach move sediment along the shore. The combined effect of these two processes is sediment transport by longshore drift.
Landforms of Coastal Erosion	Coasts that have experienced uplift within the past several million years tend to have irregular shapes and are dominated by erosional processes. Wave paths are bent where the coast is irregular and wave energy is focused on headlands. Rocky headlands are eroded into sea caves, arches, stacks, and sea cliffs, and the areas around these features are eroded into wave-cut platforms. Over the long term (millions of years), irregular coasts are straightened.

Landforms of Coastal Deposition	Coasts that have not been uplifted for tens of millions of years tend to be relatively straight, and are dominated by depositional features, although deposition is also important on irregular coasts. Waves and longshore drift are important in controlling the formation of beaches, as well as spits, tombolos, baymouth bars, and barrier islands. Beaches can be divided into zones, such as foreshore and backshore, and beach shapes typically change from season to season. Carbonate reefs and carbonate sediments form in tropical regions where there is little input of clastic sediments.
Sea-Level Change	The relative levels of the land and sea have significant implications for coastal processes and landforms, and they have been constantly changing over geological time. Eustatic sea-level changes are global in effect, and are typically related to glacial ice formation or melting. Isostatic sea-level changes are local effects caused by uplift or subsidence of continental crust, typically because of the gain or loss of glacial ice. Tectonic sea-level changes are related to plate interactions. Net sea-level rise leads to development of estuaries and fiords, while net sea-level drop creates uplifted marine terraces and beaches.
Human Interference with Shorelines	Humans have a strong urge to alter coasts for their own convenience by building seawalls, breakwaters, groynes, and other barriers. Although these types of features may have economic and other benefits, they can have both geological and ecological implications that must be considered.

CHAPTER

GEOLOGY OF THE OCEANS 18

Figure 1: Oceanic crust (pillow basalt) from the Paleogene Metchosin Igneous Complex, near Sooke, on Vancouver Island. The view is about 1.5 m across.

Oceans cover 71% of Earth's surface and hold 97% of Earth's water. The water they contain is critical to plate tectonics, to volcanism, and of course, to life on Earth. It is said that we know more about the surface of the Moon than the floor of the oceans. Whether this is true or not, the important point is that the ocean floor is covered with an average of nearly 4,000 m of water, and it's pitch black below a few hundred metres so it's not easy to discover what is down there. We know a lot more about the oceans than we used to, but there is still a great deal more to discover.

Earth has had oceans for a very long time, dating back to the point where the surface had cooled enough to allow liquid water, only a few hundred million years after Earth's formation. At that time there were no continental rocks, so the water that was here was likely spread out over the surface in one giant (but relatively shallow) ocean.

18.1 TOPOGRAPHY OF THE SEA FLOOR

We examined the topography of the sea floor from the perspective of plate tectonics in Chapter 10, but here we are going to take another look at the important features from an oceanographic perspective. The topography of the northern Atlantic Ocean is shown in Figure 2. The important features are the extensive continental shelves less than 250 m deep (pink); the vast deep ocean plains between 4,000 and 6,000 m deep (light and dark blue); the mid-Atlantic ridge, in many areas shallower than 3,000 m; and the deep ocean trench north of Puerto Rico (8,600 m).

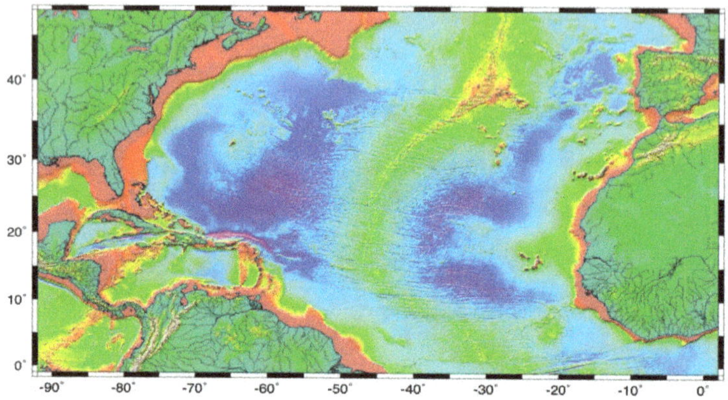

Figure 2: The topography of the Atlantic Ocean sea floor between 0° and 50° north. Red and yellow colours indicate less than 2,000 m depth; green less than 3,000 m; blue 4,000 m to 5,000 m; and purple greater than 6,000 m.

A topographic profile of the Pacific Ocean floor between Japan and British Columbia is shown in Figure 3. Be careful when interpreting this diagram (and others like it), because in order to show the various features clearly the vertical axis is exaggerated, in this case by about 200 times. The floor of the Pacific, like those of the other oceans, is actually very flat, even in areas with seamounts or deep trenches. The vast sediment-covered abyssal plains of the oceans are much flatter than any similar-sized areas on the continents.

The main features of the Pacific Ocean floor are the continental slopes, which drop from about 200 m to several thousand metres over a distance of a few hundred kilometres; the abyssal plains — exceedingly flat and from 4,000 m to 6,000 m deep; volcanic seamounts and islands; and trenches at subduction zones that are up to 11,000 m deep.

The ocean floor is almost entirely underlain by mafic oceanic crust (mostly basalt and gabbro, as described in more detail below), while the continental slopes are underlain by felsic continental crust (mostly granitic and sedimentary rocks). And, as you'll remember from Chapter 10, the heavier oceanic crust floats lower on the mantle than continental crust does, and that's why oceans are oceans.

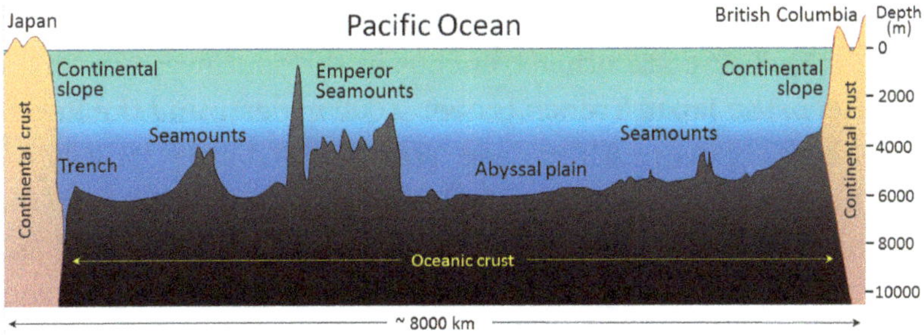

Figure 3: The generalized topography of the Pacific Ocean sea floor between Japan and British Columbia. The vertical exaggeration is approximately 200 times.

The continental shelf and slope offshore from Nova Scotia is shown in Figure 4. In this passive-margin area (no subduction zone), the shelf is over 150 km wide. On the Pacific coast of Canada, the shelf is less than half as wide. Continental shelves are typically less than 200 m in depth; 200 m is also the limit of the photic zone, the maximum depth to which sufficient light penetrates to allow photosynthesis to take place. As a result of that photosynthesis, the photic zone is oxygenated, and therefore suitable for animal life. Approximately 90% of marine life is restricted to the photic zone. The photic zone is also known as the epipelagic zone. The mesopelagic zone extends from 200 m to 1,000 m; the bathypelagic zone from 1,000 m to 4,000 m; and abyssalpelagic zone is deeper than 4,000 m. (Pelagic refers to the open ocean, and thus excludes areas that are near to the shores or the ocean floor.)

Although the temperature of the ocean surface varies widely, from a few degrees either side of freezing in polar regions to over 25°C in the tropics, in most parts of the ocean, the water temperature is around 10°C at 1,000 m depth and about 4°C from 2,000 m depth all the way to the bottom.

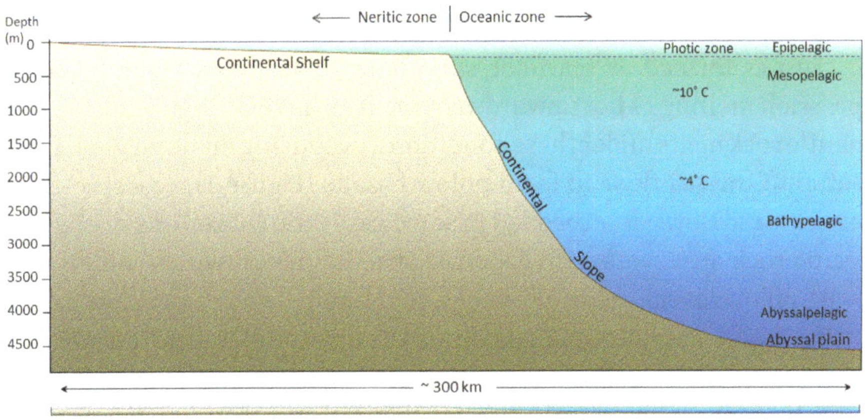

Figure 4: The generalized topography of the Atlantic Ocean floor within 300 km of Nova Scotia. The vertical exaggeration is approximately 25 times. The panel at the bottom shows the same profile without vertical exaggeration.

The deepest parts of the ocean are within the subduction trenches, and the deepest of these is the Marianas Trench in the southwestern Pacific (near Guam) at 11,000 m (Figure 5). There are other trenches in the southwestern Pacific that are over 10,000 m deep; the Japan Trench is over 9,000 m deep; and the Puerto Rico and Chile-Peru Trenches are over 8,000 m deep. Trenches that are relatively shallow tend to be that way because they have significant sediment infill. There is no recognizable trench along the subduction zone of the Juan de Fuca Plate because it has been filled with sediments from the Fraser and Columbia Rivers (or their ancient equivalents).

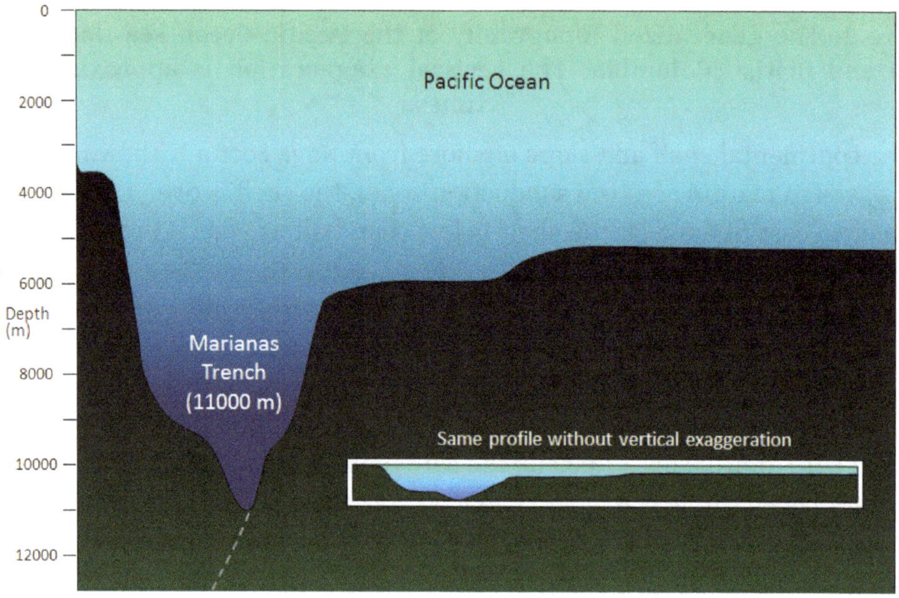

Figure 5: The generalized topography of the Pacific Ocean floor in the area of the Marianas Trench, near Guam. The dashed grey line represents the subduction of the Pacific Plate (to the right) beneath the Philippine Plate (to the left). [SE]

18.2 OCEANIC CRUST AND ITS GEOLOGY

Oceanic crust is formed at sea-floor spreading ridges from magma generated by decompression melting of hot upward-moving mantle rock (Figure 10.18). About 10% of the mantle rock melts under these conditions, producing mafic magma. This magma oozes out onto the sea floor to form pillow basalts (Figure 1), breccias (fragmented basaltic rock), and flows, interbedded in some cases with limestone or chert. Beneath the volcanic rock are layers with gabbroic sheeted dykes (which sometimes extend up into the pillow layer), gabbroic stocks, and finally layered peridotite (ultramafic rock) at the base. The ultramafic rock of the mantle lies below that. Over time, the igneous rock of the oceanic crust gets covered with layers of sediment, which eventually become sedimentary rock, including limestone, mudstone, chert, and turbidites. The lithologies of the layers of the oceanic crust are shown in Figure 6.

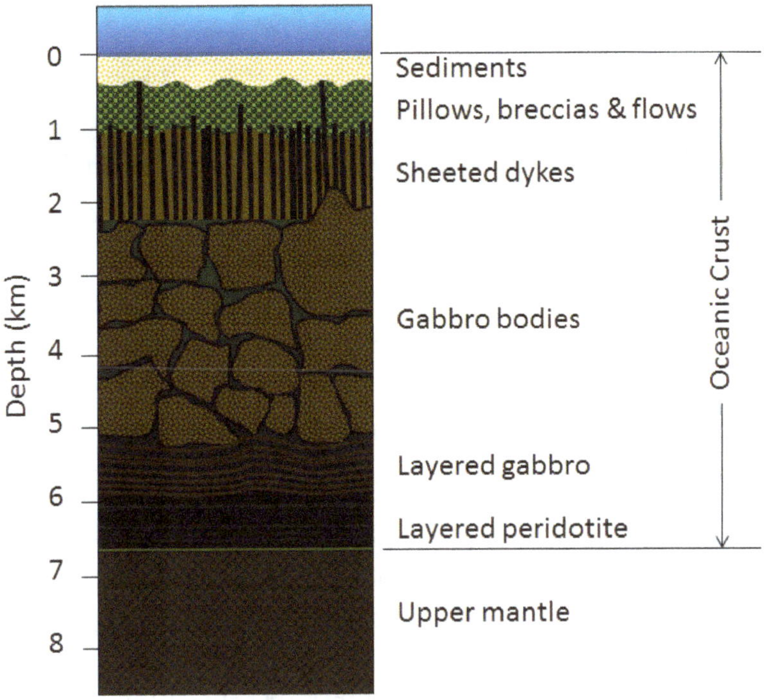

Figure 6: Schematic representation of the lithologic layers of typical oceanic crust

The age of the oceanic crust has been determined by systematic mapping variations in the strength of the Earth's magnetic field across the sea floor and comparing the results with our understanding of the record of Earth's magnetic field reversal chronology for the past few hundred million years. The ages of different parts of the crust are shown in Figure 7. The oldest oceanic crust is around 280 Ma in the eastern Mediterranean, and the oldest parts of the open ocean are around 180 Ma on either side of the north Atlantic. It may be surprising, considering that parts of the continental crust are close to 4,000 Ma old, that the oldest sea floor is less than 300 Ma. Of course, the reason for this is that all sea floor older than that has been either subducted or pushed up to become part of the continental crust. For example, there are fragments of sea floor in British Columbia that date back to around 380 and 220 Ma, and there are similar rocks in the Canadian Shield that are older than 3 Ga.

As one would expect, the oceanic crust is very young near the spreading ridges (Figure 7), and there are obvious differences in the rate of sea-floor spreading along different ridges. The ridges in the Pacific and southeastern Indian Oceans have wide age bands, indicating rapid spreading (approaching 10 cm/y on each side in some areas), while those in the Atlantic and western Indian Oceans are spreading much more slowly (less than 2 cm/y on each side in some areas).

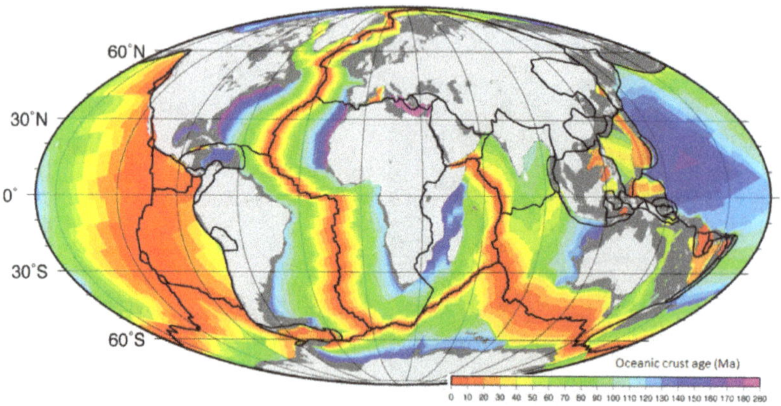

Figure 7: The age of the oceanic crust

18.3 SEA-FLOOR SEDIMENTS

Except within a few kilometres of a ridge crest, where the volcanic rock is still relatively young, most parts of the sea floor are covered in sediments. This material comes from several different sources and is highly variable in composition, depending on proximity to a continent, water depth, ocean currents, biological activity, and climate. Sea-floor sediments (and sedimentary rocks) can range in thickness from a few millimetres to several tens of kilometres. Near the surface, the sea-floor sediments remain unconsolidated, but at depths of hundreds to thousands of metres (depending on the type of sediment and other factors) the sediment becomes lithified.

The various sources of sea-floor sediment can be summarized as follows:

- Terrigenous sediment is derived from continental sources transported by rivers, wind, ocean currents, and glaciers. It is dominated by quartz, feldspar, clay minerals, iron oxides, and terrestrial organic matter.
- Pelagic carbonate sediment is derived from organisms (e.g., foraminifera) living in the ocean water (at various depths, but mostly near surface) that make their shells (a.k.a. tests) out of carbonate minerals such as calcite.
- Pelagic silica sediment is derived from marine organisms (e.g., diatoms and radiolaria) that make their tests out of silica (microcrystalline quartz).
- Volcanic ash and other volcanic materials are derived from both terrestrial and submarine eruptions.
- Iron and manganese nodules form as direct precipitates from ocean-bottom water.

The distributions of some of these materials around the seas are shown in Figure 9. Terrigenous sediments predominate near the continents and within inland seas and large lakes. These sediments tend to be relatively coarse, typically containing sand

and silt, but in some cases even pebbles and cobbles. Clay settles slowly in nearshore environments, but much of the clay is dispersed far from its source areas by ocean currents. Clay minerals are predominant over wide areas in the deepest parts of the ocean, and most of this clay is terrestrial in origin. Siliceous oozes (derived from radiolaria and diatoms) are common in the south polar region, along the equator in the Pacific, south of the Aleutian Islands, and within large parts of the Indian Ocean. Carbonate oozes are widely distributed in all of the oceans within equatorial and mid-latitude regions. In fact, clay settles everywhere in the oceans, but in areas where silica- and carbonate-producing organisms are prolific, they produce enough silica or carbonate sediment to dominate over clay.

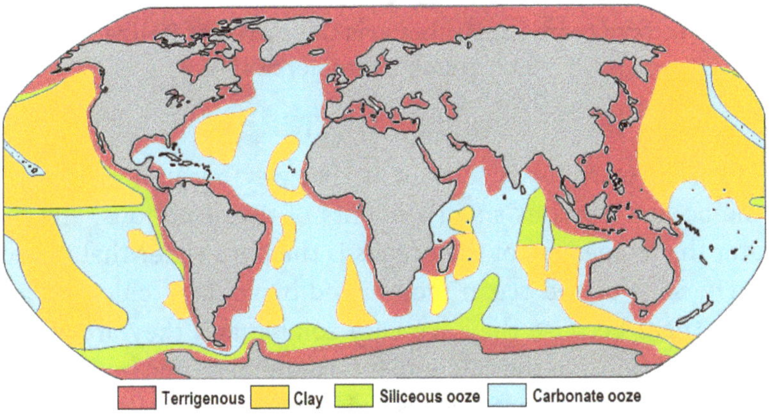

Figure 8: The distribution of sediment types on the sea floor. Within each coloured area, the type of material shown is what dominates, although other materials are also likely to be present.

Carbonate sediments are derived from a wide range of near-surface pelagic organisms that make their shells out of carbonate (Figure 9). These tiny shells, and the even tinier fragments that form when they break into pieces, settle slowly through the water column, but they don't necessarily make it to the bottom. While calcite is insoluble in surface water, its solubility increases with depth (and pressure) and at around 4,000 m, the carbonate fragments dissolve. This depth, which varies with latitude and water temperature, is known as the carbonate compensation depth, or CCD. As a result, carbonate oozes are absent from the deepest parts of the ocean (deeper than 4,000 m), but they are common in shallower areas such as the mid-Atlantic ridge, the East Pacific Rise (west of South America), along the trend of the Hawaiian/Emperor Seamounts (in the northern Pacific), and on the tops of many isolated seamounts.

Figure 9: Foraminifera from the Ambergris Caye area of Belize. Most of the shells are about 1 mm across.

18.4 SEAWATER

As everyone knows, seawater is salty. It is that way because the river water that flows into the oceans contains small amounts of dissolved ions, and for the most part, the water that comes out of the oceans is the pure water that evaporates from the surface. The salts of the ocean (dominated by sodium, chlorine, and sulphur) (Figure 10) are there because they are very soluble and they aren't consumed by biological processes (most of the calcium, for example, is used by organisms to make carbonate minerals). If salts are always going into the ocean, and never coming out, one might assume that the oceans have been continuously getting saltier over geological time. In fact this appears not to be the case. There is geological evidence that Earth's oceans became salty early during the Archaean, and that at times in the past, they have been at least half again as salty as they are now. This implies that there must be a mechanism to remove salt from the oceans, and that mechanism is the isolation of some parts of the ocean into seas (such as the Mediterranean) and the eventual evaporation of those seas to create salt beds that become part of the crust. The Middle Devonian Prairie Evaporite Formation of Saskatchewan and Manitoba is a good example of this.

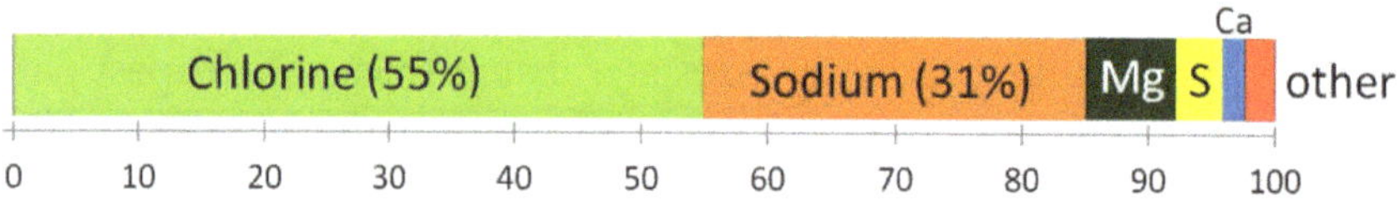

Figure 10: The proportions (by weight) of the major dissolved elements in ocean water

The average salinity of the oceans is 35 g of salt per litre of water, but there are significant regional variations in this value, as shown in Figure 11. Ocean water is least salty (around 31 g/L) in the Arctic, and also in several places where large rivers flow in (e.g., the Ganges/Brahmaputra and Mekong Rivers in southeast Asia, and the Yellow and Yangtze Rivers in China). Ocean water is most salty (over 37 g/L) in some restricted seas in hot dry regions, such as the Mediterranean and Red Seas. You might be surprised to know that, in spite of some massive rivers flowing into it (such as the Nile and the Danube), water does not flow out of the Mediterranean Sea into the Atlantic. There is so much evaporation happening in the Mediterranean basin that water flows into it from the Atlantic, through the Strait of Gibraltar.

In the open ocean, salinities are elevated at lower latitudes because this is where most evaporation takes place. The highest salinities are in the subtropical parts of the Atlantic, especially north of the equator. The northern Atlantic is much more saline than the north Pacific because the Gulf Stream current brings a massive amount of salty water from the tropical Atlantic and the Caribbean to the region around Britain, Iceland, and Scandinavia. The salinity in the Norwegian Sea (between Norway and Iceland) is substantially higher than that in other polar areas.

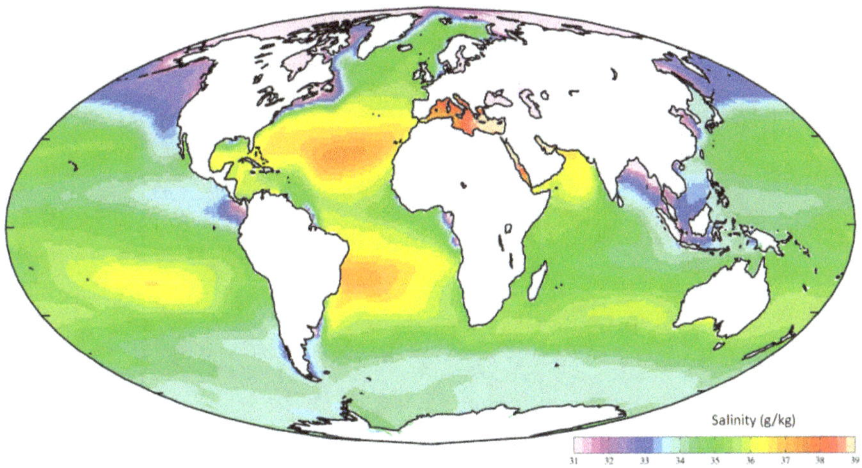

Figure 11: The distribution of salinity in Earth's oceans and major seas

AT A GLANCE

The Topography of the Sea Floor	The oceans are about 4,000 m deep on average, but they also have a wide range of topographical features, including shallow continental shelves, continental slopes, continuous ridges related to plate divergence, numerous isolated seamounts, and deep submarine canyons at subduction zones.
The Geology of the Oceanic Crust	Most oceanic crust forms during sea-floor spreading and is characterized by pillow basalts, sheeted dykes, gabbro bodies, layered gabbro, and layered ultramafic rock. The oldest parts of the sea floor are older than 200 Ma, but most of the sea floor is younger than 100 Ma. Seamounts are common and almost all are volcanoes, related to mantle plumes, subduction, or other processes. In tropical regions, ocean islands tend to be surrounded by carbonate reefs.

Sea-Floor Sediments	Almost all of the sea floor is covered by young sediments and sedimentary rocks, derived either from erosion of continents or from marine biological processes. Clastic sediments, some quite coarse, predominate on shelves and slopes. Terrigenous clays are distributed across the sea floor, but in areas where either carbonate- or silica-forming organisms thrive, the sediments are likely to be dominated by carbonate or silica oozes. Methane hydrates, derived from bacterial decomposition of organic matter, form within sediments on shelves and slopes.
Ocean Water	Average ocean water has about 35 g/L of salt, mostly made up of chlorine and sodium, but also including magnesium, sulphur, and calcium. Salinity levels are highest in the tropics where evaporation is greatest. Sea-surface temperatures range from less than 0°C at the poles to over 25°C in equatorial regions. Open-ocean currents, which generally rotate clockwise in the northern hemisphere and counter-clockwise in the south, are critically important in redistributing heat on Earth. Deep-ocean currents, driven by density differences, are another key part of the heat redistribution system. Changes to current patterns or intensity have significant implications for global climate.

CHAPTER 19
CLIMATE CHANGE AND ITS IMPACTS

If one thing has been constant about Earth's climate over geological time, it is its constant change. In the geological record, we can see this in the evidence of glaciations in the distant past and we can also detect periods of extreme warmth by looking at the isotope composition of sea-floor sediments, such as those in the core shown in Figure 1. Not only has the climate changed frequently, the temperature fluctuations have been very significant. Today's mean global temperature is about 15°C. During Snowball Earth times, the global mean was as cold as -50°C, while at various times during the Paleozoic and Mesozoic and during the Paleocene-Eocene thermal maximum, it was close to +30°C.

But in spite of these dramatic climate changes, Earth has been habitable from very early in its history — as soon as liquid water was present — right through to the present day. That continuous habitability is perhaps a little more surprising than you might think, as we'll see below.

Figure 1: Core from Ocean Drilling Program hole 1220b (southeast of Hawaii) showing the boundary between the Paleocene and the Eocene (at 55.8 Ma). Marine life was decimated during the 100,000 years of the Paleocene-Eocene thermal maximum, and the dark part of the core represents the absence of carbonate sediment from planktonic organisms. The scale is in centimetres.

A significant part of this chapter is about the natural processes of climate change and how they work. It's critically important to be aware of those natural climate change processes if we want to understand anthropogenic climate change. First, this awareness helps us to understand why our activities are causing the present-day climate to change, and second, it allows us to distinguish between natural and anthropogenic processes in the climate record of the past 250 years.

19.1 CLIMATE CHANGE

There are two parts to climate change, the first one is known as climate forcing, which is when conditions change to give the climate a little nudge in one direction or the other. The second part of climate change, and the one that typically does most of the work, is what we call a feedback. When a climate forcing changes the climate a little, a whole series of environmental changes take place, many of which either exaggerate the initial change (positive feedbacks), or suppress the change (negative feedbacks).

An example of a climate-forcing mechanism is the increase in the amount of carbon dioxide (CO_2) in the atmosphere that results from our use of fossil fuels. CO_2 traps heat in the atmosphere and leads to climate warming. Warming changes vegetation patterns; contributes to the melting of snow, ice, and permafrost; causes sea level to rise; reduces the solubility of CO_2 in sea water; and has a number of other minor effects. Most of these changes contribute to more warming. Melting of permafrost, for example, is a strong positive feedback because frozen soil contains trapped organic matter that is converted to CO_2 and methane (CH_4) when the soil thaws. Both these gases accumulate in the atmosphere and add to the warming effect. On the other hand, if warming causes more vegetation growth, that vegetation should absorb CO_2, thus reducing the warming effect, which would be a negative feedback. Under our current conditions — a planet that still has lots of glacial ice and permafrost — most of the feedbacks that result from a warming climate are positive feedbacks and so the climate changes that we cause get naturally amplified by natural processes.

Greenhouse gas

Throughout this chapter we'll be talking about the role of greenhouse gases (GHGs) in controlling the climate, so it's important to understand what greenhouse gases are and how they work. As you know, the dominant gases of the atmosphere are nitrogen (as N_2) and oxygen (as O_2). These gas molecules have only two atoms each and are not GHGs. Some of the other important gases of the atmosphere are water vapour (H_2O), carbon dioxide (CO_2), and methane (CH_4). All of these have more than two atoms, and they are GHGs

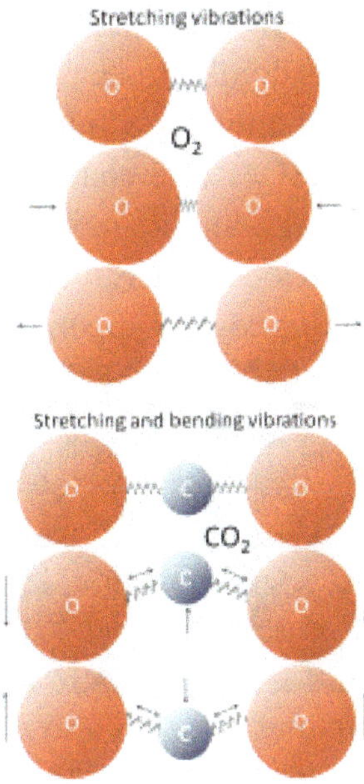

All molecules vibrate at various frequencies and in various ways, and some of those vibrations take place at frequencies within the range of the infrared (IR) radiation that is emitted by Earth's surface. Gases with two atoms, such as O_2, can only vibrate by stretching (back and forth), and those vibrations are much faster than the IR radiation. Gases with three or more atoms (such as CO_2) vibrate by stretching as well, but they can also vibrate in other ways, such as by bending. Those vibrations are slower and match IR radiation frequencies.

When IR radiation interacts with CO_2 or with one of the other GHGs, the molecular vibrations are enhanced because there is a match between the wavelength of the light and the vibrational frequency of the molecule. This makes the molecule vibrate more vigorously, heating the surrounding air in the process. These molecules also emit IR radiation in all directions, some of which reaches Earth's surface and causes the greenhouse effect.

Natural Climate Forcing

Natural climate forcing has been going on throughout geological time. A wide range of processes has been operating at widely different time scales, from a few years to billions of years.

The longest-term natural forcing variation is related to the evolution of the Sun. Like most other stars of a similar mass, our Sun is evolving. For the past 4.57 billion years, its rate of nuclear fusion has been increasing, and it is now emitting about 40% more energy (as light) than it did at the beginning of geological time (Figure 2). A difference of 40% is big, so it's a little surprising that the temperature on Earth has remained at a reasonable and habitable temperature for all of this time. The mechanism for that relative climate stability has been the evolution of our atmosphere from one that was dominated by CO_2, and also had significant levels of CH_4 — both GHGs — to one with only a few hundred parts per million of CO_2 and just under 1 part per million of CH_4. Those changes to our atmosphere have been no accident; over geological time, life and its metabolic processes have evolved and changed the atmosphere to conditions that remained cool enough to be habitable. A scientific explanation for how this could happen is known as the Gaia hypothesis.

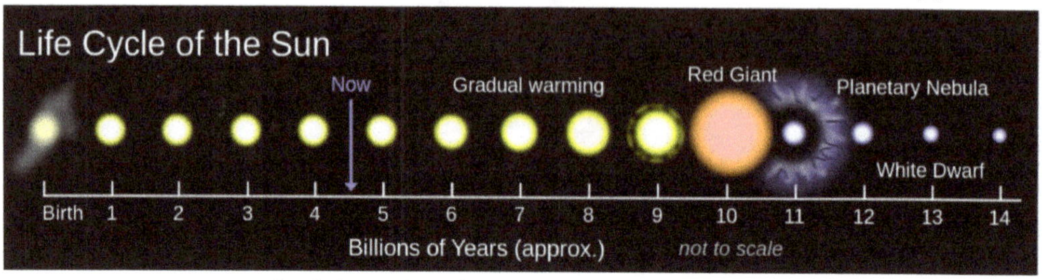

Figure 2: The life cycle of our Sun and of other similar stars

The Gaia hypothesis, developed by British scientist and environmentalist James Lovelock in the 1960s, is the theory that organisms evolve in ways that contribute to ensuring that their environment remains habitable. It does not include any sort of coordination of effort among organisms or any consciousness of a need to make changes. Gaia is not a superorganism. A way of understanding Gaia is through Lovelock's simple Daisyworld model. A planet with a warming star is populated only by two types of daisies, white ones and black ones. The black ones contribute to warming because they absorb solar energy, while the white ones reflect light and contribute to cooling. As the star's luminosity gradually increases, the white daisies have better outcomes because their reflectivity cools their local environment, while the black daisies, suffering from the heat, do not reproduce as well. Over time white daisies gradually dominate the population, but eventually the star becomes so bright that even white daisies cannot compensate, and all of the daisies perish. Obviously Earth is not Daisyworld, but similar processes — such as the evolution of photosynthetic bacteria that consume CO_2 — have taken place that influence the atmosphere and moderate the climate.

You can read more about Lovelock, Gaia, and Daisyworld by searching the Internet using any one of those terms. [SE]

Plate tectonic processes contribute to climate forcing in several different ways, and on time scales ranging from tens of millions to hundreds of millions of years. One mechanism is related to continental position. For example, we know that Gondwana (South America + Africa + Antarctica + Australia) was positioned over the South Pole between about 450 and 250 Ma, during which time there were two major glaciations (Andean-Saharan and Karoo) affecting the South polar regions (Figure 2) and cooling the rest of the planet at the same time. Another mechanism is related to continental collisions. As described in Chapter 16, the collision between India and Asia, which started at around 50 Ma, resulted in massive tectonic uplift. The consequent accelerated weathering of this rugged terrain consumed CO_2 from the atmosphere and contributed to gradual cooling over the remainder of the Cenozoic. Also, as described in Chapter 16, the opening of the Drake Passage — due to plate-tectonic separation of South America from Antarctica — led to the development of the Antarctic Circumpolar Current, which isolated Antarctica from the warmer water in the rest of the ocean and thus contributed to Antarctic glaciation starting at around 35 Ma.

Volcanic eruptions don't just involve lava flows and exploding rock fragments; various particulates and gases are also released, the important ones being sulphur dioxide and CO_2. Sulphur dioxide is an aerosol that reflects incoming solar radiation and has a net cooling effect that is short lived (a few years in most cases, as the particulates settle out of the atmosphere within a couple of years), and doesn't typically contribute to longer-term climate change. Volcanic CO_2 emissions can contribute to climate warming but only if a greater-than-average level of volcanism is sustained over a long time (at least tens of thousands of years). It is widely believed that the catastrophic end-Permian extinction (at 250 Ma) resulted from warming initiated by the eruption of the massive Siberian Traps over a period of at least a million years.

Climate Feedbacks

As already stated, climate feedbacks are critically important in amplifying weak climate forcings into full-blown climate changes. When Milankovitch published his theory in 1924, it was widely ignored, partly because it was evident to climate scientists that the forcing produced by the orbital variations was not strong enough to drive the significant climate changes of the glacial cycles. Those scientists did not recognize the power of positive feedbacks. It wasn't until 1973, 15 years after Milankovitch's death, that sufficiently high-resolution data were available to show that the Pleistocene glaciations were indeed driven by the orbital cycles, and it became evident that the orbital cycles were just the forcing that initiated a range of feedback mechanisms that made the climate change.

Since Earth still has a very large volume of ice — mostly in the continental ice sheets of Antarctica and Greenland, but also in alpine glaciers and permafrost — melting is one of the key feedback mechanisms. Melting of ice and snow leads to several different types of feedbacks, an important one being a change in albedo. Albedo is a measure of the reflectivity of a surface. Earth's various surfaces have widely differing albedos, expressed as the percentage of light that reflects off a given material. This is important because most solar energy that hits a very reflective surface is not absorbed and therefore does little to warm Earth. Water in the oceans or on a lake is one of the darkest surfaces, reflecting less than 10% of the incident light, while clouds and snow or ice are among the brightest surfaces, reflecting 70% to 90% of the incident light (Figure 3).

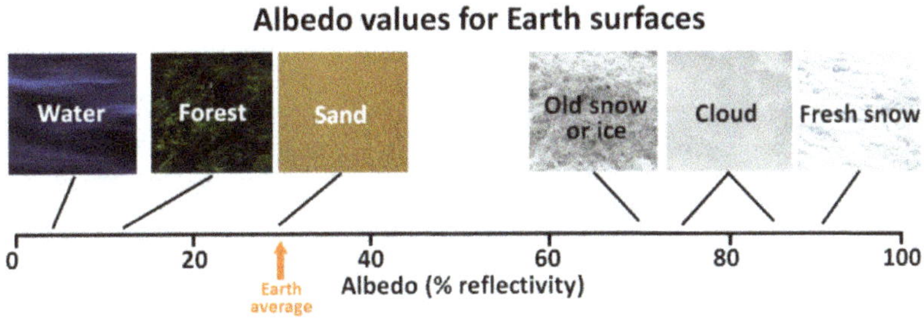

Figure 3: Typical albedo values for Earth surfaces [SE]

19.2 ANTHROPOGENIC CLIMATE CHANGE

When we talk about anthropogenic climate change, we are generally thinking of the industrial era, which really got going when we started using fossil fuels (coal to begin with) to drive machinery and trains. That was around the middle of the 18th century. The issue with fossil fuels is that they involve burning carbon that was naturally stored in the crust over hundreds of millions of years as part of Earth's process of counteracting the warming Sun.

Some climate scientists argue that anthropogenic climate change actually goes back much further than the industrial era, and that humans began to impact the climate by clearing land to grow grains in Europe and the Middle East around 8,000 years BCE and by creating wetlands to grow rice in Asia around 5,000 years BCE. Clearing forests for crops is a type of climate-forcing because the CO_2 storage capacity of the crops is generally lower than that of the trees they replace, and creating wetlands is a type of climate forcing because the anaerobic bacterial decay of organic matter within wetlands produces CH_4.

In fact, whether anthropogenic climate change started with the agricultural revolution or the industrial revolution is not important, because the really significant climate changes didn't start until the early part of the 20th century, and although our activities are a major part of the problem, our increasing numbers are a big issue as well. Figure 4 shows the growth of the world population from around 5 million, when we first started growing crops, to about 18 million when wetland rice cultivation began, to over 800 million at the start of the industrial revolution, to over 7,300 million today. A big part of the incredible growth in our population is related to the availability of the cheap and abundant energy embodied in fossil fuels, which we use for transportation, heating and cooling, industry, and food production. It will be hard to support a population of this size without fossil fuels, but we have to find a way to do it.

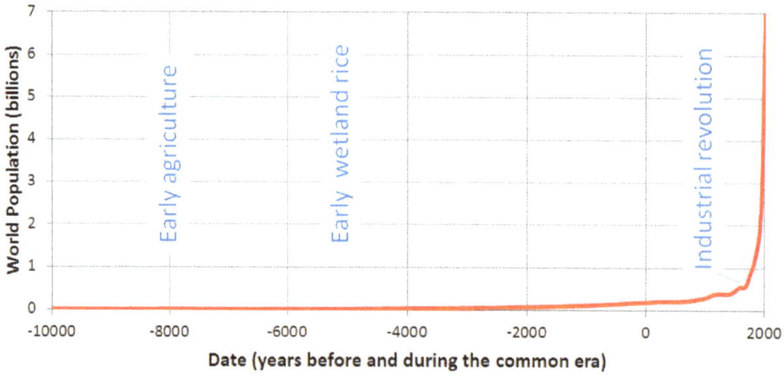

Figure 4: World population growth over the past 12,000 years

A rapidly rising population, the escalating level of industrialization and mechanization of our lives, and an increasing dependence on fossil fuels have driven the anthropogenic climate change of the past century. The trend of mean global temperatures since 1880 is shown in Figure 5. For approximately the past 55 years, the temperature has increased at a relatively steady and disturbingly rapid rate, especially compared to past changes. The average temperature now is approximately 0.8°C higher than before industrialization, and two-thirds of this warming has occurred since 1975. One of the driving factors of the recent increase in the rate of climate change has been the migration of North Americans from city centres to the suburbs, and the resulting need for virtually every household to own at least one car, when previously they were able to get around on foot or public transit.

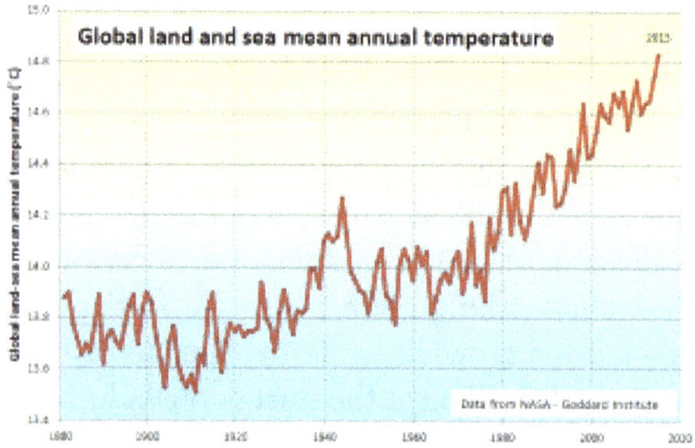

Figure 5: Global mean annual temperatures for the period from 1880 to 2015

The Intergovernmental Panel on Climate Change (IPCC), established by the United Nations in 1988, is responsible for reviewing the scientific literature on climate change and issuing periodic reports on several topics, including the scientific basis for understanding climate change, our vulnerability to observed and predicted climate changes, and what we can do to limit climate change and minimize its impacts. Figure 6, from the fifth report of the IPCC, issued in 2014, shows the relative contributions of various GHGs and other factors to current climate forcing, based on the changes from levels that existed in 1750. Figure 7 shows the IPCC's projections for temperature increases over the next 100 years.

The biggest anthropogenic contributor to warming is the emission of CO_2, which accounts for 50% of positive forcing. CH_4 and its atmospheric derivatives (CO_2, H_2O, and O_3) account for 29%, and the halocarbon gases (mostly leaked from air-conditioning appliances) and nitrous oxide (N_2O) (from burning fossils fuels) account for 5% each. Carbon monoxide (CO) (also produced by burning fossil fuels) accounts for 7%, and the volatile organic compounds other than methane (NMVOC) account for 3%.

CO_2 emissions come mostly from coal- and gas-fired power stations, motorized vehicles (cars, trucks, and aircraft), and industrial operations (e.g., smelting), and indirectly from forestry. CH_4 emissions come from production of fossil fuels (escape from coal mining and from gas and oil production), livestock farming (mostly beef), landfills, and wetland rice farming. N_2O and CO come mostly from the combustion of fossil fuels. In summary, close to 70% of our current GHG emissions come from fossil fuel production and use, while most of the rest comes from agriculture and landfills.

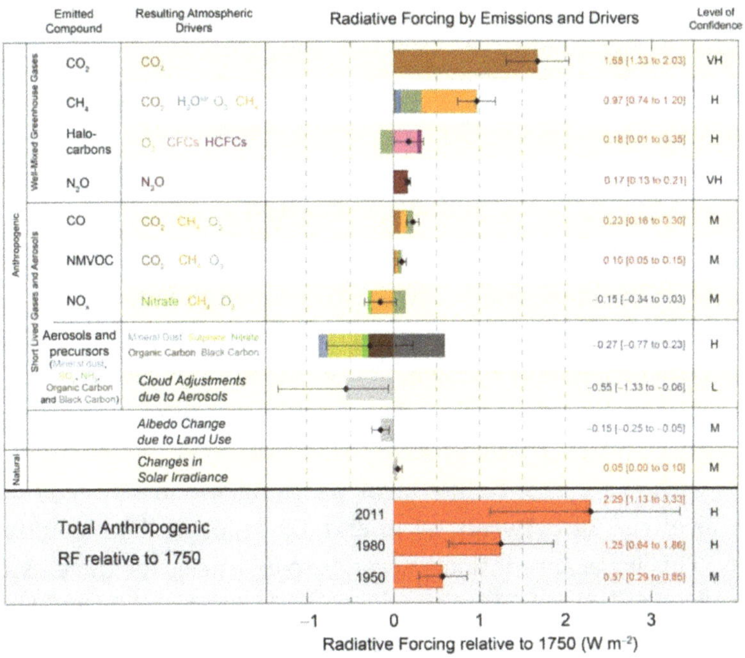

Figure 6: The relative importance of factors that are contributing to anthropogenic warming

19. 3 THE EFFECTS OF CLIMATE CHANGE

We've all experienced the effects of climate change over the past decade. However, it's not straightforward for climatologists to make the connection between a warming climate and specific weather events, and most are justifiably reluctant to ascribe any specific event to climate change. In this respect, the best measures of climate change are those that we can detect over several decades, such as the temperature changes shown in Figure 7, or the sea-level rise shown in Figure 13. As already stated, sea level has risen approximately 20 cm since 1750, and that rise is attributed to both warming (and therefore expanding) seawater and melting glaciers and other land-based snow and ice (melting of sea ice does not contribute directly to sea-level rise as it is already floating in the ocean).

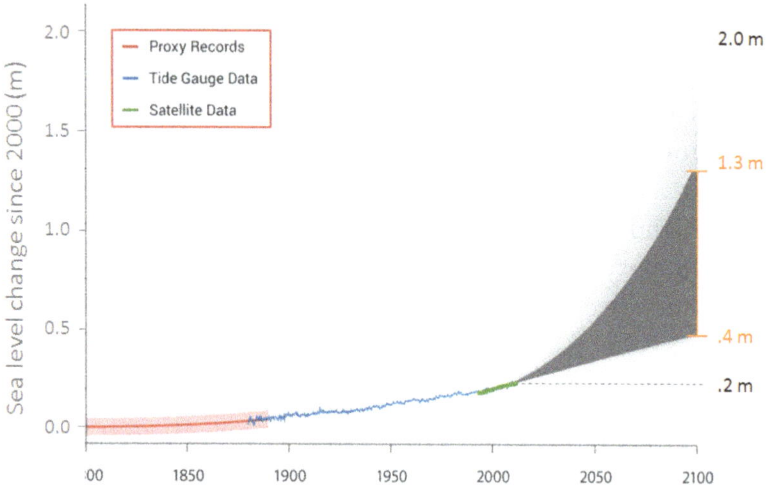

Figure 7: Projected sea-level increases to 2100, showing likely range (grey) and possible maximum

Projections for sea-level rise to the end of this century vary widely. This is in large part because we do not know which of the above climate change scenarios (Figure 12) we will most closely follow, but many are in the range from 0.5 m to 2.0 m. One of the problems in predicting sea-level rise is that we do not have a strong understanding of how large ice sheets, such as Greenland and Antarctica, will respond to future warming. Another issue is that the oceans don't respond immediately to warming. For example, with the current amount of warming, we are already committed to a future sea-level rise of between 1.3 m and 1.9 m, even if we could stop climate change today. This is because it takes decades to centuries for the existing warming of the atmosphere to be transmitted to depth within the oceans and to exert its full impact on large glaciers. Most of that committed rise would take place over the next century, but some would be delayed longer. And for every decade that the current rates of climate change continue, that number increases by another 0.3 m. In other words, if we don't make changes quickly, by the end of this century we'll be locked into 3 m of future sea-level rise.

In a 2008 report, the Organisation for Economic Co-operation and Development (OECD) estimated that by 2070 approximately 150 million people living in coastal areas could be at risk of flooding due to the combined effects of sea-level rise, increased storm intensity, and land subsidence. The assets at risk (buildings, roads, bridges, ports, etc.) are in the order of $35 trillion ($35,000,000,000,000). Countries with the greatest exposure of population to flooding are China, India, Bangladesh, Vietnam, U.S.A., Japan, and Thailand. Some of the major cities at risk include Shanghai, Guangzhou, Mumbai, Kolkata, Dhaka, Ho Chi Minh City, Tokyo, Miami, and New York.

One of the other risks for coastal populations, besides sea-level rise, is that climate warming is also associated with an increase in the intensity of tropical storms (e.g., hurricanes or typhoons), which almost always bring serious flooding from intense rain and storm surges. Some recent examples are New Orleans in 2005 with Hurricane Katrina, and New Jersey and New York in 2012 with Hurricane Sandy (Figure 8).

Figure 8: Damage to the Casino Pier, Seaside Heights, New Jersey, from Hurricane Sandy, November 2012

Tropical storms get their energy from the evaporation of warm seawater in tropical regions. In the Atlantic Ocean, this takes place between 8° and 20° N in the summer. Figure 9 shows the variations in the sea-surface temperature (SST) of the tropical Atlantic Ocean (in blue) versus the amount of power represented by Atlantic hurricanes between 1950 and 2008 (in red). Not only has the overall intensity of Atlantic hurricanes increased with the warming since 1975, but the correlation between hurricanes and sea-surface temperatures is very strong over that time period.

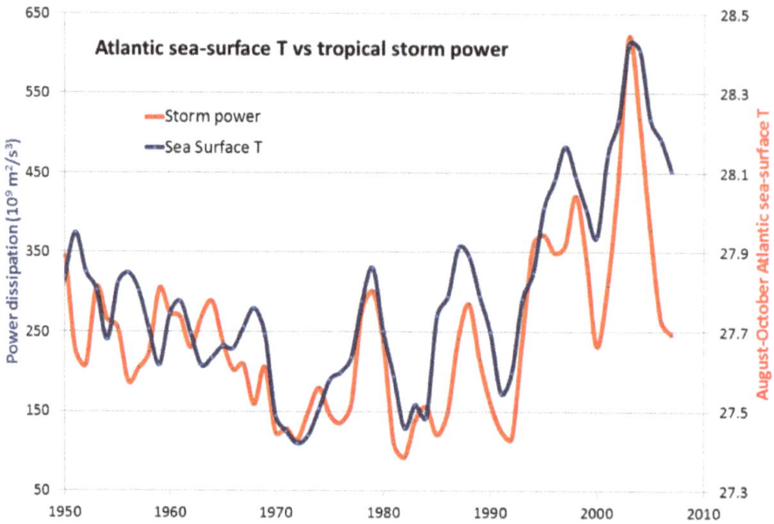

Figure 9: Relationship between Atlantic tropical storm cumulative annual intensity and Atlantic sea-surface temperatures

Because warm air is able to hold more water than cold air, the general global trend over the past century has been one of increasing precipitation (Figure 10).

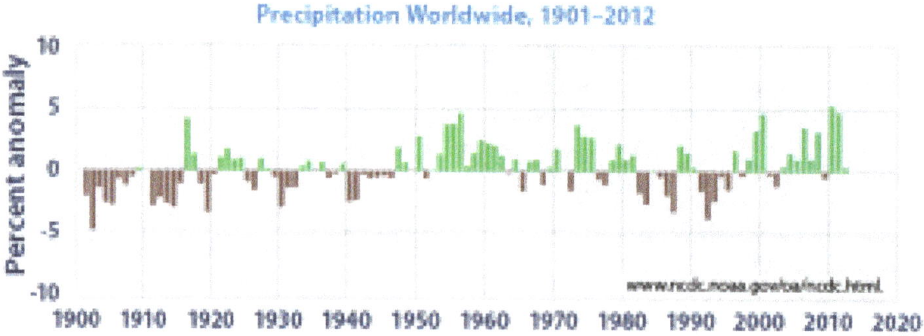

Figure 10: Global precipitation anomalies compared with the average over the period from 1901 to 2000

A similar trend is evident for British Columbia, based on weather data from 1945 to 2005 for 29 stations distributed around the province (Figure 11). Of those stations, 19 show an increase in precipitation and 10 show a decrease; although the decreases are all less than 12%, some of the increases are greater than 48%. Based on the data from these stations, it is estimated that approximately 60 mm/year more precipitation fell on British Columbia in 2005 compared with 1945. That is equivalent to about six months of the average flow of the Fraser River.

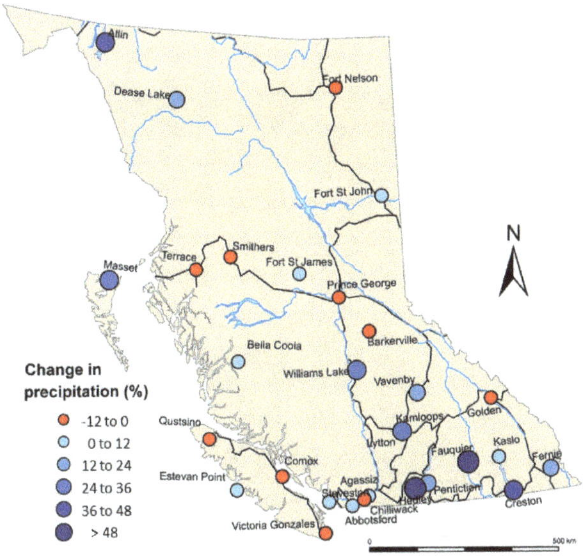

Figure 11: Change in precipitation amounts over the period 1945 to 2005 for 29 stations in British Columbia]

While the overall amount of precipitation (total volume of rain plus snow) increased at 19 out of 29 stations between 1945 and 2005, the amount of snowfall decreased at every single station. This is a disturbing trend for operators of winter resorts and hydroelectric dams, the Wildfire Management Branch, people who drink water from reservoirs that are replenished by snow, and people who eat food that is grown across western Canada and is irrigated with water derived from melting snow.

AT A GLANCE

What Makes the Climate Change?	The two components of climate change are forcings and feedbacks. Natural climate forcings, which have operated throughout geological time, include solar evolution and cycles, continental drift, continental collisions and mountain building, volcanism, orbital variations, and ocean current cycles. Feedbacks include melting of ice, snow, and permafrost (changing albedo and releasing GHGs); temperature-related changes to solubility of CO2; and vegetation growth.
Anthro-pogenic Climate Change	The key contributors to anthropogenic climate change are our use of fossil fuels and our increasing numbers, although other important factors include what we eat and how we produce it.
Implications of Climate Change	The most reliable indicators of climate change are those that we can detect by looking at records going back for decades. These include temperature and other climate parameters, of course, but also sea-level rise and the incidence of major storms. Some of the implications of climate change include changes to the distribution of disease vectors and pests, and an increase in the incidence and severity of heat waves.

CHAPTER

GEOLOGICAL RESOURCES 20

It has been said that "if you can't grow it, you have to mine it," meaning that anything we can't grow we have to extract from Earth in one way or another. This includes water, of course, our most important resource, but it also includes all the other materials that we need to construct things like roads, dams, and bridges, or manufacture things like plates, toasters, and telephones. Even most of our energy resources come from Earth, including uranium and fossil fuels, and much of the infrastructure of this electrical age depends on copper (Figure 1).

Virtually everything we use every day is made from resources from Earth. For example, let's look at a tablet computer (Figure 2). Most of the case is made of a plastic known as ABS, which is made from either gas or petroleum. Some tablets have a case made from aluminum. The glass of a touch screen is made mostly from quartz combined with smaller amounts of sodium oxide (Na_2O), sodium carbonate (Na_2CO_3), and calcium oxide (CaO). To make it work as a touch screen, the upper surface is coated with indium tin oxide. When you touch the screen you're actually pushing a thin layer of polycarbonate plastic (made from petroleum) against the coated glass — completing an electrical circuit. The computer is then able to figure out exactly where you touched the screen. Computer processors are made from silica wafers (more quartz) and also include a significant amount of copper and gold. Gold is used because it is a better conductor than copper and doesn't tarnish the way silver or copper does. Most computers have nickel-metal-hydride (NiMH) batteries, which contain nickel, of course, along with cadmium, cobalt, manganese, aluminum, and the rare-earth elements lanthanum, cerium, neodymium, and praseodymium. The processor and other electronic components are secured to a circuit board, which is a thin layer of fibreglass sandwiched between copper sheets coated with small amounts of tin and lead. Various parts are put together with steel screws that are made of iron and molybdenum.

Figure 1: The open pit (background) and waste-rock piles (middle) of the Highland Valley Copper Mine at Logan Lake, British Columbia

Figure 2: The main components of a tablet computer

That's not everything that goes into a tablet computer, but to make just those components we need a pure-silica sand deposit, a salt mine for sodium, a rock quarry for calcium, an oil well, a gas well, an aluminum mine, an iron mine, a manganese mine, a copper-molybdenum-gold mine, a cobalt-nickel mine, a rare-earth element and indium mine, and a source of energy to transport all of the materials, process them, put them together, and finally transport the computer to your house or the store where you bought it.

Exercise 1 Where Does It Come From?

Look around you and find at least five objects (other than a computer or a phone) that have been made from materials that had to be mined, quarried, or extracted from an oil or gas well. Try to identify the materials involved, and think about where they might have come from. This pen is just an example.

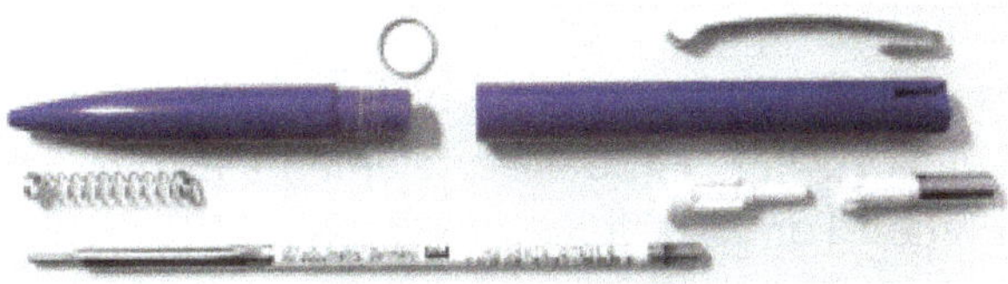

[https://upload.wikimedia.org/wikipedia/commons/f/fd/Ballpoint-pen-parts.jpg]

20.1 METAL DEPOSITS

Mining has always been a major part of Canada's economy. Canada has some of the largest mining districts and deposits in the world, and for the past 150 years, we have been one of the most important suppliers of metals. Extraction of Earth's resources goes back a long way in Canada. For example, the First Nations of British Columbia extracted obsidian from volcanic regions for tools and traded it up and down the coast. In the 1850s, gold was discovered in central British Columbia, and in the 1890s, even more gold was discovered in the Klondike area of Yukon. These two events were critical to the early development of British Columbia, Yukon, and Alaska.

Canada's mining sector had revenues in the order of $37 billion in 2013. The majority of that was split roughly equally among gold, iron, copper, and potash, with important but lesser amounts from nickel and diamonds (Figure 3). Revenues from the petroleum sector are significantly higher, at over $100 billion per year.

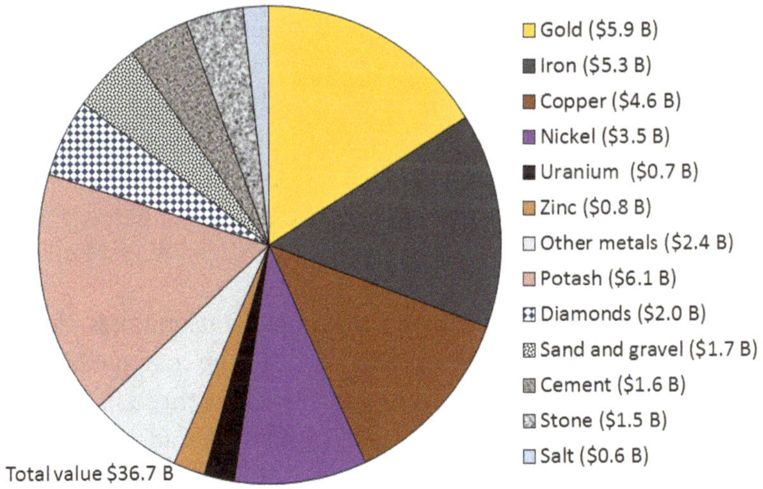

Figure 3: The value of various Canadian mining sectors in 2013

A metal deposit is a body of rock in which one or more metals have been concentrated to the point of being economically viable for recovery. Some background levels of important metals in average rocks are shown on Table 1, along with the typical grades necessary to make a viable deposit, and the corresponding concentration factors. Looking at copper, for example, we can see that while average rock has around 40 ppm (parts per million) of copper, a grade of around 10,000 ppm or 1% is necessary to make a viable copper deposit. In other words, copper ore has about 250 times as much copper as typical rock. For the other elements in the list, the concentration factors are much higher. For gold, it's 2,000 times and for silver it's around 10,000 times.

Table 1: Typical background and ore levels of some important metals

Metal	Typical Background Level	Typical Economic Grade*	Concentration Factor
Copper	40 ppm	10,000 ppm (1%)	250 times
Gold	0.003 ppm	6 ppm (0.006%)	2,000 times
Lead	10 ppm	50,000 ppm (5%	5,000 times
Molybdenum	1 ppm	1,000 ppm (0.1%)	1,000 times
Nickel	25 ppm	20,000 ppm (2%)	800 times
Silver	0.1 ppm	1,000 ppm (0.1%)	10,000 times
Uranium	2 ppm	10,000 ppm (1%)	5,000 times
Zinc	50 ppm	50,000 ppm (5%)	1,000 times

*It's important to note that the economic viability of any deposit depends on a wide range of factors including its grade, size, shape, depth below the surface, and proximity to infrastructure, the current price of the metal, the labour and environmental regulations in the area, and many other factors.

It is clear that some very significant concentration must take place to form a mineable deposit. This concentration may occur during the formation of the host rock, or after the rock forms, through a number of different types of processes. There is a very wide variety of ore-forming processes, and there are hundreds of types of mineral deposits. The origins of a few of them are described below.

Magmatic Nickel Deposits

A magmatic deposit is one in which the metal concentration takes place primarily at the same time as the formation and emplacement of the magma. Most of the nickel mined in Canada comes from magmatic deposits such as those in Sudbury (Ontario), Thompson (Manitoba) (Figure 4), and Voisey's Bay (Labrador). The magmas from which these deposits form are of mafic or ultramafic composition (derived from the mantle), and therefore they have relatively high nickel and copper contents to begin with (as much as 100 times more than normal rocks in the case of nickel). These elements may be further concentrated within the magma as a result of the addition of sulphur from partial melting of the surrounding rocks. The heavy nickel and copper sulphide minerals are then concentrated further still by gravity segregation (i.e., crystals settling toward the bottom of the magma chamber). In some cases, there are significant concentrations of platinum-bearing minerals.

Most of these types of deposits around the world are Precambrian in age – probably because the mantle was significantly hotter at that time, and the necessary mafic and ultramafic magmas were more likely to be emplaced in the continental crust.

Figure 4: The nickel smelter at Thompson, Manitoba

Volcanogenic Massive Sulphide Deposits

Much of the copper, zinc, lead, silver, and gold mined in Canada is mined from volcanic-hosted massive sulphide (VHMS) deposits associated with submarine volcanism (VMS deposits). Examples are the deposits at Kidd Creek, Ontario, Flin Flon on the Manitoba-Saskatchewan border, Britannia on Howe Sound, and Myra Falls (within Strathcona Park) on Vancouver Island.

VMS deposits are formed from the water discharged at high temperature (250° to 300°C) at ocean-floor hydrothermal vents, primarily in areas of subduction-zone volcanism. The environment is comparable to that of modern-day black smokers (Figure 5), which form where hot metal- and sulphide-rich water issues from the sea floor. They are called massive sulphide deposits because the sulphide minerals (including pyrite (FeS2) , sphalerite (ZnS), chalcopyrite ($CuFeS_2$), and galena (PbS)) are generally present in very high concentrations (making up the majority of the rock in some cases). The metals and the sulphur are leached out of the sea-floor rocks by convecting groundwater driven by the volcanic heat, and then quickly precipitated where that hot water enters the cold seawater, causing it to cool suddenly and change chemically. The volcanic rock that hosts the deposits is formed in the same area and at the same general time as the accumulation of the ore minerals.

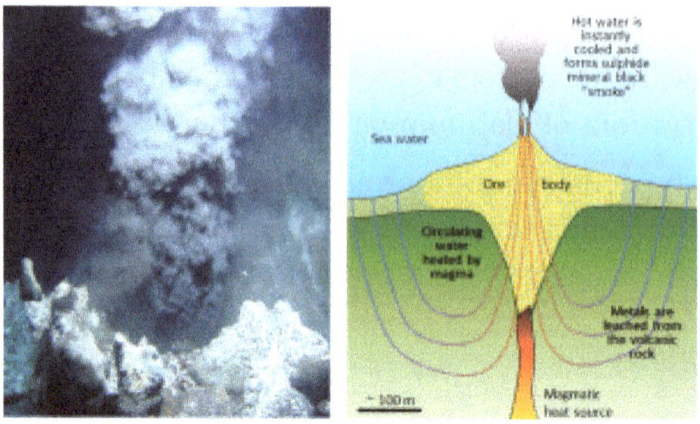

Figure 5: Left: A black smoker on the Juan de Fuca Ridge off the west coast of Vancouver Island. Right: A model of the formation of a volcanogenic massive sulphide deposit on the sea floor.

Porphyry Deposits

Porphyry deposits are the most important source of copper and molybdenum in British Columbia, the western United States, and Central and South America. Most porphyry deposits also host some gold, which may be, in rare cases, the primary commodity. B.C. examples include several large deposits within the Highland Valley mine (Figure 1) and numerous other deposits scattered around the central part of the province.

A porphyry deposit forms around a cooling felsicstock in the upper part of the crust. They are called "porphyry" because upper crustal stocks are typically porphyritic in texture, the result of a two-stage cooling process. Metal enrichment results in part from convection of groundwater related to the heat of the stock, and also from metal-rich hot water expelled by the cooling magma (Figure 6). The host rocks, which commonly include the stock itself and the surrounding country rocks, are normally highly fractured and brecciated. During the ore-forming process, some of the original minerals in these rocks are altered to potassium feldspar, biotite, epidote, and various clay minerals. The important ore minerals include chalcopyrite ($CuFeS_2$), bornite (Cu_5FeS_4), and pyrite in copper porphyry deposits, or molybdenite (MoS_2) and pyrite in molybdenum porphyry deposits. Gold is present as minute flakes of native gold.

This type of environment (i.e., around and above an intrusive body) is also favourable for the formation of other types of deposits — particularly vein-type gold deposits (a.k.a. epithermal deposits). Some of the gold deposits of British Columbia (such as in the Eskay Creek area adjacent to the Alaska panhandle), and many of the other gold deposits situated along the western edge of both South and North America are of the vein type shown in Figure 6, and are related to nearby magma bodies.

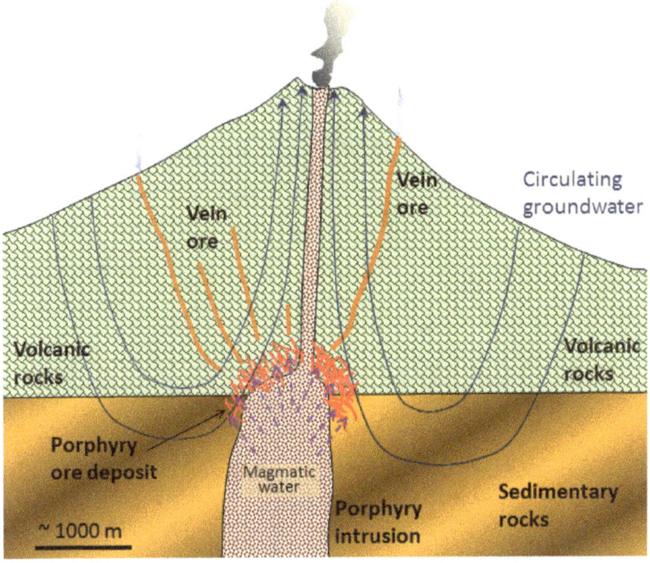

Figure 6: A model for the formation of a porphyry deposit around an upper-crustal porphyritic stock and associated vein deposits.

Banded Iron Formation

Most of the world's major iron deposits are of the banded iron formation type, and most of these formed during the initial oxygenation of Earth's atmosphere between 2,400 and 1,800 Ma. At that time, iron that was present in dissolved form in the ocean (as Fe^{2+}) became oxidized to its insoluble form (Fe^{3+}) and accumulated on the sea floor, mostly as hematite interbedded with chert (Figure 7). Unlike many other metals, which are economically viable at grades of around 1% or even much less, iron deposits are only viable if the grades are in the order of 50% iron and if they are very large.

Figure 7: Banded iron formation from an unknown location in North America on display at a museum in Germany. The rock is about 2 m across. The dark

grey layers are magnetite and the red layers are hematite. Chert is also present. [https://upload.wikimedia.org/wikipedia/commons/5/5f/Black-band_ironstone_%28aka%29.jpg]

Unconformity-Type Uranium Deposits

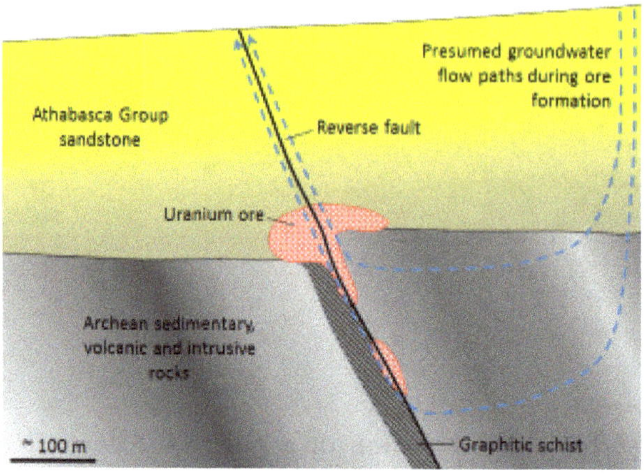

Figure 8: Model of the formation of unconformity-type uranium deposits of the Athabasca Basin, Saskatchewan

There are several different types of uranium deposits, but some of the largest and richest are those within the Athabasca Basin of northern Saskatchewan. These are called unconformity-type uranium deposits because they are all situated very close to the unconformity between the Proterozoic Athabasca Group sandstone and the much older Archean sedimentary, volcanic, and intrusive igneous rock (Figure 8). The origin of unconformity-type U deposits is not perfectly understood, but it is thought that two particular features are important: (1) the relative permeability of the Athabasca Group sandstone, and (2) the presence of graphitic schist within the underlying Archean rocks. The permeability of the sandstone allowed groundwater to flow through it and leach out small amounts of U, which stayed in solution in the oxidized form U6+. The graphite (C) created a reducing environment (non-oxidizing) that converted the U from U6+ to insoluble U4+, at which point it was precipitated as the mineral uraninite (UO2).

Mining and Mineral Processing

Metal deposits are mined in a variety of different ways depending on their depth, shape, size, and grade. Relatively large deposits that are quite close to the surface and somewhat regular in shape are mined using open-pit mine methods (Figure 1). Creating a giant hole in the ground is generally cheaper than making an underground mine, but it is also less precise, so it is necessary to mine a lot of waste rock along with the ore. Relatively deep deposits or those with elongated or irregular shapes are typically mined from underground with deep vertical shafts, declines (sloped tunnels), and levels (horizontal tunnels) (Figures 9 and 11). In this way, it is possible to focus the mining on the ore body itself. However, with relatively large ore bodies, it may be necessary to leave some pillars to hold up the roof.

Figure 9: Underground at the Myra Falls Mine, Vancouver Island.

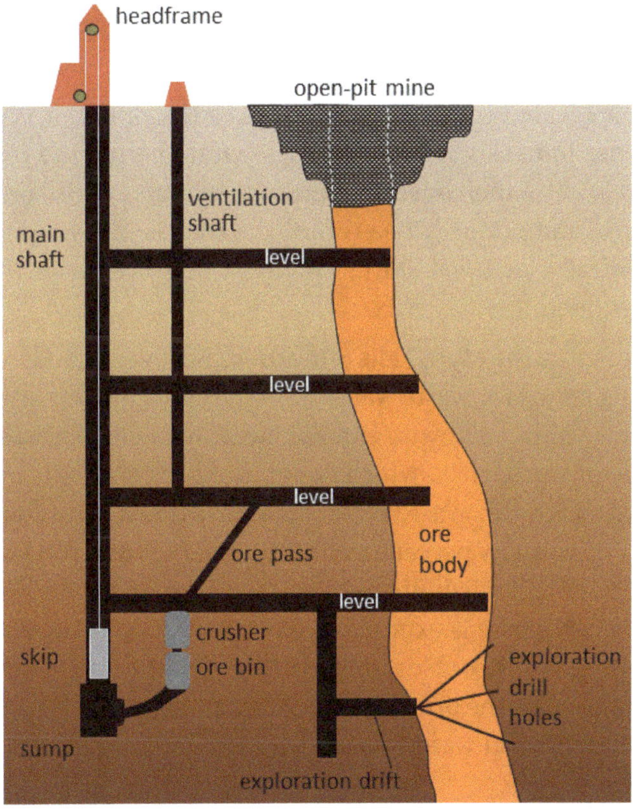

Figure 10: Schematic cross-section of a typical underground mine.

In many cases, the near-surface part of an ore body is mined with an open pit, while the deeper parts are mined underground (Figures 10 and 11).

Figure 11: Entrance to an exploratory decline (white arrow) for the New Afton Mine situated in the side of the open pit of the old Afton Mine, near Kamloops, B.C.

A typical metal deposit might contain a few percent of ore minerals (e.g., chalcopyrite or sphalerite), mixed with the minerals of the original rock (e.g., quartz or feldspar). Other sulphide minerals are commonly present within the ore, especially pyrite.

When ore is processed (typically very close to the mine), it is ground to a fine powder and the ore minerals are physically separated from the rest of the rock to make a concentrate. At a molybdenum mine, for example, this concentrate may be almost pure molybdenite (MoS_2). The rest of the rock is known as tailings. It comes out of the concentrator as a wet slurry and must be stored near the mine, in most cases, in a tailings pond.

The tailings pond at the Myra Falls Mine on Vancouver Island is shown in Figure 12, and the settling ponds for waste water from the concentrator are shown in Figure 13. The tailings are contained by an embankment. Also visible in the foreground of Figure 12 is a pile of waste rock, which is non-ore rock that was mined in order to access the ore. Although this waste rock contains little or no ore minerals, at many mines it contains up to a few percent pyrite. The tailings and the waste rock at most mines are an environmental liability because they contain pyrite plus small amounts of ore minerals. When pyrite is exposed to oxygen and water, it generates sulphuric acid — also known as acid rock drainage (ARD). Acidity itself is a problem to the environment, but because the ore elements, such as copper or lead, are more soluble in acidic water than neutral water, ARD is also typically quite rich in metals, many of which are toxic.

Figure 12: The tailings pond at the Myra Falls Mine on Vancouver Island. The dry rock in the middle of the image is waste rock. The structure on the right is the headframe for the mine shaft. Myra Creek flows between the tailings pond and the headframe.

Figure 13: The tailings pond (lower left) at Myra Falls Mine with settling ponds (right) for processing water from the concentrator.

Tailings ponds and waste-rock storage piles must be carefully maintained to ensure their integrity and monitored to ensure that acidic and metal-rich water is not leaking out. In August 2014, the tailings pond at the Mt. Polley Mine in central B.C. failed and 10 million cubic metres of waste water along with 4.5 million cubic metres of tailings slurry was released into Polley Lake, Hazeltine Creek, and Quesnel Lake (Figure 14, a and b). As of July 2015, the environmental implications of this event are still not fully understood.

Figure 14a: The Mt. Polley Mine area prior to the dam breach of August 2014. The tailings were stored in the area labelled "retention basin."

Figure 14b: The Mt. Polley Mine area after the tailings dam breach of August 2014. The water and tailings released flowed into Hazeltine Creek, and Polley and Quesnel Lakes.

Most mines have concentrators on site because it is relatively simple to separate ore minerals from non-ore minerals and thus significantly reduce the costs and other implications of transportation. But separation of ore minerals is only the preliminary stage of metal refinement, for most metals the second stage involves separating the actual elements within the ore minerals. For example, the most common ore of copper is chalcopyrite ($CuFeS_2$). The copper needs to be separated from the iron and sulphur to make copper metal and that involves complicated and very energy-intensive processes that are done at smelters or other types of refineries. Because of their cost and the economies of scale, there are far fewer refineries than there are mines.

There are several metal refineries (including smelters) in Canada; some examples are the aluminum refinery in Kitimat, B.C. (which uses ore from overseas); the lead-zinc smelter in Trail, B.C.; the nickel smelter at Thompson, Manitoba; numerous steel smelters in Ontario, along with several other refining operations for nickel, copper, zinc, and uranium; aluminum refineries in Quebec; and a lead smelter in New Brunswick.

20.2 INDUSTRIAL MINERALS

Metals are critical for our technological age, but there are a lot of other not-so-shiny materials that are needed to facilitate our way of life. For everything made out of concrete or asphalt, we need sand and gravel. To make the cement that holds concrete together, we also need limestone. For the glass in our computer screens and for glass-sided buildings, we need silica sand plus sodium oxide (Na_2O), sodium carbonate (Na_2CO_3), and calcium oxide (CaO). Potassium is an essential nutrient for farming in many areas, and for a wide range of applications (e.g., ceramics and many industrial processes), we also need various types of clay.

The best types of aggregate (sand and gravel) resources are those that have been sorted by streams, and in Canada the most abundant and accessible fluvial deposits are associated with glaciation. That doesn't include till of course, because it has too much silt and clay, but it does include glaciofluvial outwash, which is present in thick deposits in many parts of the country, similar to the one shown in Figure 15. In a typical gravel pit, these materials are graded on-site according to size and then used in a wide range of applications from constructing huge concrete dams to filling children's sandboxes. Sand is also used to make glass, but for most types of glass, it has to be at least 95% quartz (which the sandy layers shown in Figure 15 are definitely not), and for high-purity glass and the silicon wafers used for electronics, the source sand has to be over 98% quartz.

Figure 15: Sand and gravel in an aggregate pit near Nanaimo, BC.

Approximately 80 million tonnes of concrete are used in Canada each year – a little over 2 tonnes per person. The cement used for concrete is made from approximately 80% calcite ($CaCO_3$) and 20% clay. This mixture is heated to 1450°C to produce the required calcium silicate compounds (e.g., Ca_2SiO_4). The calcite typically comes from limestone quarries like the one on Texada Island, B.C. (Figure 16). Limestone is also used as the source material for many other products that require calcium compounds, including steel and glass, pulp and paper, and plaster products for construction.

Figure 16: Triassic Quatsino Formation limestone being quarried on Texada Island, B.C.

Sodium is required for a wide range of industrial processes, and the most convenient source is sodium chloride (rock salt), which is mined from evaporite beds in various parts of Canada. The largest salt mine in the world is at Goderich, Ontario, where salt is recovered from the 100 m thick Silurian Salina Formation. The same formation is mined in the Windsor area. Rock salt is also used as a source of sodium and chlorine in the chemical industry to melt ice on roads, as part of the process of softening water, and as a seasoning. Under certain conditions, the mineral sylvite (KCl) accumulates in evaporite beds, and this rock is called potash. This happened across the Canadian prairies during the Devonian, creating the Prairie evaporite formation (Figure 16). Potassium is used as a crop fertilizer, and Canada is the world's leading supplier, with most of that production coming from Saskatchewan.

Another evaporite mineral, gypsum ($CaS0_4.2H_20$), is the main component of plasterboard (drywall) that is widely used in the construction industry. One of the main mining areas for gypsum in Canada is in the Milford Station area of Nova Scotia, site of the world's largest gypsum mine.

Rocks are quarried or mined for many different uses, such as building facades (Figure 17), countertops, stone floors, and headstones. In most of these cases, the favoured rock types are granitic rocks, slate, and marble. Quarried rock is also used in some applications where rounded gravel isn't suitable, such as the ballast (road bed) for railways, where crushed angular rock is needed.

Figure 17: Slate used as a facing material on a concrete building column in Vancouver

20.3 FOSSIL FUELS

There are numerous types of fossil fuels, but all of them involve the storage of organic matter in sediments or sedimentary rocks. Fossil fuels are rich in carbon and almost all of that carbon ultimately originates from CO_2 taken out of the atmosphere during photosynthesis. That process, driven by solar energy, involves reduction (the opposite of oxidation) of the carbon, resulting in it being combined with hydrogen instead of oxygen. The resulting organic matter is made up of complex and varied carbohydrate molecules.

Most organic matter is oxidized back to CO_2 relatively quickly (within weeks or years in most cases), but any of it that gets isolated from the oxygen of the atmosphere (for example, deep in the ocean or in a stagnant bog) may last long enough to be buried by sediments and, if so, may be preserved for tens to hundreds of millions of years. Under natural conditions, that means it will be stored until those rocks are eventually exposed at the surface and weathered.

In this section, we'll discuss the origins and extraction of the important fossils fuels, including coal, oil, and gas. Coal, the first fossil fuel to be widely used, forms mostly on land in swampy areas adjacent to rivers and deltas in areas with humid tropical to temperate climates. The vigorous growth of vegetation leads to an abundance of organic matter that accumulates within stagnant water, and thus does not decay and oxidize. This situation, where the dead organic matter is submerged in oxygen-poor water, must be maintained for centuries to millennia in order for enough material to accumulate to form a thick layer (Figure 18a). At some point, the swamp deposit is covered with more sediment — typically because a river changes its course or sea level rises (Figure 18b). As more sediments are added, the organic matter starts to become compressed and heated. Low-grade lignite coal forms at depths between a few 100 m and 1,500 m and temperatures up to about 50°C (Figure 18c). At between 1,000 m to 5,000 m depth and temperatures up to 150°C m, bituminous coal forms (Figure 18d). At depths beyond 5,000 m and temperatures over 150°C, anthracite coal forms.

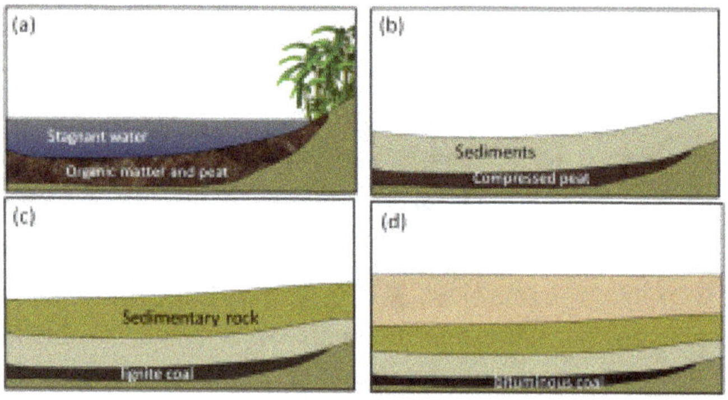

Figure 18 Formation of coal: (a) accumulation of organic matter within a swampy area; (b) the organic matter is covered and compressed by deposition of a new layer of clastic sediments; (c) with greater burial, lignite coal forms; and (d) at even greater depths, bituminous and eventually anthracite coal form.

There are significant coal deposits in many parts of Canada, including the Maritimes, Ontario, Saskatchewan, Alberta, and British Columbia. In Alberta and Saskatchewan, much of the coal is used for electricity generation. Coal from the Highvale Mine (Figure 19), Canada's largest, is used to feed the Sundance and Keephills power stations west of Edmonton. Almost all of the coal mined in British Columbia is exported for use in manufacturing steel.

Figure 19: The Highvale Mine (background) and the Sundance (right) and Keephills (left) generating stations on the southern shore of Wabamun Lake, Alberta

While almost all coal forms on land from terrestrial vegetation, most oil and gas is derived primarily from marine micro-organisms that accumulate within sea-floor sediments. In areas where marine productivity is high, dead organic matter is delivered to the sea floor fast enough that some of it escapes oxidation. This material accumulates in the muddy sediments, which become buried to significant depth beneath other sediments.

As the depth of burial increases, so does the temperature — due to the geothermal gradient — and gradually the organic matter within the sediments is converted to hydrocarbons (Figure 20). The first stage is the biological production (involving anaerobic bacteria) of methane. Most of this escapes back to the surface, but some is trapped in methane hydrates near the sea floor. At depths beyond about 2 km, and at temperatures ranging from 60° to 120°C, the organic matter is converted by chemical processes to oil. This depth and temperature range is known as the oil window. Beyond 120°C most of the organic matter is chemically converted to methane.

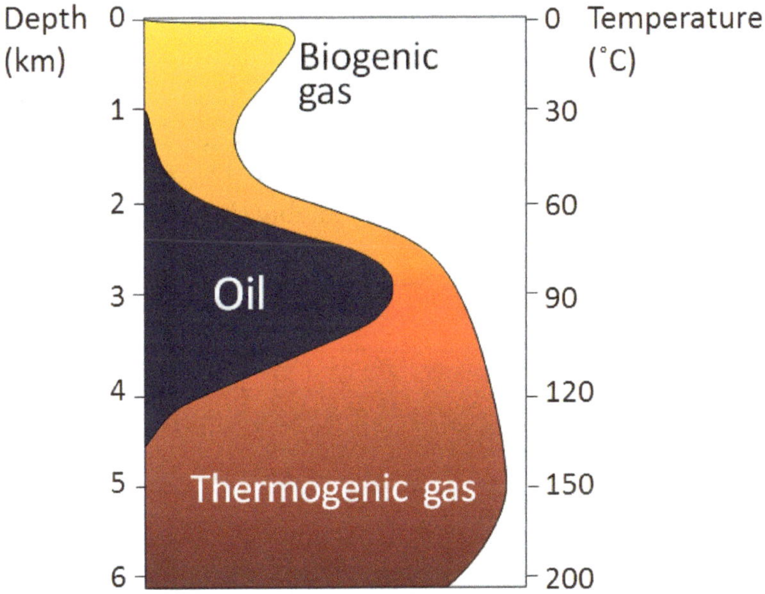

Figure 20: The depth and temperature limits for biogenic gas, oil, and thermogenic gas

The organic matter-bearing rock within which the formation of gas and oil takes place is known to petroleum geologists as the source rock. Both liquid oil and gaseous methane are lighter than water, so as liquids and gases form, they tend to move slowly toward the surface, out of the source rock and into reservoir rocks. Reservoir rocks are typically relatively permeable because that allows migration of the fluids from the source rocks, and also facilitates recovery of the oil or gas. In some cases, the liquids and gases make it all the way to the surface, where they are oxidized, and the carbon is returned to the atmosphere. But in other cases, they are contained by overlying impermeable rocks (e.g., mudrock) in situations where anticlines, faults, stratigraphy changes, and reefs or salt domes create traps (Figure 21).

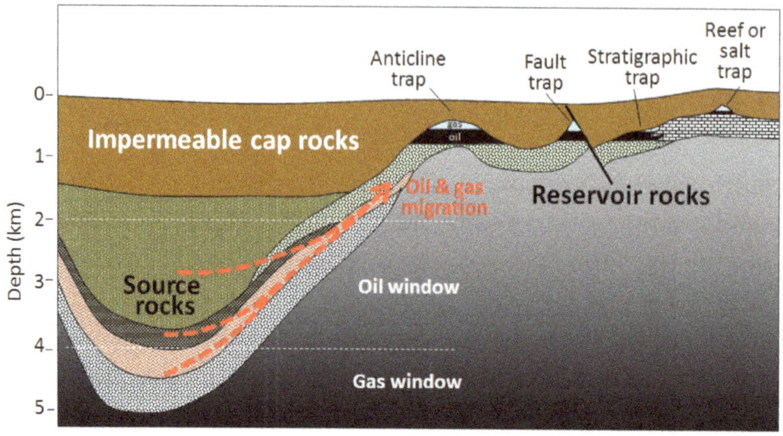

Figure 21: Migration of oil and gas from source rocks into traps in reservoir rocks

The liquids and gases that are trapped within reservoirs become separated into layers based on their density, with gas rising to the top, oil below it, and water underneath the oil. The proportions of oil and gas depend primarily on the temperature in the source rocks. Some petroleum fields, such as many of those in Alberta, are dominated by oil, while others, notably those in northeastern B.C., are dominated by gas.

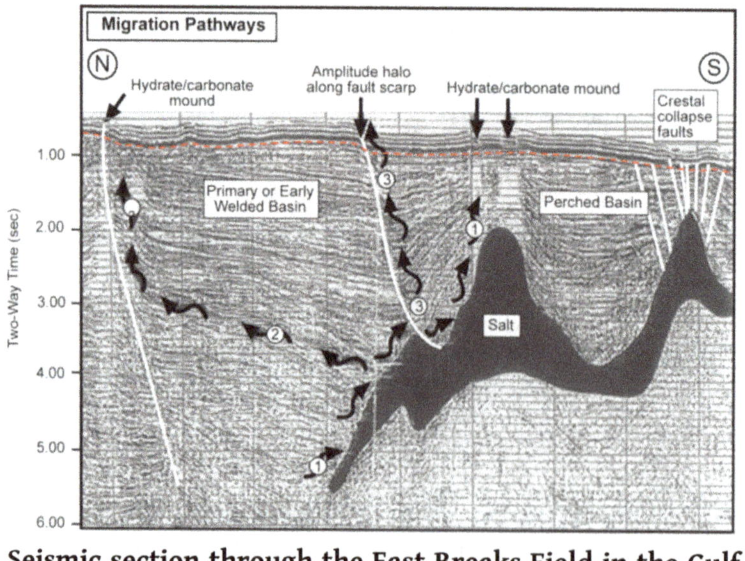

Figure 22: Seismic section through the East Breaks Field in the Gulf of Mexico. The dashed red line marks the approximate boundary between deformed rocks and younger undeformed rocks. The wiggly arrows are interpreted migration paths. The total thickness of this section is approximately 5 km.

In general, petroleum fields are not visible from the surface, and their discovery involves the search for structures in the subsurface that have the potential to form traps. Seismic surveys are the most commonly used tool for early-stage petroleum exploration, as they can reveal important information about the stratigraphy and structural geology of subsurface sedimentary rocks. An example from the Gulf of Mexico south of Texas is shown in Figure 22. In this area, a thick evaporite deposit ("salt") has formed domes because salt is lighter than other sediments and tends to rise slowly toward the surface; this has created traps. The sequence of deformed rocks is capped with a layer of undeformed rock.

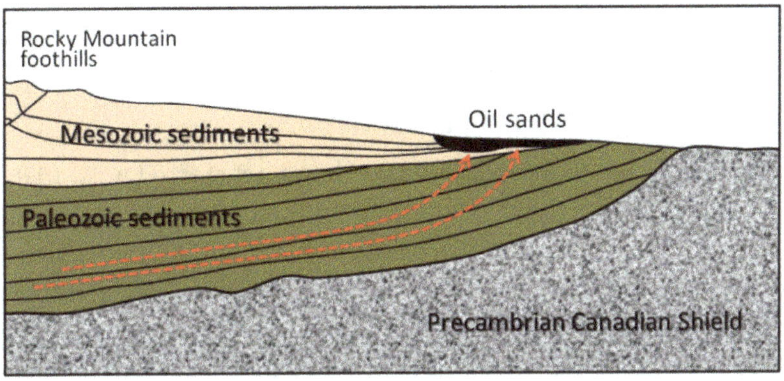

Figure 23: Schematic cross-section of northern Alberta showing the source rocks and location of the Athabasca Oil Sands

Shale gas is gas that is trapped within rock that is too impermeable for the gas to escape under normal conditions, and it can only be extracted by fracturing the reservoir rock using water and chemicals under extremely high pressure. This procedure is known as hydraulic fracturing or "fracking." Fracking is controversial because of the volume of water used, and because, in some jurisdictions, the fracking companies are not required to disclose the nature of the chemicals used. Although fracking is typically done at significant depths, there is always the risk that overlying water-supply aquifers could be contaminated (Figure 24). Fracking also induces low-level seismicity.

During the process that converts organic matter to coal, some methane is produced, which is stored within the pores of the coal. When coal is mined, methane is released into the mine where it can become a serious explosion hazard. Modern coal-mining machines have methane detectors on them and actually stop operating if the methane levels are dangerous. It is possible to extract the methane from coal beds without mining the coal; gas recovered this way is known as coal-bed methane.

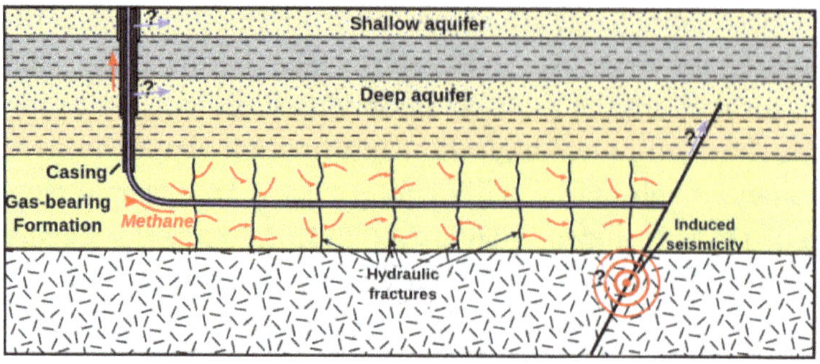

Figure 24: Depiction of the process of directional drilling and fracking to recover gas from impermeable rocks. The light blue arrows represent the potential for release of fracking chemicals to aquifers.

20.4 DIAMONDS

Although Canada's diamond mining industry didn't get started until 1998, diamonds are currently the sixth most valuable product mined in the country (Figure 3), and Canada ranks sixth in the world in diamond production. Diamonds form deep in the mantle (approximately 200 km to 250 km depth) under very specific pressure and temperature conditions, from carbon that is naturally present in mantle rock (not from coal). The diamond-bearing rock is brought to the surface coincidentally via a type of volcanism that is extremely rare (the most recent kimberlite eruption is thought to have taken place 10,000 years ago and prior to that at around 30 Ma). There is more on the volcanology of kimberlites in section 4.3. All of the world's kimberlite diamond deposits are situated within ancient shield areas (cratons) in Africa, Australia, Russia, South America, and North America.

It has long been known that diamonds could exist within the Canadian Shield, but up until 1991, exploration efforts had been unsuccessful. In 1980 two geologists, Chuck Fipke and Stu Blusson, started searching in the Northwest Territories by sampling glacial sediments looking for some of the minerals that are normally quite abundant within kimberlites: chromium-bearing garnet, chromium-bearing pyroxene, chromite (Cr2O3), and ilmenite (FeTiO3). These distinctive minerals are used for this type of exploration because they are many times more abundant in kimberlite than diamond is. After more than a decade of exploration, Fipke and Blusson finally focused their search on an area 250 km northeast of Yellowknife, and, in 1991, they announced the discovery of a diamond-bearing kimberlite body at Lac de Gras. That discovery is now the Diavik Mine, and there is another diamond mine — Ekati — 25 km to the northwest (Figure 25). There are two separate mines at Diavik accessing three different kimberlite bodies, and there are five at Ekati. See Figure 22 for a close-up view of the Ekati Mine. There are six operating diamond mines in Canada, four in the Northwest Territories (including Diavik and Ekati), and one each in Nunavut and Ontario.

Figure 25: Diamond mines in the Lac de Gras region, Nunavut. The twin pits of the Diavik Mine are visible in the lower right on an island within Lac de Gras. The five pits of the Ekati mine are also visible, on the left and the upper right. The two main mine centres are 25 km apart.

AT A GLANCE

Metal Deposits	Geological resources are critical to our way of life and important to the Canadian economy. Gold, iron, copper, nickel, and potash are Canada's most valuable mined commodities. The proportions of metals in mineral deposits are typically several thousand times higher than those in average rocks, and special processes are required to extract the valuable content. Some deposits form through processes within a magma chamber, others during volcanism or adjacent to a stock, and some are related to sedimentary processes. Mining involves both surface and underground methods, but in either case, rock is brought to surface that can react with water and oxygen to produce acid rock drainage and metal contamination.
Industrial Minerals	Non-metallic materials are very important to infrastructure and agriculture. Some of the major industrial minerals include sand and gravel, limestone for cement and agriculture, salt for a range of applications, potash fertilizer, and decorative stone.
Fossil Fuels	The main fossil fuels are coal, oil, and gas. Coal forms on land in wet environments where organic matter can remain submerged and isolated from oxygen for millennia before it is buried by more sediments. The depth of that burial influences the grade of coal produced. Oil and gas originate from organisms living in marine environments, and again, fairly rapid burial is required to preserve the organic matter on the sea floor. At moderate burial depth (2 km to 4 km), oil is produced, and at greater depth, gas is produced. Both oil and gas migrate toward the surface and can be trapped beneath impermeable rock layers in structural features, such as anticlines or faults. Some unconventional fossil fuel resources include oil sands, shale gas, and coal-bed methane.

Diamonds	Diamonds originate in the mantle and are only brought to the surface by the very rare eruption of kimberlitic volcanoes. The relatively recent discovery of diamonds in Canada was based on the exhaustive search for diamond indicator minerals in glacial sediments. There are now six diamond mines in Canada.

CHAPTER

21 WESTERN CANADA'S GEOLOGICAL HISTORY

Western Canada has a fascinating geological history with rocks ranging in age from the Archean to the Holocene. Over that time, almost every conceivable geological process has taken place here, resulting in the formation of a wide array of rock types, and some of the most important fossil deposits in the world. The region is also endowed with a range of geological resources, spanning the periodic table from beryllium to uranium, and the geological processes have produced awe-inspiring scenery and world-class recreational opportunities.

This chapter focuses on the important geological history and geological features of western Canada, but includes an overview of Canadian geology as a whole, starting with the development and journey of the ancient continent of Laurentia.

Figure 1: Crowsnest Mountain in the southern Alberta Rockies is made up of Paleozoic rocks that were uplifted by continental convergence during the Mesozoic, and then eroded by glaciation during the Cenozoic

21.1 CANADA: GEOLOGICAL HISTORY

Laurentia, which makes up the core of North America, is the largest and arguably the oldest of Earth's cratons (regions of stable ancient crust). Some of the rocks are over 4 billion years old, and Laurentia has been together in its present form for the last billion years. Over the past 650 million years, Laurentia has moved along a zigzag path from deep in the southern hemisphere to close to the North Pole (Figure 2). During that time, it collided several times with other continents and was temporarily part of two supercontinents (Pannotia and Pangea).

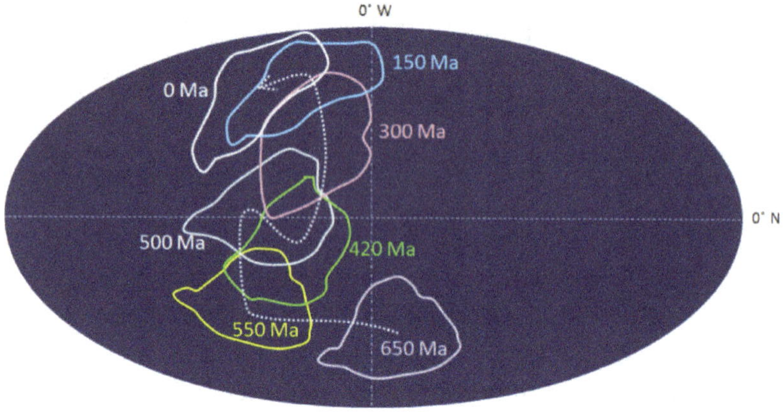

Figure 2: The path of Laurentia over the past 650 Ma

Bodies of rock tend to be eroded and recycled through the processes of plate tectonics, including uplift leading to erosion and burial leading to melting, and thus there are very few areas of truly ancient rocks on Earth. The oldest undisputed rocks are those of the Acasta Gneiss from north of Yellowknife, Northwest Territories, aged 4.03 Ga. But there are some rocks that could be even older within the Nuvvuagittuq greenstone belt on the east coast of Hudson Bay, in Quebec. These have been isotopically dated at 4.28 Ga, although the reliability of that date has been questioned. Based on other data, it is acknowledged that the Nuvvuagittuq rocks are at least as old as 3.75 Ga. The Acasta and Nuvvuagittuq rocks are situated within the Slave and Superior Cratons respectively, the oldest parts of Laurentia (Figure 3). Although these ancient cratons are not consistently that old, they are generally older than 3 Ga, as is part of the Wyoming Craton. The Hearne and Rae Cratons are older than 2 Ga, while most of the other parts of Laurentia are aged between 1 Ga and 2 Ga. The various provinces of Laurentia were assembled by plate-tectonic processes between 1 Ga and 3 Ga.

The areas of Figure 3 that are left uncoloured — the Appalachian, Innuitian, and Cordilleran fold belts — are geological regions that have been added to North America since 500 Ma. These are at least partly made up of sedimentary rocks that were deposited along the coasts and then folded, faulted, and uplifted during continental collisions.

The term Laurentia is geologically equivalent to the term Canadian Shield, although the latter is generally considered to be the area where the ancient Laurentian rocks are exposed at the surface and not covered with younger rocks. That applies to most of the region to the north and east of the red dotted line in Figure 3.

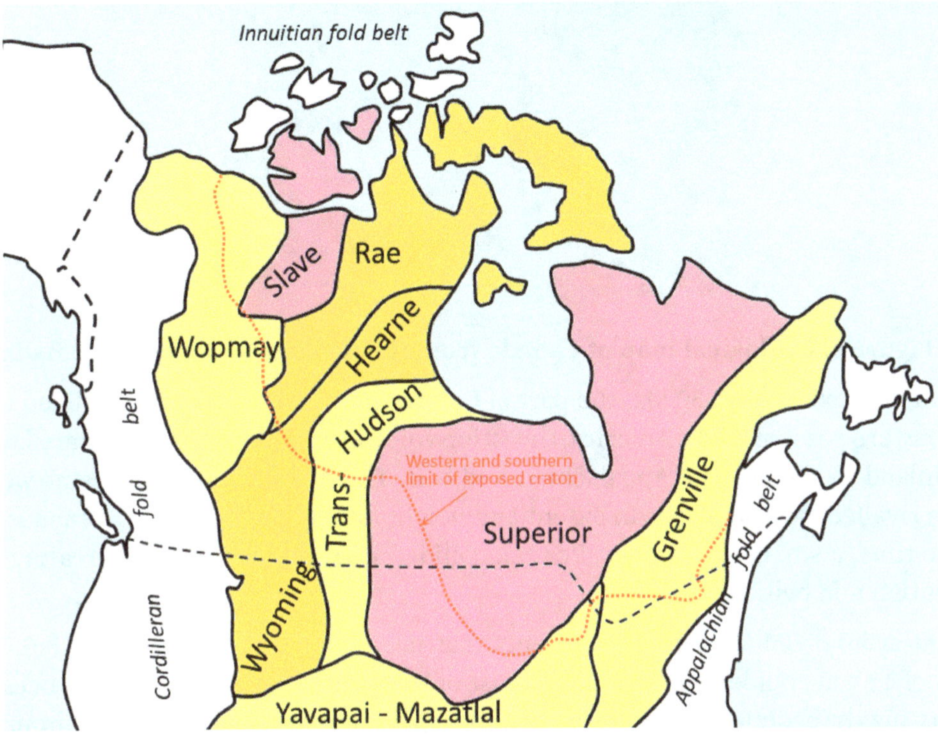

Figure 3: The main provinces of Laurentia. The pink areas are the oldest; light yellow are the youngest. All of the areas south and west of the dotted red line are now covered with younger rocks. The white areas represent rocks that were added to North America since 700 Ma.

Laurentia was part of the supercontinent Rodinia during the period between 1,100 Ma and 700 Ma. As Rodinia started to break up after 700 Ma, sediments derived from the erosion of the interior of the continent began to accumulate along its coasts, initially along the west coast, then the east coast at around 600 Ma, and finally on the north coast by around 550 Ma. This process continued for several hundred million years. By around 450 Ma, large areas of the interior of Laurentia were depressed below sea level — probably because of the downward pull of an underlying subducting plate — and marine sediments were deposited over parts of Quebec, Ontario, Manitoba, Saskatchewan, Alberta, and the Northwest Territories during the Ordovician, Silurian, and Devonian Periods (450 Ma to 350 Ma). These sediments are coloured various shades of blue on the geological map of Canada (Figure 4).

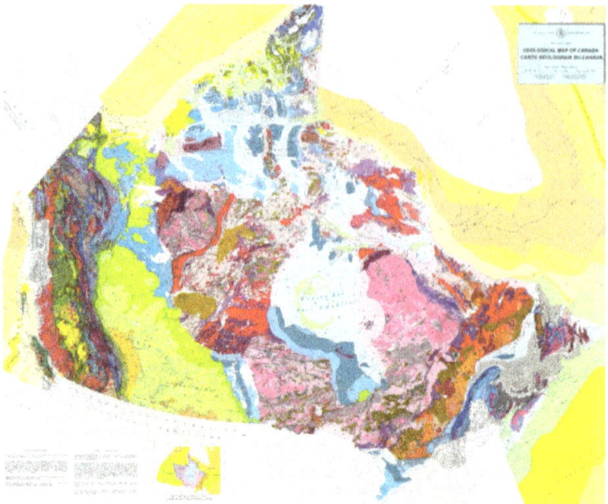

Figure 4: Geological map of Canada from the Geological Survey of Canada

At approximately 350 Ma, the part of Gondwana that is now Africa collided with the eastern coast of North America, thrusting volcanic islands and sedimentary layers far inland to become the Appalachian fold belt. The Appalachian Mountains would have rivalled the Himalayas in extent and height during the Devonian. At about the same time, a smaller continent, Pearya, collided with the north coast, creating the Innuitian fold belt.

At around 200 Ma, small continents that now make up the interior of B.C. and part of Yukon collided with the west coast of North America, starting the process of thrusting the sedimentary rocks inland and upward to form the Rocky Mountains.

The west-central part of North America subsided once again at around 150 Ma, due to an underlying subducting plate, and this led to the deposition of more marine rocks across Manitoba, Saskatchewan, and Alberta, and north into the Northwest Territories and Yukon (the green areas in Figure 3).

Finally, at around 90 Ma, more small continents, which now comprise Vancouver Island and Haida Gwaii, collided with the west coast, leading to further uplift of the Rocky Mountains.

21.2 IN THE PRECAMBRIAN PERIOD WESTERN CANADA

Laurentia extends as far west as eastern B.C. (Figure 3), but the ancient rocks of the craton are almost completely covered by younger rocks in B.C., Yukon, and all of Alberta except the far northeast corner. Laurentia is well represented in northern Saskatchewan and across large parts of Manitoba, the Northwest Territories, and Nunavut (Figure 5). Where they are exposed, the rocks of the Canadian Shield are highly varied lithologically, typically strongly metamorphosed due to their deep burial at some time in the past, and in some cases, quite different from what could be expected to occur on Earth today.

Starting from the south, in eastern Manitoba and adjacent Ontario, we have the ancient rocks of the Superior Province. On the map the Superior Province, rocks are mostly pink, representing granitic and gneissic rocks, with strips and blotches of green, representing metamorphosed sea-floor basalt and sediments, also known as greenstone belts. These rocks are widely interpreted to have deep crustal origins, and include large areas of granulite facies metamorphic rock formed at high temperatures and moderate to high pressures (see Figure 19). Superior Province greenstone belts in Ontario and Quebec host some of the world's largest volcanogenic massive sulphide deposits. As described in Chapter 20, the Superior Province in northern Manitoba is host to important nickel deposits at Thompson. These formed from mantle-derived mafic magma that interacted with sulphur-bearing crustal rocks, and within which heavy-metal sulphide minerals formed.

The Trans-Hudson Orogen (THO), as its name implies, extends through Saskatchewan and Manitoba and over to the eastern side of Hudson Bay. It represents the continent-continent collision zone between the Superior Craton to the south and the Churchill Craton (including the Wyoming, Hearne, and Rae Cratons) to the north; thus it's a remnant of the initial formation of Laurentia at around 1.9 Ga. At the time of the collision, the THO would have been a major mountain range, and the rocks that we see there now — which evolved deep beneath those mountains — are highly metamorphosed sedimentary and volcanic rocks intruded by large granitic bodies. The important volcanogenic massive sulphide deposits around Flin Flon are within the THO.

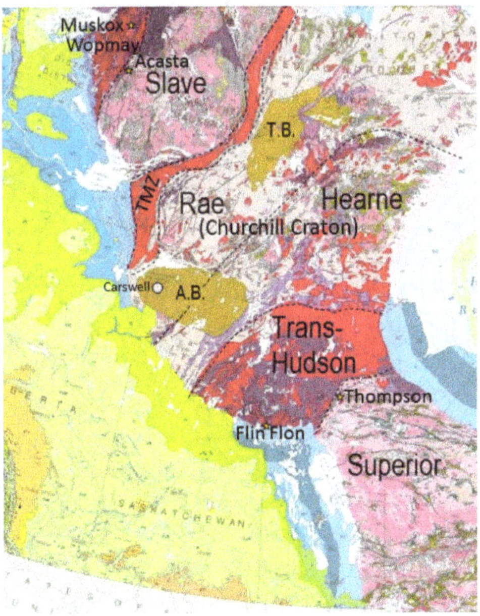

Figure 5: Geological features of the Canadian Shield of western Canada. A.B.: Athabasca Basin, T.B.: Thelon Basin, and TMZ: Taltson Magmatic Zone

The Churchill Craton is lithologically similar to the Superior Craton, although not generally as old. It includes two important sedimentary basins: the Athabasca Basin in Saskatchewan and the Thelon Basin in Nunavut, both filled with rocks aged around 1.7 Ga. These consist primarily of sandstones and minor mudstones that are only weakly metamorphosed and essentially undeformed (not folded) because they are situated within a stable craton and so have not been subjected to significant tectonic forces. The Athabasca Basin is economically important for its large and rich unconformity-type uranium deposits (see Chapter 20). At its western end, there is the remnant of a large extraterrestrial impact, the 40 km diameter Carswell Crater. When the meteor struck at this location, at around 115 Ma, the impact and subsequent rebound of the crust was enough to bring metamorphic rock up to surface from beneath about 2,000 m of Athabasca Group sandstone. There is no connection between the Carswell Crater and the much older (~1.2 Ga) uranium deposits.

The Taltson Magmatic Zone (TMZ), which forms the boundary between the Churchill and Slave Cratons, consists primarily of granitic rock. One interpretation is that the TMZ formed along a convergent boundary, although this is not universally accepted.

The Slave Craton is dominated by granitic rocks and metamorphosed clastic sedimentary rocks. On its western edge, there is a large area of very old gneissic rock that includes the Acasta Gneiss, dated at 4.03 Ga, which, for the time being at least, is the oldest rock in the world (Figure 6).

Figure 6: A sample of the Acasta Gneiss on display at the Natural History Museum in Vienna

The Wopmay Orogen, interpreted as the site of another ancient continent-continent collision, lies to the west of the Slave Craton. Although mostly composed of felsic igneous rocks and gneisses, the Wopmay Orogen includes a body of mafic and ultramafic igneous rock called the Muskox Intrusion. Derived from a mantle plume and dated at about 1.1 Ga, the Muskox is comparable to a handful of other mafic and ultramafic intrusions around the world in that it has distinctive repetitive layering caused by settling of heavy metal-rich minerals within the low-viscosity magma. Muskox has high levels of nickel, copper, and chromium, and has the potential to have platinum and palladium like a similar body in South Africa. Ultramafic intrusions like Muskox do not take place on Earth today because the mantle is no longer hot enough.

The oldest rocks in British Columbia are the strongly metamorphosed sedimentary, volcanic, and intrusive rocks of the Monashee Complex, situated to the west of the Columbia River near Revelstoke (Figure 7). Aged around 2 Ga, these may actually be part of Laurentia.

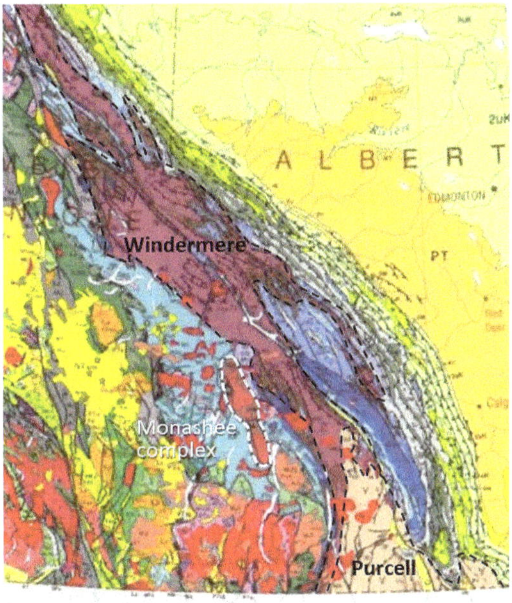

Figure 7: Precambrian rocks in southern B.C. and Alberta

There are much more extensive Precambrian rocks within the Columbia and Rocky Mountains of southeastern B.C. and the southwestern corner of Alberta. The rocks of the Purcell Supergroup (a supergroup comprises more than one group) are present in the extreme southeastern corner of B.C. and adjacent Alberta, and extend well into the United States (as the Belt Supergroup). These are mostly unmetamorphosed clastic rocks deposited in rivers and lakes during the middle Proterozoic, at around 1,400 Ma, while Laurentia was still part of the supercontinent

Columbia. When Columbia rifted apart, the division happened within the area of the Purcell/Belt rocks. Similar rocks of the same age are present in Tasmania and Siberia, and it is postulated that they were once part of the same depositional basin.

21.3 IN THE PALEOZOIC PERIOD WESTERN CANADA

At the beginning of the Paleozoic (542 Ma), Laurentia was near the equator (Figure 2) and sedimentation was continuing on all of Laurentia's marine margins, including the passive margin (not tectonically active) on what is now the west coast. The clastic sediments of the Windermere Group are succeeded by mostly limestone beds (represented by the blue areas in Figure 7) interbedded in some areas with mudstone and sandstone. The most famous Cambrian rocks in the Rockies are those around Field, B.C., within Yoho and Kootenay National Parks. The Burgess Shale of the Stephen Formation is considered by some to be the most important fossil bed in the world because of its spectacular preservation of detail in a wide array of organisms that are ancestors to many of today's organisms and are not present in earlier rocks. The Walcott Quarry, on the pass between Mt. Field and Wapta Mountain has been known and studied for over 100 years (Figure 8). In 2012 a new Burgess Shale discovery was made at Marble Canyon, about 30 km to the southeast, by a team led by the Royal Ontario Museum (ROM). Fossils with similar levels of preservation are present, and several previously unknown organisms have been found. The Paleozoic strata of the Rockies also include Ordovician, Devonian, Carboniferous, and Permian sedimentary rocks. For example, Carboniferous limestone makes up most of the upper part of Crowsnest Mountain in southern Alberta (Figure 1).

Figure 8: The Cambrian Burgess Shale at the Walcott Quarry, Yoho Park, B.C., with Wapta Mountain in the upper right.

While clastic and carbonate sediments were accumulating along the western edge of Laurentia, much of the interior of the continent was submerged under inland seas that were connected to ocean most of the time. This region is known as the Western Canada Sedimentary Basin (WCSB). The Paleozoic sediments that accumulated within this basin show up as the blue areas in Figures 4 and 5; however, their extent is much wider than that because Paleozoic sedimentary rocks also underly the Mesozoic rocks within most of the areas that are light green on those maps. By way of example, a schematic cross-section through the Paleozoic and Mesozoic rocks of southern Manitoba is given in Figure 10. The section extends from the Saskatchewan-Manitoba border on the left to just east of Winnipeg on the right, and shows the Paleozoic rocks overlain on the rocks of the Precambrian Superior Craton.

Fifteen different Paleozoic formations, ranging in age from Ordovician to Carboniferous, are shown in Figure 9. Of these, 11 are dominated by carbonate rocks (limestone or dolomite) that very likely formed in an ocean-connected marine environment. The non-carbonate formations are the lowermost one (resting on Precambrian rocks), which is sandstone of marine origin; the Devonian Prairie Evaporite Formation (in red) — the same formation from which potash is mined in Saskatchewan; and the upper two Devonian formations (in yellow), which are shale. When the Prairie Evaporite formed, the basin was isolated from the open ocean, and the rate of evaporation was greater than the rate of input from precipitation and river inflow. During that time, probably at least several million years, there were numerous changes in sea level or land level that allowed additional ocean water — and therefore additional salt — into the basin.

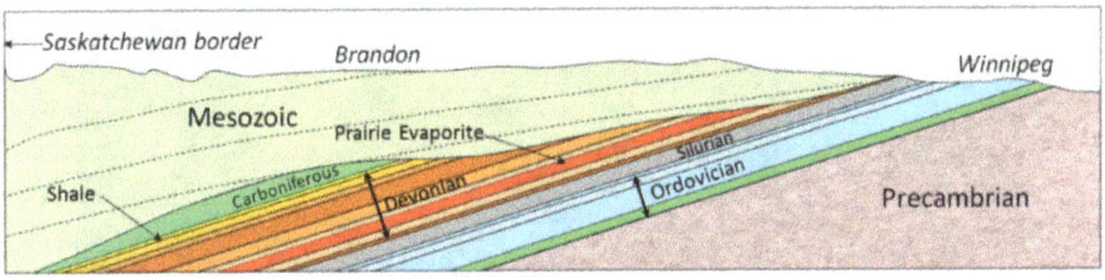

Figure 9: The Paleozoic sedimentary rocks of southern Manitoba along a section extending from the Saskatchewan border on the left to the Winnipeg area on the right. The section is 400 km wide and 1,800 m high, and the vertical exaggeration is about 100 times. The dip of the beds is also exaggerated by 100 times; their original and current attitudes are close to horizontal.

There are Paleozoic rocks in the central and western parts of British Columbia and Yukon, but they formed far away and did not become part of North America until the Mesozoic. Subduction started along the western edge of Laurentia by the middle Paleozoic. That meant that oceanic crust was moving toward the continent, bringing small segments of exotic continental crust with it (Figure 10). These crustal blocks along western North America are called terranes, indicating that they are sections of the continent that have an exotic origin (Figure 11). Most of British Columbia is made up of terranes that include sedimentary rocks with fossils that imply an origin south of the equator, or volcanic rocks with magnetic orientations that indicate a southern-hemisphere origin.

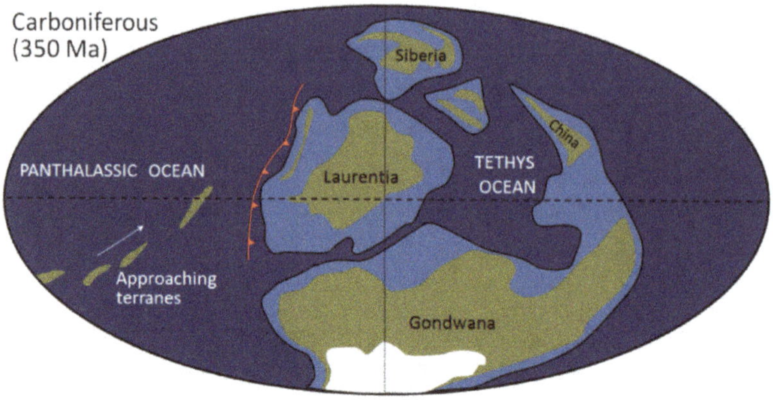

Figure 10: The distribution of continents in the early Carboniferous, showing the terranes that later became attached to the west coast of North America. The light blue areas are continental shelves, the white is ice of the Karoo Glaciation, and the red line shows subduction of oceanic crust beneath Laurentia. Panthalassic is the name for the huge ocean that preceded the Pacific Ocean.

Figure 11: The Carboniferous Mt. Mark Formation on Vancouver Island is part of the Wrangellia Terrane, which arrived on the edge of North America during the Cretaceous.

21.4 IN THE MESOZOIC PERIOD WESTERN CANADA

The Mesozoic extends over 187 million years from the beginning of the Triassic (252 Ma) to the end of the Cretaceous (65 Ma). It was a particularly important period for the geology of western Canada. During this time, several continental collisions occurred along the west coast, resulting in the formation of the Rocky Mountains and the accretion (addition) of much of the land mass of British Columbia, and continuing deposition within the WCSB.

Terrane Accretion in British Columbia and Yukon

Continued subduction along the western edge of North America carried a number of continental terranes toward the coast, with the first collisions taking place in the early part of the Triassic, as the Quesnel, Cache Creek, and Stikine Terranes combined to form the Intermontane Superterrane, so named because it forms the interior plateau of British Columbia, between the Rockies to the east and the Coast Range to the west (Figure 12).

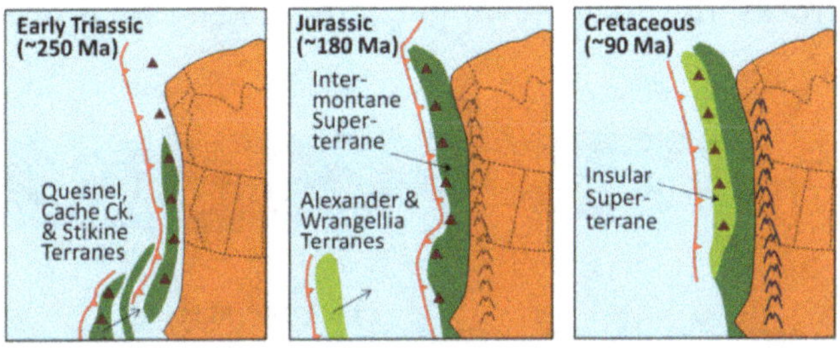

Figure 12: Model of the accretion of the Intermontane and Insular Superterranes to the west coast of North America during the Mesozoic. Subduction zones are the red-toothed lines. The dark-red triangles represent volcanoes.

Approximately 100 million years later, another pair of terranes — Alexander and Wrangellia — collided to form most of Vancouver Island and Haida Gwaii, plus a significant part of Alaska. During the Cenozoic, additional terranes (the Outboard terranes) were added to the western edge of North America. An overview of the accreted terranes of B.C., Yukon, and Alaska is given in Figure 13.

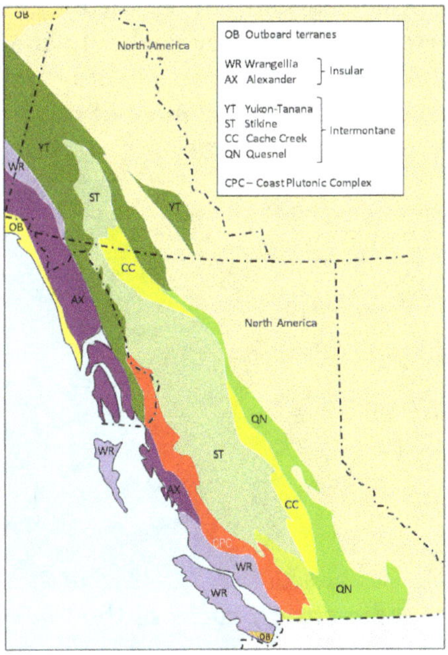

Figure 13: A generalized overview of the accreted terranes of B.C., Yukon, and Alaska. The Intermontane terranes are in green, the Insular terranes in purple,

and the Outboard terranes in yellow. The Coast Plutonic Complex (CPC) formed in situ and is not a terrane.

During the Jurassic, the Intermontane Superterrane acted like a giant bulldozer, pushing, folding, and thrusting the existing Proterozoic and Paleozoic west coast sediments eastward and upward to form the Rocky Mountains (Figure 15). The same process continued into the Cretaceous as the Insular Superterrane collided with North America and pushed the Intermontane Superterrane farther east. Folding in Rocky Mountain rocks, like that shown in Figures 15 and 16, is one of the results of this process.

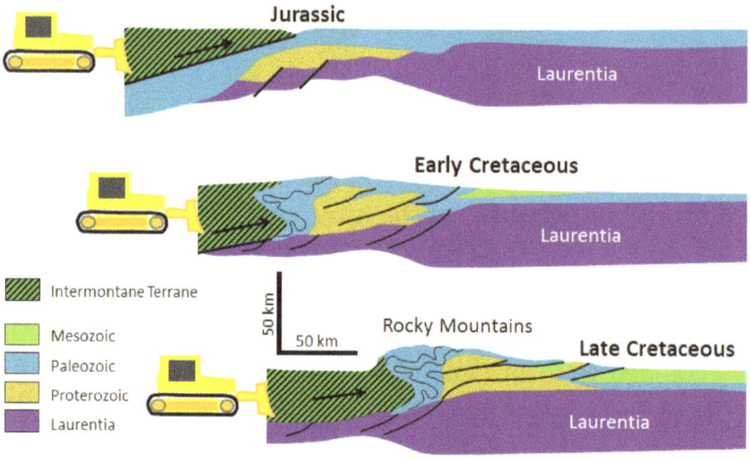

Figure 14: Cross-section of the accretion of the Intermontane Superterrane to the west coast of North America and the resulting compression, folding, and thrusting of North American sedimentary rocks. In the Late Cretaceous, it was the accretion of the Insular Superterrane pushing against the Intermontane Superterrane that did most of the work.

Figure 15: Tight folds in sedimentary rocks of the Rocky Mountains near Field, B.C.

Figure 16: Folding of sedimentary rocks at Mt. Rae, Alberta.

Thrusting is another important process in the formation of fold-belt mountains, as described in Chapter 12. During plate convergence, entire sheets of sedimentary rock are slowly pushed over top of other sheets, resulting in situations where older rocks lie on top of younger ones. One of the best known examples of this is at Mt. Yamnuska, near Exshaw, Alberta (Figure 17), where the older Cambrian rocks were pushed east by a total of 40 km, over top of younger rocks.

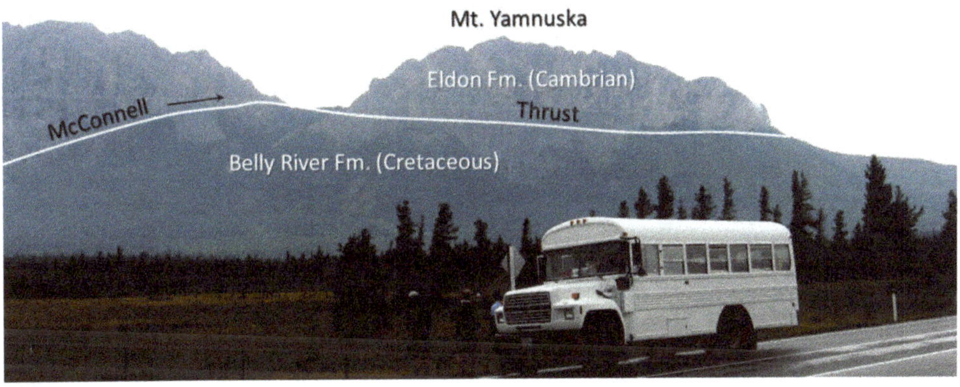

Figure 17: The McConnell Thrust at Mt. Yamnuska near Exshaw, Alberta. Carbonate rocks of Cambrian age have been thrust over top of Cretaceous mudstone.

In the area near the U.S. border, within B.C., Alberta, and Montana, a sheet of Paleozoic rocks has been thrust about 80 km east over top of Cretaceous rocks along the Lewis Fault (Figure 18).

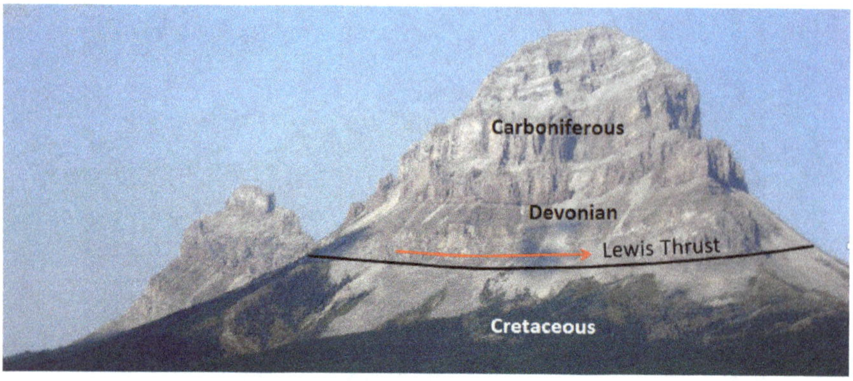

Figure 18: The Lewis Thrust at Crowsnest Mountain near Frank, Alberta. Carbonate rocks of Devonian and Carboniferous age have been pushed 80 km to the east and thrust over top of Cretaceous mudstone.

Not only did the subduction of oceanic crust beneath North America during the Mesozoic deliver geologically exotic terranes to the western edge of the continent, it also resulted in massive amounts of volcanism along the boundary (Figure 13). The upper-crustal magma chambers that fed those now-eroded volcanoes slowly cooled into granitic and dioritic stocks, and those stocks gradually coalesced into batholiths that extend from the southwest corner of B.C. all the way into Yukon and Alaska (Figures 19 and 18). Most of the granitic rocks of this region fall into two main age ranges: many are middle-Jurassic to early Cretaceous in age (~ 170 Ma to 140 Ma), while others are late Cretaceous to Paleogene (~50 Ma to 90 Ma). Many of the older bodies intruded into the terranes they are on before they arrived on the North America coast. This applies to those on Vancouver Island and Haida Gwaii. Some of the older ones formed in situ when the subduction zone was farther east (see Figure 13). Most of the younger bodies formed in situ when the subduction zone was close to where it is now (west of Vancouver Island) or slightly to the east.

Although these intrusive igneous rocks cooled at depth in the upper crust, they now form some of the highest peaks in Canada, many of them hundreds of metres higher than those of the Rocky Mountains (including Mt. Waddington, the highest peak entirely within British Columbia at 4,091 m). It is estimated that over the past 100 million years some of these igneous bodies have been uplifted in the order of 8,000 m. Much of that uplift is a result of the relative low density of the granitic rocks compared with the surrounding rocks.

Figure 19: Jurassic (J) and Cretaceous (K) granitic bodies in south-central B.C.

Figure 20: Cretaceous granite at The Lions, north of Vancouver, with Howe Sound in the background [Isaac Earle, used with permission]

The Western Canada Sedimentary Basin during the Mesozoic

The construction of the Rocky Mountains during the Jurassic and Cretaceous — and their ensuing erosion — created a significant new source of sediments for the WCSB. Based on the ages, distributions, and thicknesses of the sedimentary layers (Figure 23), it is evident that the greatest volumes of sediment were produced in the Upper Cretaceous (100 Ma to 65 Ma) and into the Paleocene (65 Ma to 55 Ma). The sediments accumulated in a basin that is thought to have been at least partly formed by the presence of a subducting slab of oceanic lithosphere underneath this part of North America. However, the ongoing uplift of the Rockies through this time period also led to isostatic depression of the crust. The western edge of the basin, which has about 4,500 m of Mesozoic rock alone, is a foreland basin. (See Chapter 6 for more on the origins of basins.)

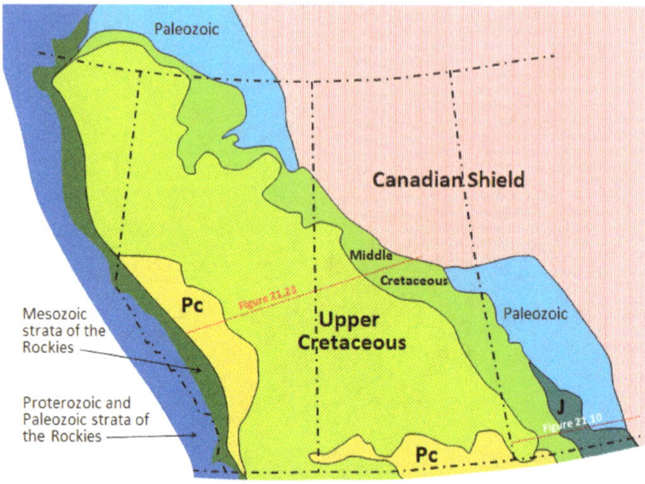

Figure 21: The distribution of Mesozoic sedimentary rocks in the Western Canada Sedimentary Basin. J stands for Jurassic and Pc for Paleocene. The lines of the cross-sections of Figures 21.10 and 21.23 are shown. [SE after Alberta Geological Survey]

From time to time during the Mesozoic, the WCSB was filled to varying degrees with marine water. The Jurassic rocks at the base of the sequence are marine in origin, although the Jurassic sequence in Manitoba also includes evaporite layers. Most of the Middle Cretaceous rocks across the basin are marine, but the majority of the Upper Cretaceous rocks are of terrestrial origin, deposited within the flood plains and deltas of rivers. Some of these terrestrial sediments include coal layers; as described in Chapter 20, there are significant coal deposits in central Alberta.

The Paleozoic sediments within the WCSB were buried deeply beneath the Mesozoic sediments and were heated enough to form both oil and gas. There are large petroleum resources in reservoir rocks of various ages extending from northeastern B.C. to southwestern Manitoba.

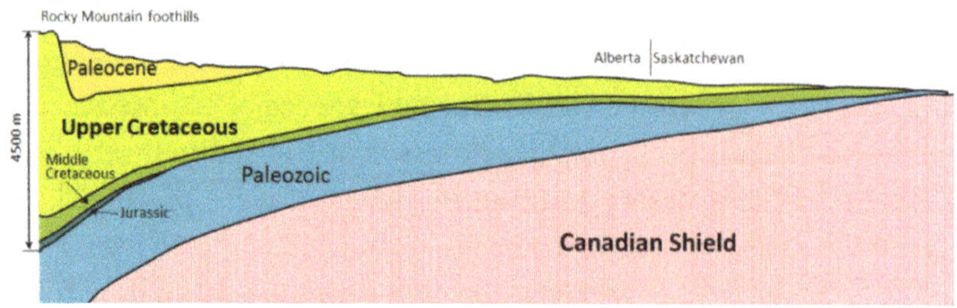

Figure 22: A cross-section showing the Mesozoic sedimentary rocks in the WCSB from the Rocky Mountain foothills to north-central Saskatchewan [SE after Alberta Geological Survey]

Several of the terrestrial Cretaceous formations in the WCSB are host to important dinosaur fossils. Some are within B.C. and Saskatchewan, but the most famous are in Alberta, including the Dinosaur Park Formation (Figure 1), the Scollard Formation, and the Horseshoe Canyon Formation (Figure 23). The Dinosaur Park Formation has one of the greatest concentrations of dinosaur fossils of any rock on Earth, with at least 50 genera of dinosaurs represented, ranging from tiny Hesperonychus to giant Albertosaurus. The Hilda Bone Bed, situated about 80 km to the east of Dinosaur Park, is estimated to have the remains of approximately 1500 ceratopsians, all of which are interpreted to have died in a flood related to a tropical storm. A few of the larger herbivorous dinosaurs found at Dinosaur Park are illustrated in Figure 25.

Figure 23: The Horseshoe Canyon Formation near Drumheller, Alberta [SE]

Figure 24: Depiction of some of the large herbivorous dinosaurs from the upper part of the Dinosaur Park Formation. Left to right: the ceratopsian Pentaceratops , the hadrosaur Lambeosaurus eating from a tall tree, the ceratopsian Styracosaurus, the ankylosuar Scolosaurus, the hadrosaur Prosaurolophus (in the distance), the ankylosaur Panoplosaurus, and a herd of Styracosaurs in the background. [by J.T. Csotonyi at https://upload.wikimedia.org/wikipedia/commons/e/e9/Dinosaur_park_formation_fauna.png]

21.5 WESTERN CANADA DURING THE CENOZOIC

Two additional relatively small terranes collided with North America early in the Cenozoic. At around 55 Ma, metamorphosed sedimentary and volcanic rocks of the Pacific Rim Terrane were forced a few tens of kilometres underneath the west coast of Vancouver Island (Figure 25). These rocks are distributed along the west coast of the island and in the area around Victoria (Figure 26). At around 42 Ma, sea-floor pillow basalt and gabbro of the Crescent Terrane accreted to the southern margin of Vancouver Island and also to the adjacent part of Washington State. These terranes are shown as Outboard terranes in Figure 14.

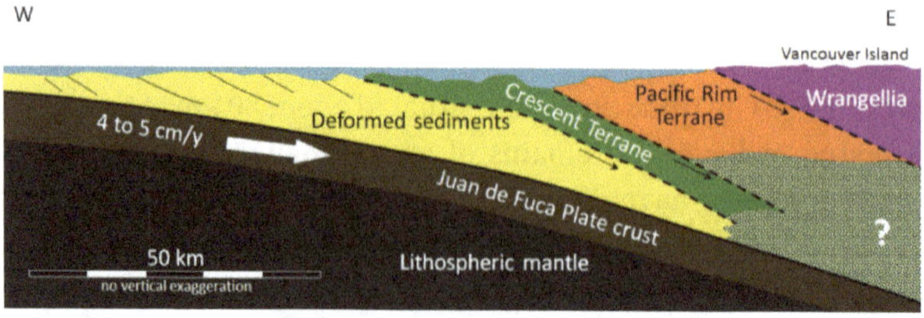

Figure 25: East-west cross-section showing the accretion of the Pacific Rim and Crescent Terranes beneath Vancouver Island, and the ongoing subduction of the Juan de Fuca Plate. The dashed lines are inactive faults.

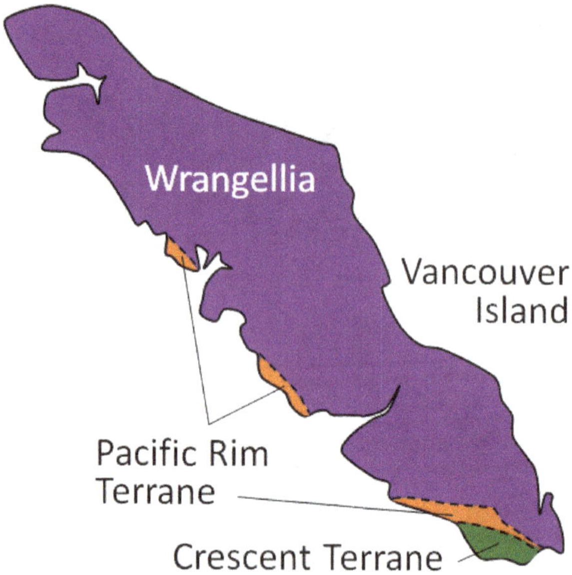

Figure 26: The distribution of Pacific Rim and Crescent Terrane rocks on Vancouver Island

The accretion of the Pacific Rim and Crescent Terranes had the effect of pushing Vancouver Island closer to the North American mainland, resulting in the uplift of the sediments deposited within the Nanaimo Basin to form islands in the Strait of Georgia (Figure 27) and mountains on Vancouver Island.

Figure 27: The Geoffrey Formation of the Nanaimo Group on Ruxton Island, B.C.

Following these events, the subduction of the Juan de Fuca Plate, which is a remnant of the former, much larger, Farallon Plate, was re-established at its current location farther to the west of Vancouver Island. This subduction, and that of the North America Plate beneath Alaska, has produced recently active volcanoes in Alaska, and all along the west coast from north of Vancouver Island to northern California (Figure 29). In southwestern B.C., there are several dormant volcanoes of Pleistocene age (including Garibaldi and Meager) that trend along a line that also passes through Mt. Baker in Washington State. About 40 km to the east is a trend of slightly older igneous complexes (Pliocene to Oligocene). The displacement between these belts could be explained by a westward shift in the position of the subduction zone over that time period.

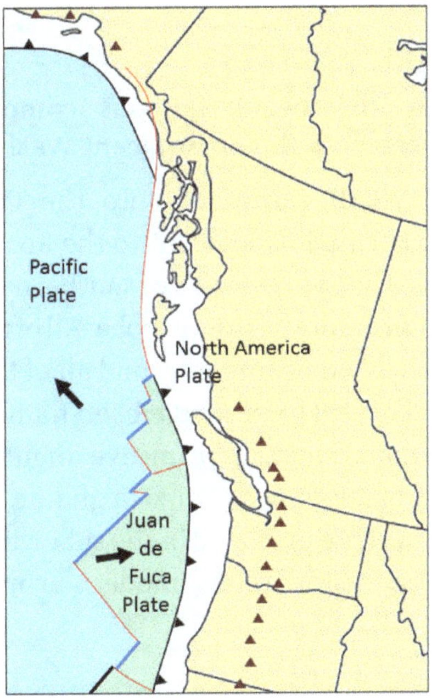

Figure 28: The current plate situation along the western edge of northern North America. Blue lines are divergent boundaries, red lines are transform boundaries,

and black lines with teeth are subduction boundaries. The dark red triangles are volcanoes.

The subduction and transform boundaries along this coast also generate relatively frequent earthquakes throughout this region, as illustrated in Exercise 1.

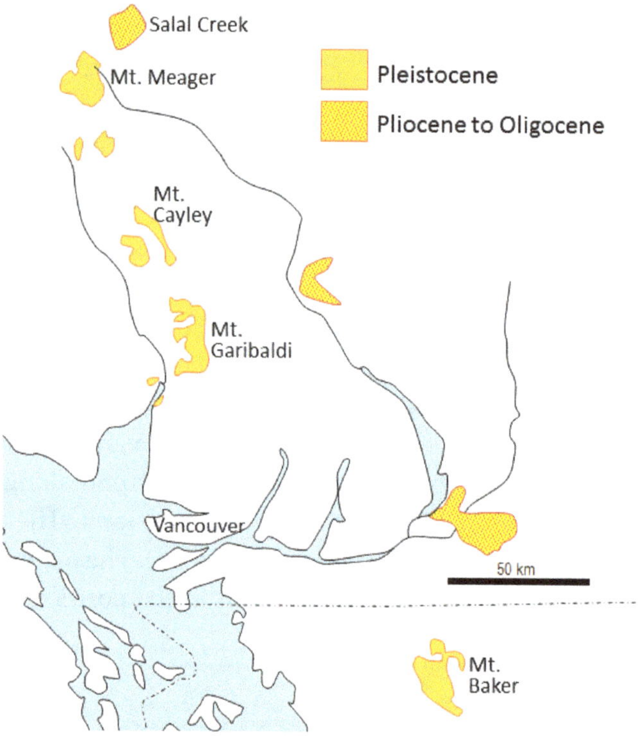

Figure 29: Oligocene to Pleistocene igneous complexes and volcanoes in southwestern B.C. and adjacent Washington

Sedimentation in the WCSB continued into the Cenozoic (Figure 22) with deposition of the Paskapoo Formation adjacent to the Rockies in Alberta (Figure 30), the Ravenscrag Formation in the Cypress Hills of southern Alberta and Saskatchewan, and the Turtle Hills Formation in southern Manitoba. All of these strata were deposited in terrestrial fluvial and deltaic environments, and all of them include coal deposits. Numerous mammalian (and other) fossils have been found in these rocks in Alberta and Saskatchewan. The mammals include primitive ungulates (ancestors to the deer and their relatives), a type of pangolin, a colugo (a gliding mammal that was possibly a primate ancestor), and some true primates in the suborder Plesiadapiformes, which became extinct and are not ancestors to any modern primates.

Figure 30: The Paskapoo Formation exposed on the banks of the Red Deer River, Alberta

AT A GLANCE

Geological History of Canada	The continent Laurentia, which includes what is now the Canadian Shield, was formed through the assembly of a number of smaller continents over the period from around 4 Ga to 1 Ga. Over the past 650 Ma, Laurentia moved north from deep within the southern hemisphere. During that time, a number of important plate-tectonic events have taken place, including formation of the Appalachian and Innuitian fold belts, sedimentation within the interior of the continent, and terrane accretion and mountain formation along the west coast.
Western Canada during the Precambrian	The two oldest parts of the Canadian Shield in western Canada are the Slave and Superior Cratons, and both include some of Earth's oldest rocks. During the formation of Laurentia, these cratons were combined with the Rae and Hearn Cratons, with the collision zones now represented by the Trans-Hudson Orogen and the Taltson Magmatic Zone. Continental sediments accumulated in Saskatchewan and Nunavut at around 1,700 Ma and in the area of the southern B.C.-Alberta border at around 1,400 Ma. This region was rifted apart during the breakup of the supercontinent Columbia, after which sedimentation continued on the western margin of Laurentia with the deposition of the Windermere Group.
Western Canada during the Paleozoic	Sedimentation on the west coast of North America continued into the early Paleozoic, and by the Ordovician, the Western Canada Sedimentary Basin (WCSB) had developed, extending from southern Manitoba to the northern Northwest Territories. Most of the Paleozoic rocks in this basin, which range in age up to the Carboniferous, have marine affinities, although there are important evaporites as well. Sedimentation continued on the west coast through this time, but by the late Paleozoic, a subduction boundary had developed along the coast and small continents were moving toward North America.

Western Canada during the Mesozoic	The various parts of the Intermontane Superterrane began colliding with the west coast of North America during the Jurassic (~180 Ma). This started the building of the Rocky Mountains by the thrusting of existing sedimentary rocks toward the east. The arrival of the Insular Superterrane during the Cretaceous (~90 Ma) contributed to further thrusting and uplift of the Rockies, creating a significant source of sediments for the WCSB. The greatest volume of Mesozoic rocks in the basin are of Upper Cretaceous age, and that likely coincides with the period of maximum collision-related uplift of the Rockies. Subduction of oceanic crust at various locations along the west coast and within the accreted terranes prior to their arrival has produced massive volumes of intrusive igneous rocks within the Coast Range.
Western Canada during the Cenozoic	The Pacific Rim and Crescent Terranes were added to the western edge of Vancouver Island during the Paleogene, pushing the island closer to the mainland and forcing recently deposited Nanaimo Group rocks onto the island. Continuing subduction along the coast has generated ongoing volcanism and earthquake activity in southwestern B.C. Sedimentation continued in the WCSB into the Cenozoic, especially in the Paleocene with deposition of the terrestrial Paskapoo Formation in Alberta and similar rocks in southern Saskatchewan.

CHAPTER

THE SOLAR SYSTEM'S AND EARTH'S HISTORY 22

The story of how Earth came to be is a fascinating contradiction. On the one hand, many, many things had to go just right for Earth to turn out the way it did and develop life. On the other hand, the formation of planets similar to Earth is an entirely predictable consequence of the laws of physics, and it seems to have happened more than once.

We will start Earth's story from the beginning — the *very* beginning — and learn why generations of stars had to be born and then die explosive deaths before Earth could exist. We will look at what it takes for a star to form, and for objects to form around it, as well as why the nature of those objects depends on how far away from the central star they form.

Earth spent its early years growing up in a very rough neighbourhood, and we will discuss how Earth's environment influenced its development, including how it got its moon from what was quite literally an Earth-shattering blow. This chapter will also discuss the hunt for Earth-like exoplanets (planets that exist outside of our solar system).

22.1 STARTING WITH A BIG BANG THEORY

According to the big bang theory, the universe blinked violently into existence 13.77 billion years ago (Figure 1). The big bang is often described as an explosion, but imagining it as an enormous fireball isn't accurate. The big bang involved a sudden expansion of matter, energy, and space from a single point. The kind of Hollywood explosion that might come to mind involves expansion of matter and energy within space, but during the big bang, space itself was created.

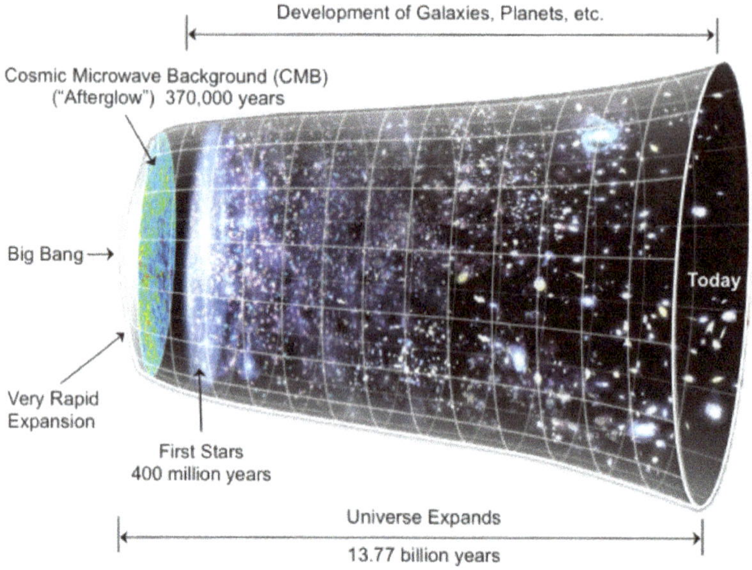

Figure 1: The big bang. The universe began 13.77 billion years ago with a sudden expansion of space, matter, and energy, and it continues to expand today.

At the start of the big bang, the universe was too hot and dense to be anything but a sizzle of particles smaller than atoms, but as it expanded, it also cooled. Eventually some of the particles collided and stuck together. Those collisions produced hydrogen and helium, the most common elements in the universe, along with a small amount of lithium.

You may wonder how a universe can be created out of nothing, or how we can know that the big bang happened at all. Creating a universe out of nothing is mostly beyond the scope of this chapter, but there is a way to think about it. The particles that make up the universe have opposites that cancel each other out, similar to the way that we can add the numbers 1 and -1 to get zero (also known as "nothing"). As far as the math goes, having zero is exactly the same as having a 1 and a -1. It is also exactly the same as having a 2 and a -2, a 3 and a -3, two -1s and a 2, and so on. In other words, nothing is really the potential for something if you divide it into its opposite parts. As for how we can know that the big bang happened at all, there are very good reasons to accept that it is indeed how our universe came to be.

The early stages of the big bang

The notion of seeing the past is often used metaphorically when we talk about ancient events, but in this case it is meant literally. In our everyday experience, when we watch an event take place, we perceive that we are watching it as it unfolds in real time. In fact, this isn't true. To see the event, light from that event must travel to our eyes. Light travels very rapidly, but it does not travel instantly. If we were watching a digital clock 1 m away from us change from 11:59 a.m. to 12:00 p.m., we would actually see it turn to 12:00 p.m. three billionths of a second after it happened. This isn't enough of a delay to cause us to be late for an appointment, but the universe is a very big place, and the "digital clock" in question is often much, much farther away. In fact, the universe is so big that it is convenient to describe distances in terms of light years, or the distance light travels in one year. What this means is that light from distant objects takes so long to get to us that we see those objects as they were at some considerable time in the past. For example, the star Proxima Centauri is 4.24 light years from the sun. If you viewed Proxima Centauri from Earth on January 1, 2015, you would actually see it as it appeared in early October 2010.

We now have tools that are powerful enough to look deep into space and see the arrival of light from early in the universe's history. Astronomers can detect light from approximately 375,000 years after the big bang is thought to have occurred. Physicists tell us that if the big bang happened, then particles within the universe would still be very close together at this time. They would be so close that light wouldn't be able to travel far without bumping into another particle and getting scattered in another direction. The effect would be to fill the sky with glowing fog, the "afterglow" from the formation of the universe (Figure 1). In fact, this is exactly what we see when we look at light from 375,000 years after the big bang. The fog is referred to as the cosmic microwave background (or CMB), and it has been carefully mapped throughout the sky (Figure 2). The map displays the cosmic microwave background as temperature variations, but these variations translate to differences in the density of matter in the early universe. The red patches are the highest density regions and the blue patches are the lowest density. Higher density regions represent the eventual beginnings of stars and planets. The map in Figure 2 has been likened to a baby picture of the universe.

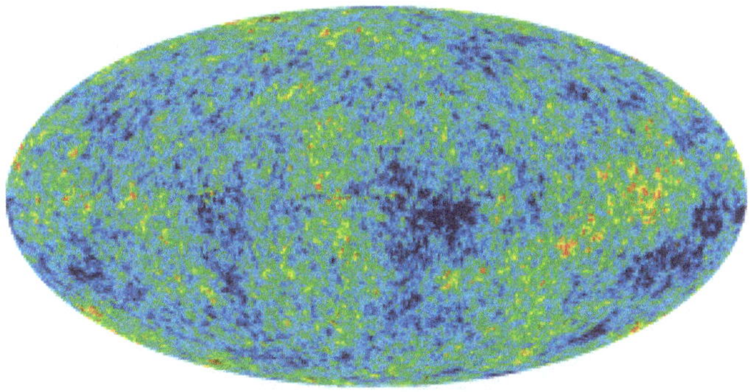

Figure 2: Cosmic microwave background (CMB) map of the sky, a baby picture of the universe. The CMB is light from 375,000 years after the big bang. The colours reveal variations in density. Red patches have the highest density and blue patches have the lowest density. Regions of higher density eventually formed the stars, planets, and other objects we see in space today. [NASA/ WMAP Science Team http://bit.ly/CMBMap]

The big bang is still happening, and we can see the universe expanding

The expansion that started with the big bang never stopped. It continues today, and we can see it happen by observing that large clusters of billions of stars, called galaxies, are moving away from us. (The exception is the Andromeda galaxy with which we are on a collision course.) The astronomer Edwin Hubble came to this conclusion when he observed that the light from other galaxies was red-shifted. The red shift is a consequence of the Doppler effect. This refers to how we see waves when the object that is creating the waves is moving toward us or away from us.

Figure 3: Duckling illustrates the Doppler effect in water. The ripples made in the direction the duckling is moving (blue lines) are closer together than the ripples behind the duckling (red lines).

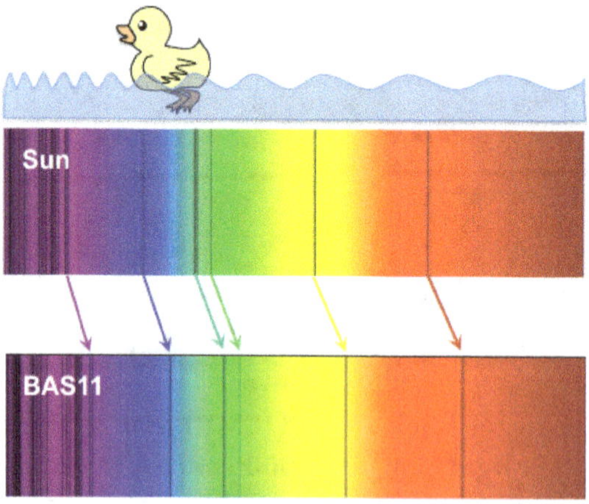

Figure 4: Red shift in light from the supercluster BAS11 compared to the sun's light. Black lines represent wavelengths absorbed by atoms (mostly H and He). For BAS11 the black lines are shifted toward the red end of the spectrum compared to the Sun.

Before we get to the Doppler effect as it pertains to the red shift, let's see how it works on something more tangible. The duckling swimming in Figure 3 is generating waves as it moves through the water. It is generating waves that move forward as well as back, but notice that the ripples ahead of the duckling are closer to each other than the ripples behind the duckling. The distance from one ripple to the next is called the wavelength. The wavelength is shorter in the direction that the duckling is moving, and longer as the duckling moves away.

When waves are in air as sound waves rather than in water as ripples, the different wavelengths manifest as sounds with different pitches — the short wavelengths have a higher pitch, and the long wavelengths have a lower pitch. This is why the pitch of a car's engine changes as the car races past you.

For light waves, wavelength translates to colour (Figure 4). In the spectrum of light that we can see, shorter wavelengths are on the blue end of the spectrum, and longer wavelengths are on the red end of the spectrum. Does this mean that galaxies look red because they are moving away from us? No, but the colour we see is shifted toward the red end of the spectrum and longer wavelengths.

Notice that the sun's spectrum in the upper part of Figure 4 has some black lines in it. The black lines are there because some colours are missing in the light we get from the Sun. Different elements absorb light of specific wavelengths, and many of the black lines in Figure 22.4 represent colours that are absorbed by hydrogen and helium within the Sun. This means the black lines are like a bar code that can tell us what a star is made of. The lower spectrum in Figure 4 is the light coming from BAS11, an enormous cluster of approximately 10,000 galaxies located 1 billion light years away. The black lines represent the same elements as in the Sun's spectrum, but they are shifted to the right toward the red end of the spectrum because BAS11 is moving away from us as the universe continues to expand. So to summarize, because almost all of the galaxies we can see have light that is red-shifted, it means they are all moving away from us. In fact, the farther away they are, the faster they are going. This is evidence that the universe is still expanding.

22.2 PLANETS FORMATION FROM THE REMNANTS OF EXPLODING STARS

If we were to take an inventory of the elements that make up Earth, we would find that 95% of Earth's mass comes from only four elements: oxygen, magnesium, silicon, and iron. Most of the remaining 5% comes from aluminum, calcium, nickel, hydrogen, and sulphur. We know that the big bang made hydrogen, helium, and lithium, but where did the rest of the elements come from?

The answer is that the other elements were made by stars. Sometimes stars are said to "burn" their fuel, but burning is not at all what is going on within stars.

The burning that happens when wood in a campfire is turned to ash and smoke is a chemical reaction — heat causes the atoms that were in the wood and in the surrounding atmosphere to exchange partners. Atoms group in different ways, but the atoms themselves do not change. What stars do is change the atoms. The heat and pressure within stars cause smaller atoms to smash together and fuse into new, larger atoms. For example, when hydrogen atoms smash together and fuse, helium is formed. Large amounts of energy are released when some atoms fuse and that energy is what causes stars to shine.

It takes larger stars to make elements as heavy as iron and nickel. Our Sun is an average star; after it uses up its hydrogen fuel to make helium, and then some of that helium is fused to make small amounts of beryllium, carbon, nitrogen, oxygen, and fluorine, it will be at the end of its life. It will stop making atoms and will cool down and bloat until its middle reaches the orbit of Mars. In contrast, large stars end their lives in spectacular fashion, exploding as supernovae and casting off newly formed atoms —including the elements heavier than iron — into space. It took many generations of stars creating heavier elements and casting them into space before heavier elements were abundant enough to form planets like Earth.

Until recently, astronomers have only been able to see stars that already contain heavier elements in small amounts, but not the first-generation stars that started out before any of the heavier elements were produced. That changed in June of 2015 when it was announced that a distant galaxy called CR7 had been found that contained stars made only of hydrogen and helium. The galaxy is so far away that it shows us a view of the universe from only 800 million years after the big bang.

22.3 BUILD A SOLAR SYSTEM

A solar system consists of a collection of objects orbiting one or more central stars. All solar systems start out the same way. They begin in a cloud of gas and dust called a nebula. Nebulae are some of the most beautiful objects that have been photographed in space, with vibrant colours from the gases and dust they contain, and brilliant twinkling from the many stars that have formed within them (Figure 22.5). The gas consists largely of hydrogen and helium, and the dust consists of tiny mineral grains, ice crystals, and organic particles.

Figure 5: Photograph of a nebula. The Pillars of Creation within the Eagle Nebula viewed in visible light (left) and near infrared light (right). Near infrared light captures heat from stars and allows us to view stars that would otherwise be hidden by dust. This is why the picture on the right appears to have more stars than the picture on the left. [NASA, ESA, and the Hubble Heritage Team (STScI/AURA) http://bit.ly/1Dm2X5a]

Step 1: Collapse a nebula

A solar system begins to form when a small patch within a nebula (small by the standards of the universe, that is) begins to collapse upon itself. Exactly how this starts isn't clear, although it might be triggered by the violent behaviour of nearby stars as they progress through their life cycles. Energy and matter released by these stars might compress the gas and dust in nearby neighbourhoods within the nebula.

Once it is triggered, the collapse of gas and dust within that patch continues for two reasons. One of those reasons is that gravitational force pulls gas molecules and dust particles together. But early in the process, those particles are very small, so the gravitational force between them isn't strong. So how do they come together? The answer is that dust first accumulates in loose clumps for the same reason dust bunnies form under your bed: static electricity. Given the role of dust bunnies in the early history of the solar system, one might speculate that an accumulation of dust bunnies poses a substantial risk to one's home (Figure 22.6). In practice, however, this is rarely the case.

Figure 6: Public service announcement. If you don't think housekeeping is important, then you don't understand the gravity of the situation. [KP, with inspiration from NASA/JPL, http://1.usa.gov/1K1dVcU]

Step 2: Make a disk and put a star at its centre

As the small patch within a nebula condenses, a star begins to form from material drawn into the centre of the patch, and the remaining dust and gas settle into a disk that rotates around the star. The disk is where planets eventually form, so it's called a protoplanetary disk. In Figure 7 the image in the upper left shows an artist's impression of a protoplanetary disk, and the image in the upper right shows an actual protoplanetary disk surrounding the star HL Tauri. Notice the dark rings in the protoplanetary disk. These are gaps where planets are beginning to form. The rings are there because incipient planets are beginning to collect the dust and gas in their orbits. There is an analogy for this in our own solar system, because the dark rings are akin to the gaps in the rings of Saturn (Figure 7, lower left), where moons can be found (Figure 7, lower right).

Figure 7: Protoplanetary disks and Saturn's rings. Upper left: An artists impression of a protoplanetary disk containing gas and dust, surrounding a new star. [NASA/ JPL-Caltech, http://1.usa.gov/1E5tFJR] Upper right: A photograph of the protoplanetary disk surrounding HL Tauri. The dark rings within the disk are thought to be gaps where newly forming planets are sweeping up dust and gas. [ALMA (ESO/NAOJ/NRAO) http://bit.ly/1KNCq0e]. Lower left: A photograph of Saturn showing similar gaps within its rings. The bright spot at the bottom is an aurora, similar to the northern lights on Earth. [NASA, ESA, J. Clarke (Boston University), and Z. Levay (STScI) http://bit.ly/1IfSCX5] Lower right: a close-up view of a gap in Saturn's rings showing a small moon as a white dot.

Step 3: Build some planets

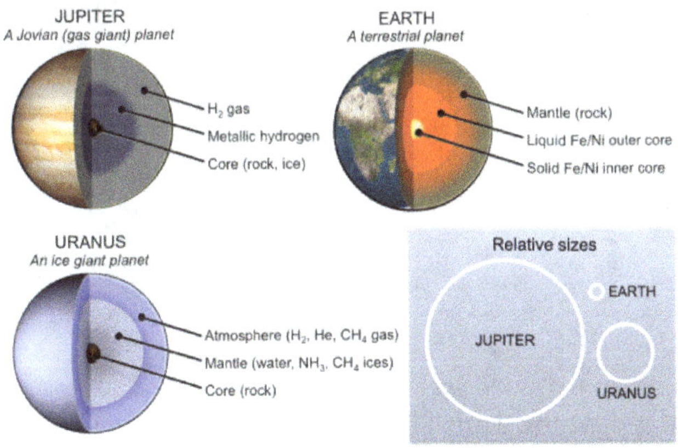

Figure 8: Three types of planets. Jovian (or gas giant) planets such as Jupiter consist mostly of hydrogen and helium. They are the largest of the three types. Ice giant planets such as Uranus are the next largest. They contain water, ammonia, and methane ice. Terrestrial planets such as Earth are the smallest, and they have metal cores covered by rocky mantles. [KP, after public domain images by FrancescoA, WolfmanSF

In general, planets can be classified into three categories based on what they are made of (Figure 8). Terrestrial planets are those planets like Earth, Mercury, Venus, and Mars that have a core of metal surrounded by rock. Jovian planets (also called gas giants) are those planets like Jupiter and Saturn that consist predominantly of hydrogen and helium. Ice giants are planets such as Uranus and Neptune that consist largely of water ice, methane (CH_4) ice, and ammonia (NH_3) ice, and have rocky cores. Often, the ice giant planets Uranus and Neptune are grouped with Jupiter and Saturn as gas giants; however, Uranus and Neptune are very different from Jupiter and Saturn.

These three types of planets are not mixed together randomly within our solar system. Instead they occur in a systematic way, with terrestrial planets closest to the sun, followed by the Jovian planets and then the ice giants (Figure 9). Smaller solar system objects follow this arrangement as well. The asteroid belt contains bodies of rock and metal. Bodies ranging from metres to hundreds of metres in diameter are classified as asteroids, and smaller bodies are referred to as meteoroids. In contrast, the Kuiper belt (Kuiper rhymes with piper), and the Oort cloud (Oort rhymes with sort), which are at the outer edge of the solar system, contain bodies composed of large amounts of ice in addition to rocky fragments and dust. (We will talk more about smaller solar system objects in a moment.)

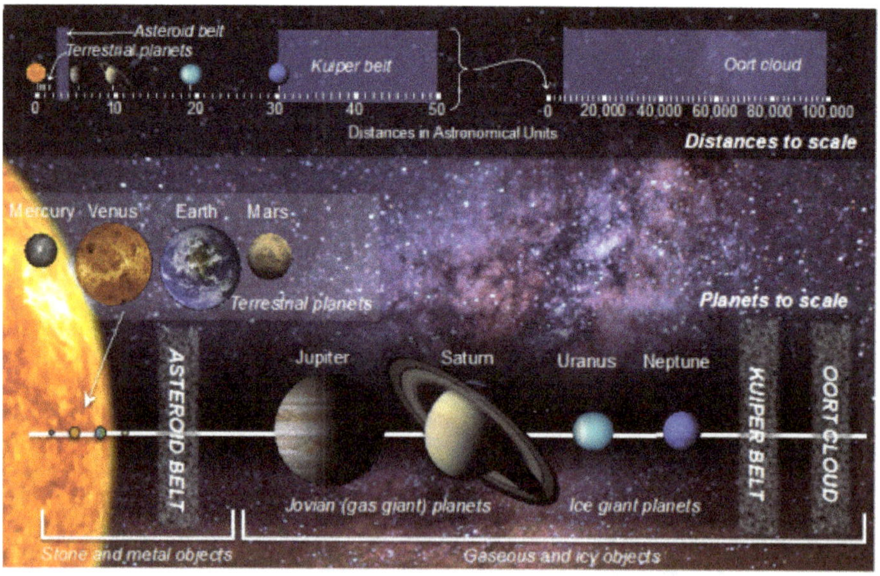

Figure 9: Our solar system. Top: The solar system shown with distances to scale. Distances are in astronomical units (AU), where 1 AU is the average distance from Earth to the Sun. The edge of the Kuiper belt extends to 50 AU (7.5 billion km), but this distance is minuscule compared to the size of the solar system as a whole, which extends to the edge of the Oort cloud, thought to be 15 trillion km away. Bottom: Solar system with the Sun and planets to scale. The gas giants are

the largest planets, followed by the ice giants, and then the terrestrial planets. Note that the planets in this diagram likely do not reflect the entire population of planets in our solar system because evidence suggests that large planets are present beyond the Kuiper belt.

Part of the reason for this arrangement is the frost line (also referred to as the snow line). The frost line separated the inner part of the protoplanetary disk closer to the sun, where it was too hot to permit anything but silicate minerals and metal to crystalize, from the outer part of the disk farther from the Sun, where it was cool enough to allow ice to form. As a result, the objects that formed in the inner part of the protoplanetary disk consist largely of rock and metal, while the objects that formed in the outer part consist largely of gas and ice. The young sun blasted the solar system with raging solar winds (winds made up of energetic particles), which helped to drive lighter molecules toward the outer part of the protoplanetary disk.

The objects in our solar system formed by accretion. Early in this process, particles collected in fluffy clumps because of static electricity. As the clumps grew larger, gravity became more important and collected clumps into solid masses, and solid masses into larger and larger bodies. If you were one of these bodies in the early solar system, and participating in the accretion game with the goal of becoming a planet, you would have to follow some key rules:

- Keep your velocity just right. If you move too fast and collide with another body, you both smash up and have to start again. If you move slowly enough, gravity will keep you from bouncing off each other and you can grow larger.
- Your distance from the Sun will determine how big you can get. If you are closer, there is less material for you to collect than if you are farther away.
- To begin with, you can only collect mineral and rock particles. You have to grow above a certain mass before your gravity is strong enough to hang onto gas molecules, because gas molecules are very light.
- As your mass increases, your gravity becomes stronger and you can grab material from farther away. The bigger you are, the faster you grow.

You would also have to watch out for some dangers:

- In the early stages of the game, the protoplanetary disk is turbulent, and you and other objects can get thrown into different orbits or at each other. This might be a good thing, or it might not, depending on how the rules above apply to you.
- If the game progresses to the point where there is no more material within your reach and you are not yet a planet, then it's game over.
- If you slow down too much (e.g., from bumping into other objects), you could spiral into the Sun (game over).

- If another planet gets big enough, it can:
- Rip you apart and then swing the pieces around so fast that for the rest of the game you collide too hard with other pieces to grow any bigger (game over)
- Fling you out of the solar system (game over)
- Grab you for itself (game over)
- Trap you in an orbit around it, turning you into a moon (game over, and incredibly humiliating)

The outcome of the game is evident in Figure 22.9. Today eight official winners are recognized, with Jupiter taking the grand prize, followed closely by Saturn. Both planets have trophy cases with more than 60 moons each, and each has a moon that is larger than Mercury. Prior to 2006, Pluto was also counted a winner, but in 2006 a controversial decision revoked Pluto's planet status. The reason was a newly formalized definition of a planet, which stated that an object can only be considered a planet if it is massive enough to have swept its orbit clean of other bodies. Pluto is situated within the icy clutter of the Kuiper belt, so it does not fit this definition. Pluto's supporters have argued that Pluto should have been grandfathered in, given that the definition came after Pluto was declared a planet, but to no avail. Pluto has not given up, and on July 13, 2015, it launched an emotional plea with the help of the NASA's New Horizons probe. New Horizons sent back images of Pluto's heart (Figure 10). On closer inspection, Pluto's heart was discovered to be broken.

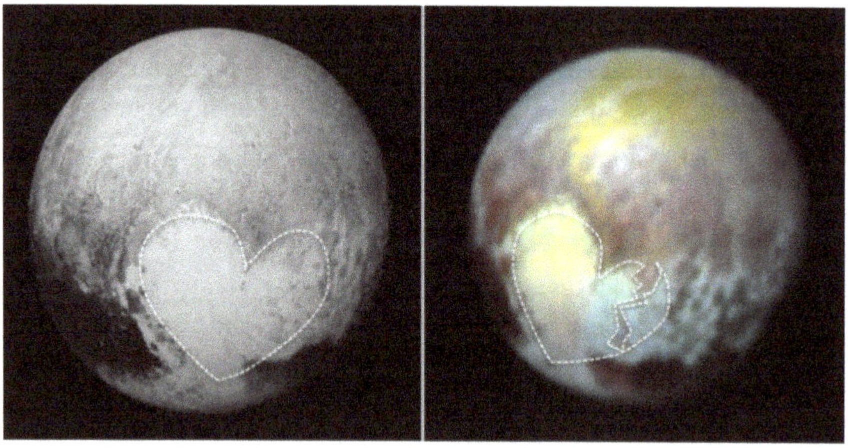

Figure 10: Photographs of Pluto. Left: The heart-shaped region called Tombaugh Regio is outlined

The rules and dangers of the planet-forming game help to explain many features of our solar system today.

- Proximity to the Sun explains why the terrestrial planets are so much smaller than the gas giant and ice giant planets.

- Mars is smaller than it should be, given the rule that distance from the Sun determines how much material a body can accumulate, and this can be explained by its proximity to Jupiter. Jupiter's immense gravity interfered with Mars' ability to accrete. Further evidence of Jupiter's interference is the debris field that forms the asteroid belt. From time to time, Jupiter still flings objects from the asteroid belt out into other parts of the solar system, some of which have collided with Earth to catastrophic effect.
- The Kuiper belt is an icy version of the asteroid belt, consisting of fragments left over from the early solar system. The material in the Kuiper belt is scattered because of Neptune's gravity. From time to time, Jupiter interferes here as well, flinging Kuiper belt objects toward the Sun and into orbit. As these objects approach the Sun, the Sun causes dust and gas to be blasted from their surface, forming tails. We know these objects as comets.
- Comets may also come from the Oort cloud where gravitational forces from outside of the solar system can hurl objects from the Oort cloud toward the Sun.

22.4 THE FIRST 2 BILLION YEARS OF EARTH

If you were to get into a time machine and visit Earth shortly after it formed (around 4.5 billion years ago), you would probably regret it. Large patches of Earth's surface would still be molten, which would make landing your time machine very dangerous indeed. If you happened to have one of the newer time-machine models with hovering capabilities and heat shields, you would still face the inconvenience of having nothing to breathe but a tenuous wisp of hydrogen and helium gas, and depending on how much volcanic activity was going on, volcanic gases such as water vapour and carbon dioxide. Some ammonia and methane might be thrown in just to make it interesting, but there would be no oxygen. Assuming you had the foresight to purchase the artificial atmosphere upgrade for your time machine, it would all be for naught if you materialized just in time to see an asteroid, or worse yet another planet, bearing down on your position. The moral of the story is that early Earth was a nasty place, and a time machine purchase is not something to take lightly.

The early Earth was hot

Chapter 9 explains that Earth's heat comes from the decay of radioactive elements within Earth, as well as from processes associated with Earth's formation. Let's look more closely at how those formation processes heated up Earth.

- Heat came from the thermal energy already contained within the objects that accreted to form the Earth.

- Heat came from collisions. When objects hit Earth, some of the energy from their motion went into deforming Earth, and some of it was transformed into heat. Clap your hands vigorously to experience this on a much smaller (and safer!) scale.
- As Earth became larger, its gravitational force became stronger. This increased Earth's ability to draw objects to it, but it also caused the material making Earth to be compressed, rather like Earth giving itself a giant gravitational hug. Compression causes materials to heat up.

Heating had a very important consequence for Earth's structure. As Earth grew, it collected a mixture of silicate mineral grains as well as iron and nickel. These materials were scattered throughout Earth. That changed when Earth began to heat up: it got so hot that both the silicate minerals and the metals melted. The metal melt was much denser than the silicate mineral melt, so the metal melt sank to Earth's centre to become its core, and the silicate melt rose upward to become Earth's crust and mantle. In other words, Earth unmixed itself. The separation of silicate minerals and metals into a rocky outer layer and a metallic core, respectively, is called differentiation. The movement of silicate and metal melts within Earth caused it to heat up even more.

Earth's high temperature early in its history also means that early tectonic processes were accelerated compared to today, and Earth's surface was more geologically active.

Earth was heavily bombarded by objects from space

Although Earth had swept up a substantial amount of the material in its orbit as it was accreting, unrest within the solar system caused by changes in the orbits of Saturn and Jupiter was still sending many large objects on cataclysmic collision courses with Earth. The energy from these collisions repeatedly melted and even vaporized minerals in the crust, and blasted gases out of Earth's atmosphere. Very old scars from these collisions are still detectable, although we have to look carefully to see them. For example, the oldest impact site discovered is the 3 billion year old Maniitsoq "crater" in west Greenland, although there is no crater to see. What is visible are rocks that were 20 km to 25 km below Earth's surface at the time of the impact, but which nevertheless display evidence of deformation that could only be produced by intense, sudden shock.

The evidence of the very worst collision that Earth experienced is not subtle at all. In fact, you have probably looked directly at it hundreds of times already, perhaps without realizing what it is. That collision was with a planet named Theia, which was approximately the size of Mars (Figure 11). Not long after Earth formed, Theia struck Earth. When Theia slammed into Earth, Theia's metal core merged with

Earth's core, and debris from the outer silicate layers was cast into space, forming a ring of rubble around Earth. The material within the ring coalesced into a new body in orbit around Earth, giving us our moon. Remarkably, the debris may have coalesced in 10 years or fewer! This scenario for the formation of the moon is called the giant impact hypothesis.

Earth's atmosphere as we know it took a long time to develop

Earth's first experiment with having an atmosphere didn't go well. It started out with a thin veil of hydrogen and helium gases that came with the material it accreted. However, hydrogen and helium are very light gases, and they bled off into space.

Earth's second experiment with having an atmosphere went much better. Volcanic eruptions built up the atmosphere by releasing gases. The most common volcanic gases are water vapour and carbon dioxide (CO_2), but volcanoes release a wide variety of gases. Other important contributions include sulphur dioxide (SO_2), carbon monoxide (CO), hydrogen sulphide (H_2S), hydrogen gas, and methane (CH_4). Meteorites and comets also brought substantial amounts of water and nitrogen to Earth. It is not clear what the exact composition of the atmosphere was after Earth's second experiment, but carbon dioxide, water vapour, and nitrogen were likely the three most abundant components.

Figure 11: Artist's impression of a collision between planets. A similar collision between Earth and the planet Theia might have given us our Moon. Fortunately for us, the collision that gave us the moon was a glancing blow rather than the direct hit shown here. Earth might not have survived a direct hit. [NASA/ JPL-Caltech, http://1.usa.gov/1IkP069]

One thing we can say for sure about Earth's second experiment is that there was effectively no free oxygen (O_2, the form of oxygen that we breathe) in the atmosphere. We know this in part because prior to 2 billion years ago, there were no sedimentary beds stained red from oxidized iron minerals. Iron minerals were present, but not in oxidized form. At that time, O_2 was produced in the atmosphere when the Sun's ultraviolet rays split water molecules apart; however, chemical reactions removed the oxygen as quickly as it was produced.

It wasn't until well into Earth's third experiment — life — that the atmosphere began to become oxygenated. Photosynthetic organisms used the abundant CO_2 in the atmosphere to manufacture their food, and released O_2 as a by-product. At first all of the oxygen was consumed by chemical reactions, but eventually the organisms released so much O_2 that it overwhelmed the chemical reactions and oxygen began to accumulate in the atmosphere, although present levels of 21% oxygen didn't occur until about 350 Ma. Today the part of our atmosphere that isn't oxygen consists largely of nitrogen (78%).

The oxygen-rich atmosphere on our planet is life's signature. If geologic process were the only processes controlling our atmosphere, it would consist mostly of carbon dioxide, like the atmosphere of Venus. It is an interesting notion (or a disconcerting one, depending on your point of view) that for the last 2 billion years the light reflected from our planet has been beaming a bar code out to the universe, similar to the ones in Figure 22.4, except ours says "oxygen." For 2 billion years, our planet has been sending out a signal that could cause an observer from another world to say, "That's odd... I wonder what's going on over there."

22.5 EXISTS ANOTHER EARTH?

If by that you mean, are there other planets where we could walk out of a spaceship with no equipment other than a picnic basket, and enjoy a pleasant afternoon on a grassy slope near a stream, then that remains to be seen. On the other hand, if you are asking if other planets exist that are rocky worlds approximately Earth's size, and orbiting within their star's habitable zone (the zone in which liquid water, and potentially life, can exist), then many planet hunters are cautiously optimistic that we have found at least 12 such worlds so far.

As of July 2015, NASA's Kepler mission has detected a total of 4,696 possible exoplanets. The Kepler spacecraft has an instrument to measure the brightness of stars, and looks for tiny variations in brightness that could be caused by a planet passing between the star it orbits and the instrument observing the star. Potential candidates are then examined in more detail to see whether they are in fact planets or not. So far 1,030 of those candidates have been confirmed as planets.

The uncertainty about the 12 possible Earth-like worlds is related to their composition. We don't yet know their composition; however, it is tempting to conclude that they are rocky because they are similar in size to Earth. Remember the rules of the accretion game: you can only begin to collect gas once you are a certain size, and how much matter you collect depends on how far away from the Sun you are. Given how large our gas giant and ice giant planets are compared to Earth, and how far away they are from the Sun, we would expect that a planet similar in size to Earth, and a similar distance from its star, should be rocky.

It isn't quite as simple as that, however. We are finding that the rules to the accretion game can result in planetary systems very different from our own, leading some people to wonder whether those planetary systems are strange, or ours is, and if ours is strange, how strange is it?

Consider that in the Kepler mission's observations thus far, it is very common to find planetary systems with planets larger than Earth orbiting closer to their star than Mercury does to the Sun. It is rare for planetary systems to have planets as large as Jupiter, and where large planets do exist, they are much closer to their star than Jupiter is to the Sun. To summarize, we need to be cautious about drawing conclusions from our own solar system, just in case we are basing those conclusions on something truly unusual.

On the other hand, the seemingly unique features of our solar system would make planetary systems like ours difficult to spot. Small planets are harder to detect because they block less of a star's light than larger planets. Larger planets farther from a star are difficult to spot because they don't go past the star as frequently. For example, Jupiter goes around the Sun once every 12 years, which means that if someone were observing our solar system, they might have to watch for 12 years to see Jupiter go past the Sun once. For Saturn, they might have to watch for 30 years.

So let's say the habitable-zone exoplanets are terrestrial. Does that mean we could live there?

The operational definition of "other Earths," which involves a terrestrial composition, a size constraint of one to two times that of Earth, and location within a star's habitable zone, does not preclude worlds incapable of supporting life as we know it. By those criteria, Venus is an "other Earth," albeit right on the edge of the habitable zone for our Sun. Venus is much too hot for us, with a constant surface temperature of 465°C (lead melts at 327°C). Its atmosphere is almost entirely carbon dioxide, and the atmospheric pressure at its surface is 92 times higher than on Earth. Any liquid water on its surface boiled off long ago. Yet the characteristics that make Venus a terrible picnic destination aren't entirely things we could predict from its distance from the sun. They depend in part on the geochemical evolution of Venus, and at one time Venus might have been a lot more like a youthful Earth. These are the kinds of things we won't know about until we can look carefully at the atmospheres and compositions of habitable-zone exoplanets.

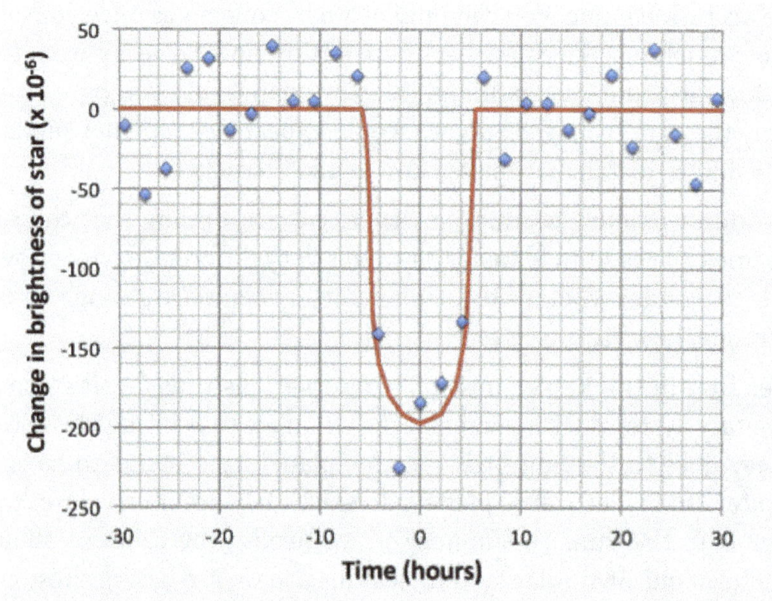

Plot showing how the star Kepler-452 dims as the planet Kepler-452b moves in front of it. [KP, after Jenkins, J. et al, 2015, Discovery and validation of Kepler-452b: a 1.6REarth super Earth exoplanet in the habitable zone of a G2 star, Astronomical Journal, V 150, DOI 10.1088/0004-6256/150/2/56.]

$$decrease\ in\ brightness = \frac{planet\ radius^2}{star\ radius^2}$$

Let's try this out for the Earth-like exoplanet called Kepler-452b. The first thing we need to know is the size of the host star Kepler-452. We can get that information by comparing its surface temperature and brightness to that of the sun. Start by calculating the ratios of the sun's temperature to the star's temperature, and the star's luminosity to the sun's luminosity using the data in Table 22.5. Record your answers in the table. Then find the star's radius using the following equation, and record your result:

$$star\ radius = sun\ radius \times \left(\frac{sun\ temperature}{star\ temperature}\right)^2 \times \sqrt{\frac{star\ luminosity}{sun\ luminosity}}$$

Table 5: Calculating the radius of star Kepler-452

Description	Sun	Kepler-452	Ratio
Temperature (degrees Kelvin)	5,778	5,757	
Luminosity (x 1026 watts)	3.846	4.615	
Radius (km)	696,300		

The second thing we need to know is how the brightness of Kepler-452 changes as planet Kepler-452b moves in front of it. Use the plot shown in this exercise box to find this information. Find the value on the y-axis where the red curve shows the most dimming from the planet and record your result in Table 6.

Table 6: Calculating the radius of planet Kepler-452b

Decrease in brightness*	Earth radius in km	Kepler-452b radius in km	Kepler-452b radius/Earth radius
$\times 10^{-6}$	6,378		

* Because we know this is a decrease, you don't need to keep the negative sign.

Use the following equation to find the radius of Kepler-452b:

$$planet\ radius = \sqrt{star\ radius^2 \times decrease\ in\ brightness}$$

To put the size of Kepler-452b in perspective, divide its radius by that of Earth and record your answer.

AT A GLANCE

Starting with a Big Bang	The universe began 13.77 billion years ago when energy, matter, and space expanded from a single point. Evidence for the big bang is the cosmic "afterglow" from when the universe was still very dense, and red-shifted light from distant galaxies, which tell us the universe is still expanding.
Forming Planets from the Remnants of Exploding Stars	The big bang produced hydrogen, helium, and lithium, but heavier elements come from nuclear fusion reactions in stars. Large stars make elements such as silicon, iron, and magnesium, which are important in forming terrestrial planets. Large stars explode as supernovae and scatter the elements into space.
How to Build a Solar System	Solar systems begin with the collapse of a cloud of gas and dust. Material drawn to the centre forms a star, and the remainder forms a disk around the star. Material within the disk clumps together to form planets. In our solar system, rocky planets are closer to the Sun, and ice and gas giants are farther away. This is because temperatures near the Sun were too high for ice to form, but silicate minerals and metals could solidify.

Earth's First 2 Billion Years	Early Earth was heated by radioactive decay, collisions with bodies from space, and gravitational compression. Heating melted Earth, causing molten metal to sink to Earth's centre and form a core, and silicate melt to float to the surface and form the mantle and crust. A collision with a planet the size of Mars knocked debris into orbit around Earth, and the debris coalesced into the moon. Earth's atmosphere is the result of volcanic degassing, contributions by comets and meteorites, and photosynthesis.
Are There Other Earths?	The search for exoplanets has identified 12 planets that are similar in size to Earth and within the habitable zone of their stars. These are thought to be rocky worlds like Earth, but the compositions of these planets are not known for certain.

GLOSSARY

Ablation melting of ice in the context of glaciation

Ablation till till that is formed when englacial and supraglacial sediments are deposited because the ice that was supporting them melts

Abyssal plain the flat surface of the deep ocean, typically beyond the limits of the continental slopes

Abyssalpelagic zone (18) the deeper parts of the ocean, between 4000 and 6000 m.

Accretion (plate tectonics) the process by which continental blocks (terranes) are added to existing continental areas

Accretion (planetary) the process by which solid celestial bodies are added to existing bodies during collisions

Acid rock drainage the production of acid from oxidation of sulphide minerals (especially pyrite) in either naturally or anthropogenically exposed rock

Aeolian processes related to transportation and deposition of sediments by wind

Aerobic processes that take place in the presence of abundant oxygen

Aerosol an aggregate of fine solid particles or a small droplet of liquid suspended in the air

Aftershock an earthquake that can be shown to have been caused by another earthquake

Aggregate unconsolidated materials (typically sediments) that are used in the construction industry

Albedo the reflectivity of a surface of a planet (expressed as the percentage of light that reflects)

Albite sodium-rich plagioclase feldspar

Alpine glacier a glacier formed in a mountainous region and confined to a valley (same as valley glacier)

Amphibole a double-chain ferromagnesian silicate mineral (e.G., Hornblende)

Amphibolite a foliated metamorphic rock in which the mineral amphibole is an important component

Amplification in the context of seismic shaking the process by which the amplitude of the seismic waves are enhanced, especially because the

Amplitude for any type of wave, the difference in height between a crest and the adjacent trough

Anaerobic processes that take place without oxygen

Andesite a volcanic rock of intermediate composition

Anion a negatively charged ion

Angular unconformity a geological boundary at the base of a sedimentary layer where the sedimentary rock beneath has been tilted or folded and then eroded

Anorthite calcium-rich plagioclase feldspar

Antarctic bottom water water at abyssal depths in the ocean that forms from the sinking of dense cold water adjacent to antarctica

Anticline an upward fold where the beds are known not to be overturned

Anthracite a high grade of coal (92 to 98% carbon) that is formed from deep burial and weak metamorphism

Anthropogenic resulting from the influence of humans

Antiform an upward fold where it is not known if the beds have been overturned

Aphanitic an igneous texture characterized by crystals that are too small to see with the naked eye

Aquifer a body of rock or sediment that has sufficient permeability to allow it to be used as a source of groundwater

Aquitard a body of rock or sediment that has insufficient permeability to allow it to be used as a source of groundwater

Arch a rock weathering remnant in the form of an arch (typically along a coast and resulting from wave erosion)

Arenite a sandstone with less than 15% silt and clay

Arete a sharp ridge that separates adjacent glacially carved valleys

Arkose a sandstone with more than 10% feldspar and more feldspar than lithic fragments

Arkosic arenite an arkose with less than 15% clay/silt matrix

Artesian well a well that is completed in a confined aquifer and in which the water level in the well rises above the top of the aquifer

Asteroid a rocky body orbiting the sun

Asteroid belt the region between the orbits of mars and jupiter that is populated with many asteroids

Asthenosphere the part of the mantle, from about 100 to 200 km below surface, within which the mantle material is close to its melting point, and therefore relatively weak

Asymmetrical in the context of folds, where the two sides of the fold make significantly different angles with respect to the axial plane

Atoll a ring-shaped carbonate (or coral) reef or series of islands

Atomic mass the total number of neutrons plus protons in an atom

Atomic number the total number of protons in an atom

Attitude the orientation of a sloping geological feature, such as a bedding plane or fracture

Auerole a zone of metamorphism around a source of heat such as a magma body

Axial plane a plane that can be traced through all of the hinge lines of a fold

Back reef the zone of shallow water on the shore-side of a reef

Background (geochemistry) the typical level of an element in average rocks or sediments

Backwash the wash of wave water down the slope of a beach

Banded iron formation an iron-bearing sedimentary rock that is rich in minerals such as hematite and magnetite, which may be interbedded with chert

Bank-full stage the water level of stream when it is in flood and just about to flow over its banks

Barrier reefa carbonate (or coral) reef that forms a barrier to waves along a coast

Basal sliding the motion of glacial ice along the base of a glacier that is warm enough to have liquid water

Basalt a volcanic rock of mafic composition

Base level in the context of a stream the base level is the lowest level that it can erode down to, as defined by the ocean, a lake or another stream that it flows into

Batholith an irregular body of intrusive igneous rock that has an exposed surface of at least 100 km2

Bathypelagic zone the moderately deep parts of the ocean, between 1000 and 4000 m.

Baymouth bar a spit that extends across the mouth of a bay

Beach face the part of the beach that is relatively steep and lies between the high and low tide levels

Bed a sedimentary layer

Bed load the fraction of a stream's sediment load that typically rests on the bottom and is moved by saltation and traction

Bedding repeated layering in a sedimentary rock

Bentonite a type of smectite clay that has strong swelling properties and is effective at absorbing dissolved ions

Berm a flat area of a beach in the backshore area (above the high tide level)

Big-bang theory the theory that the universe started with a giant explosion approximately 13.77 Billion years ago

Biotite a sheet silicate mineral (mica) that includes iron and or magnesium, and is therefore a ferromagnesian silicate

Biozone a stratigraphic interval that can be defined on the basis of a specific fossil

Bituminous a medium-grade type of coal with 70 to 92% carbon

Blueschist a metamorphic facies characterized by relatively low temperatures and high pressures, such as can exist within a subduction zone

Body wave a seismic wave that travels through rock (e.G., A p-wave or an s-wave)

Boulder a sediment clast with a diameter of at least 256 mm

Bowen reaction series the scheme that defines the typical order of crystallization of minerals from magma

Braided a stream pattern which is characterized by abundant sediment and numerous intertwining channels around bars

Breakwater a structure built offshore in order to deflect the energy of waves

Breccia a sedimentary- or volcanic-rock texture characterized by angular clasts

Brunisol a relatively immature forest soil, lacking in well-defined horizons

Caldera a volcanic depression that is many times larger than the volcanic vents within it

Caliche a white calcium-carbonate rich layer within soils in arid regions

Calving the loss of ice from the front of a glacier by collapse into water

Canadian shield the exposed part of the continent laurentia

Carbonate a mineral in which the anion is co3-2

Carbonate compensation depth the depth in the ocean (typically around 4000 m) below which carbonate minerals are soluble

Cation a positively charged ion

Cementation the process by which minerals are precipitated between grains in sediments

Cenozoic the most recent of the eras, representing the past 65.5 Ma of geological time

Chemical sedimentary rock a sedimentary rock comprised of material that was transported as ions in solution

Chernozem a black soil typical of grasslands in cold climates such as the canadian prairies

Chert a very fine grained sedimentary rock formed almost entirely of silica

Chlorite a ferromagnesian sheet silicate mineral, typically present as fine crystals and forming from the low-temperature metamorphism of mafic rock

Cinder cone a steep-sided volcano comprised almost entirely of loose rock fragments and typically formed during a single eruptive event

Cirque a steep-sided semi-circular basin eroded by a glacier at the head of its valley

Clast a sedimentary fragment of mineral or rock

Clastic sedimentary rock a sedimentary rock comprised of material that was transported as clasts or fragments

Clay sediment particle that is less than 1/256 mm in diameter

Clay mineral a hydrous sheet silicate mineral that typically exists as clay-sized grains

Claystone a sedimentary rock comprised mostly of clay-sized grains

Cleavage the tendency for a mineral to break along smooth planes that are predetermined by its lattice structure

Climate feedback a process by which the physical effects of a climate forcing can have other effects (either negative or positive) on the climate

Climate forcing a mechanism, such as a change in greenhouse gas levels, that forces the climate to change

Coal-bed methane methane that is trapped within the porosity of coal

Coastal straightening the tendency for an irregular coast to be straightened over time by coastal erosion processes

Cobble sediment particle that is between 64 and 256 mm in diameter

Col the low point or pass along a ridge between two glacial valleys

Columnar jointing the fracturing of rock or sediment (but typically volcanic rock) into columns that are typically 6-sided

Composite volcano (or stratovolcano) a volcano that is constructed of alternating layers of pyroclastic debris and lava flows

Concentrate (mining) a product of ore processing that includes a specific ore mineral, separated from the rest of the rock

Concordant parallel to pre-existing layering or foliation within a rock

Cone of depression the depression of the water table around a well that is heavily pumped

Confined aquifer an aquifer that lies below a confining layer

Confining layer an aquitard that overlies an aquifer and restricts the flow of water down from the surface

Conglomerate a sedimentary rock that is comprised predominantly of rounded grains that are larger than 2 mm

Contact metamorphism metamorphism that takes place adjacent to a source of heat, such as a body of magma

Continental drift the concept that tectonic plates can move across the surface of the earth

Continental glacier a glacier that covers a significant part of a continent and has an area of at least 50,000 km2

Continental shelf the shallow (typically less than 200 m) and flat sub-marine extension of a continent

Continental slope the steeper part of a continental margin, that slopes down from a continental shelf towards the abyssal plain

Contractionism the now discredited theory that mountain ranges formed as a result of the contraction of the earth

Convergent boundary a plate boundary at which the two plates are moving towards each other

Cordilleran ice sheet the continental glacier that covered part of western north america, including almost all of british columbia, part of the yukon, and part of northern washington, during the pleistocene glaciations

Core (1) the metallic interior part of the earth, extending from a depth of 2900 km to the centre

Core-mantle boundary the boundary, at 2900 km depth, between the mantle and the core

Coriolis effect the tendency for moving bodies (e.G., Ocean currents) to rotate on the surface of the earth, clockwise in the northern hemisphere and counter-clockwise in the southern hemisphere

Cosmic microwave background radiation left over from the an early stage in the development of the universe at the time when protons and neutrons were recombining to form atoms

Country rock the original rock of a region, into which younger rock (typically igneous) rock has been intruded

Covalent bond a bond between two atoms in which electrons are shared

Crater a volcanic depression that is related to a specific volcanic vent

Craton a region of ancient (typically precambrian) crystalline rock (equivalent to a shield)

Creep the very slow (mm to cm per year) flow of unconsolidated material on a gentle slope

Crest the highest point on a wave

Crevasse an open fissure on the surface of a glacier

Cross bedding small-scale inclined bedding within larger horizontal beds

Crust the uppermost layer of the earth, ranging in thickness from about 5 km (in the oceans) to over 50 km (on the continents)

Cyanobacteria photosynthetic bacteria that evolved in the early archean

D" layer (d-double-prime layer) a low seismic velocity zone within the basal 200 km of the mantle

Debris flow a gravity-driven flow of water and sediment that includes a significant proportion of coarse (cobble to boulder) material

Decline in mining a decline is a sloped tunnel used to access lower parts of a mine with wheeled equipment

Decompression melting melting (or partial melting) of rock resulting from a reduction in pressure without a significant reduction in temperature

Dendritic a pattern of drainage channels that resembles the branches in a tree

Density weight per volume of a substance (e.G., G/cm3) used widely in the context of minerals or rocks

Deranged a pattern of drainage channels that is chaotic

Detrital referring to fragments of rocks or minerals

Diatom photosynthetic algae that make their tests (shells) from silica

Differentiation the un-mixing of a magma, typically by the physical separation of minerals that crystallize early and settle towards the bottom

Diorite an intermediate intrusive igneous rock

Dip the angle below horizontal at which a sedimentary bed or other feature slopes

Discharge the volume of water flow in a stream expressed in terms of volume per unit time (e.G., M3/s)

Discharge area the part of an aquifer where groundwater discharge takes place

Disconformity a boundary between parallel sedimentary layers where some erosion of the lower layer has taken place

Discordant a geological feature that is not parallel to any existing layering in the country rock

Divalent an ion with a charge or +2 or -2

Divergent a plate boundary at which the two plates are moving towards away from each other

Dodecahedron an object with twelve surfaces, such as a garnet crystal

Dolomite a calcium-magnesium carbonate mineral (ca,mg)co3

Dolomitization the addition of magnesium to limestone during which some or all of the calcium carbonate is converted to dolomite

Dolostone a carbonate rock made up primarily of the mineral dolomite

Drainage basin the catchment area of a stream, including the area where all surface water drains into the stream

Drop stone a fragment of rock within otherwise fine-grained sediment that has been dropped from floating ice on a body of water

Drumlin a streamlined glacial erosional feature comprised of sediments and/or bedrock

Dyke a tabular intrusive igneous body that is discordant to any existing layering in the country rock

Eccentricity in the context of milankovitch cycles, the degree to which the sun is offset from the geometric centre of the earth's orbit

Eclogite a garnet-pyroxene-glaucophane bearing rock that is the product of high-pressure metamorphism of oceanic crustal rock, typically within a subduction zone

Effusive a volcanic eruption dominated by the relatively gentle flow of lava

El niño a periodic climatic situation in which warm water extends all or most of the way to the eastern edge of the equatorial pacific

Elastic deformation the deformation of material (including rock) from which it can fully recover if the stress is removed

Electron a sub-atomic particle of essentially no mass and a single negative charge

End moraine a deposit of sediment that accumulates at the front of a glacier

Englacial within a glacier, referring especially to sediment carried within the glacial ice

Epicentre the location on the surface vertically above the location (i.E., "Hypocentre" or "focus") where an earthquake takes place

Epipelagic zone the upper layer of water (0 to 200 m) in areas of the open ocean

Epithermal deposit a mineral deposit formed near to surface in an area of hydrothermal activity, typically associated with a body of magma

Equilibrium line on a glacier, the line between the zone of accumulation and the zone of ablation (in late summer the equilibrium line is the boundary between snow-covered ice and bare ice)

Equipotential lines in the context of groundwater an equipotential line connects locations with equal hydraulic head or water pressure

Esker a ridge of sediment deposited by a sub-glacial stream

Eustatic sea level change sea level change related to a change in the volume of the oceans, typically because of an increase or decrease in the amount of glacial ice on land

Exfoliation the fracturing of rock that results from a reduction in the pressure when overlying rock is eroded away

Exoplanet a planet that orbits a star other than the sun

Extrusive igneous rock that cooled at surface

Fall in mass wasting, the vertical or near-vertical fall of rock

Fault a boundary in rock or sediment along which displacement has taken place

Feedback a process by which the physical effects of a climate forcing can have other effects (either negative or positive) on the climate

Feldspar a very common framework silicate mineral

Felsic silica rich (>65% sio2) in the context of magma or igneous rock

Ferric the oxidized form of an ion of iron (fe3+)

Ferromagnesian referring to a silicate mineral that contains iron and or magnesium

Ferrous the reduced (non-oxidized) form of an ion of iron (fe2+)

Fetch the distance over which wind blows to form waves

Finger lake a lake that occupies a glacial valley

Firn the granular transitional state between snow and ice within a glacier

Flood plain the area that is occupied by water when a stream floods and overtops its banks

Flow the fluid-like motion of material during mass-wasting

Flow path the path that groundwater flows along between a recharge area and a discharge area

Flowing artesian well an artesian well in which the water level naturally rises above the surface of the ground

Flux melting melting of rock that is facilitated by the addition of a flux (typically water) which lowers the rocks melting point

Focus (earthquake) the actual point below surface at which an earthquake takes place (equivalent to hypocentre)

Foliated the existence of foliation in a metamorphic rock

Foliation the alignment of mineralogical or structural features of a rock – especially a metamorphic rock

Footwall the lower surface of a non-vertical fault

Foraminifera a single-celled protist with a shell that is typically made of caco3

Fore-reef the zone on the ocean side of a reef

Formation a unit of sedimentary rock that is lithologically consistent and sufficiently thick and extensive to be shown on a geological map at the scale that is typically used in the area in question

Fracking fracturing rock by injecting water and chemicals down a well at very high pressure (equivalent to hydraulic fracturing)

Fractional crystallization the sequential crystallization of minerals from magma, and the physical separation of early-forming crystals from the magma in the area where they crystallized

Fracture a break within a body of rock in which the rock on either side is not displaced

Fringing reef a reef adjacent to a shoreline where there is either a very narrow back reef area or none at all (in which case the reef is effectively attached to the shore)

Frost line in the context of planetary systems the boundary beyond which volatile components (e.G., Water, carbon dioxide, methane, ammonia etc.) Are frozen

Frost wedging the situation where the expansion of freezing water pries rock apart

Ga (gigaannus) billions of years before the present

Gabbro a mafic intrusive igneous rock

Gaia hypothesis the hypothesis advanced by james lovelock that the organisms have affected the atmosphere and oceans such that conditions on earth have been kept habitable, in spite of significantly changing energy received from the sun

Galaxy a gravitationally-bound system of stars and interstellar matter

Gas giant a large planet composed mostly of hydrogen and helium (e.G. Jupiter)

Geosyncline a kilometres thick deposit of sediments that has accumulated along the edge of a continent and is sufficient mass to depress the crust beneath it

Geothermal gradient the rate of increase of temperature with depth in the earth (typically around 30° c/km within the crust)

Giant impact hypothesis the theory that the moon formed when a mars-sized planet (theia) collided with the earth at 4.5 Ga

Glacial period a period of earth's history during which glacial ice was present over a sufficient extent to have left recognizable evidence

Glacial groove a straight line created on a rock surface by erosion by a rock fragment embedded in the base of glacial ice (larger and deeper than a glacial striation)

Glacial striation a straight line created on a rock surface by erosion by a rock fragment embedded in the base of glacial ice (finer than a glacial groove - typically less than 1 cm wide)

Glacier a long lasting (centuries or more) body of ice on land that moves under its own weight

Glaciofluvial referring to sediments deposited from a stream that is derived from a glacier

Glaciolacustrine referring to sediments deposited within a lake in a glacial environment

Glaciomarine referring to sediments deposited within the ocean in a glacial environment

Glaucophane a blue-coloured sodium-magnesium bearing amphibole mineral that forms during metamorphism at high pressures and relatively low pressures, typically within a subduction zone

Gneiss high-grade metamorphic rock in which the mineral components are separated into bands

Graben a down-dropped fault block, bounded on either side by normal faults

Grade in the context of a mineral deposit, the amount of a specific metal or mineral expressed as a proportion of the whole rock

Graded bedding an individual sedimentary layer that shows a distinctive gradation in grain size (normal graded bedding is finer towards the top, reverse graded bedding is coarser towards the top)

Gradient the slope of a stream bed over a specific distance, typically expressed in m per km

Granite a felsic intrusive igneous rock

Granule a sedimentary particle ranging in size from 2 to 4 mm in diameter

Greenhouse gas a gaseous molecule with 3 or more atoms that is able to absorb infrared radiation

Greenhouse effect in the context of climate, the ability of an atmosphere to absorb infrared radiation due to the presence of greenhouse gases

Greenschist a foliated metamorphosed rock (typically derived from basalt) in which the green colouration is derived from either chlorite, epidote or green amphibole

Greenstone a non-foliated metamorphosed rock (typically derived from basalt) in which the green colouration is derived from either chlorite, epidote or green amphibole

Greywacke a sandstone with more than 15% silt and clay, and with a significant proportion of sand-sized rock fragments

Groundwater water that lies beneath the surface of the ground

Group a stratigraphically-continuous series of related formations

Groyne a man-made structure extending from the shore built to deflect the energy of waves

Gyre a closed circular ocean current

Habi a characteristic crustal form or combination of forms of a mineral

Habitable zone the region around a star that is considered to be suitable for a life-bearing planet

Hadean the first eon of earth history, extending from 4.57 To 3.80 Ga

Halide a mineral in which the anion is one of the halide elements (e.G., Halite - nacl or fluorite - caf2)

Halite nacl, a halide mineral also known as table salt

Halogen an element in the second-last column of the periodic table that forms anions with a negative-1 charge

Hanging valley a glacial valley created by a tributary glacier which does not erode as deeply as the main-valley glacier that it joins

Hanging wall the upper surface of a non-vertical fault

Headland a point extending out to sea

Horn a peak that has been eroded on at least three sides by glaciers

Hornfels a fine-grained metamorphic rock that is not foliated

Horst an uplifted fault block, bounded on either side by normal faults

Hot spot the surface area of volcanism and high heat flow above a mantle plume

Hydrated mineral a mineral that includes either hydroxyl (oh) or water (h2o) in its chemical formula (e.G., Gypsum caso4.2H2o)

Hydraulic conductivity an expression of the rate at which a liquid will flow through a porous medium, as determined by the permeability of the medium and the viscosity of the liquid

Hydraulic fracturing fracturing rock by injecting water and chemicals down a well at very high pressure (equivalent to fracking)

Hydrolysis a reaction between a mineral and water in which h+ ions are added to the mineral and a chemically equivalent amount of cations are released into solution

Hydroxide the anion oh- or an mineral that includes that anion

Hydrothermal alteration chemical alteration of minerals by hot water solutions

Hypocentre the actual point below surface at which an earthquake takes place (equivalent to focus)

Ice giant a planet that is comprised mainly of gases heavier than hydrogen and helium, including oxygen, carbon, nitrogen, and sulfur (e.G., Uranus and neptune)

Igneous a rock formed from the cooling of magma

Illite a clay mineral with a composition similar to that of muscovite mica

Imbricate aligned and overlapping, like the tiles on a roof

Index fossil a fossil with a distinctive appearance and a wide geographic range but from a relatively restricted time range, thus making it useful for dating a correlating rocks from different regions (the most useful index fossils are from organisms that lived for less than a million years)

Inert in chemistry, an element that does not readily react with other elements (e.G., Neon)

Infiltration the recharge of groundwater from the downward percolation of surface water

Insolation a measure of the intensity of solar energy at a specific location or time (expressed in w/m2)

Intensity in seismology, a qualitative measure of the amount of shaking at specific location, based on what was felt by observers, or the amount of damage done

Intergovernmental panel on climate change (ipcc) an international body established in 1988 by the un's world meteorological organization and the un environment program to prepare periodic reports on the status of global climate change and its mitigation

Intrusive an igneous rock that has cooled slowly beneath the surface

Ionic bond a bond in which electrons are transferred from one atom to another, thus forming ions

Ion an atom that has either gained or lost electrons and has thus become charged (or a group of atoms that also has a charge – e.G., Hco3-)

Isoclinal fold a tight fold in which the limbs are parallel to each other

Isostasy the equilibrium between a block of crust floating on the underlying plastic mantle

Isostatic sea level change the effect on relative sea level of a vertical adjustment of the crust resulting from a change in the mass of the crust (e.G., From losing or gaining ice)

Isotope an form of an element that differs from other forms because it has a different number of neutrons (e.G., 16O has 8 protons and 8 neutrons while 18o has 8 protons and 10 neutrons)

Joint a fracture in rock

Jovian planet a gas giant

Ka (kiloannus) thousands of years before the present

Kaolinite a clay mineral that does not have cations other than al and si

Karst the solutional erosion of an area with soluble rock (typically limestone) to form depressions and caves

Kettle a depression formed at the front of a large glacier when a stranded ice block that was surrounded by sediment eventually melts

Kettle lake a lake that forms within a kettle

Kimberlite an ultramafic volcanic rock that originates at significant depth (> 200 m) in the mantle (some kimberlites include diamonds)

Kuiper belt a region of the solar system beyond the orbit of neptune that is populated by small objects and dwarf planets (including pluto)

Laccolith concordant intrusion in which the central part has formed an upward dome

Lahar a mudflow or debris flow that is either caused by a volcanic eruption, or forms on the flank of a volcano as a result of flooding not related to an eruption

Landfill gas gases produced within a landfill during the microbial breakdown of landfill components (most are dominated by carbon dioxide and methane)

Large igneous province a very large area of mafic volcanic rock produced by a massive eruption typically related to a mantle plume

Lateral moraine a deposit of rocky material that forms along the margin of a valley or alpine glacier, mostly from the freeze-thaw release of material from the steep slopes above

Lattice the regular and repeating three-dimensional structure of a mineral

Laurentide ice sheet the continental glacier that extended across central eastern north america during the pleistocene, covering most of canada and a significant part of the united states

Lava levée a ridge that forms along the edge of a lava flow because the magma at the edge cools faster than that in the middle

Lava tube a tube that forms as mafic lava flows along a channel and lava leveés build up on either side, eventually forming a roof (once a lava tube forms it insulates the flowing magma, allowing it to stay hot a liquid for longer and therefore flow much further)

Leachate in the context of landfills, the liquid (rainwater) that passes through the waste and becomes contaminated with soluble components from the waste

Levée on a stream, the ridge that naturally forms along the edge of the channel during flood events

Level in mining, a horizontal mine opening

Light year the distance that light can travel in one year (9.4607 X 1012 km)

Lignite a low-grade type of coal with less than 70% carbon

Limbs the layers of rock on either side of a fold

Limestone (6) a sedimentary rock that is comprised mostly of calcite

Liquefaction the tendency for unconsolidated and water saturated sediments to lose strength during seismic shaking

Lithic arenite an arenite in which there is more than 10% lithic clasts and in which there are more lithic clasts than feldspar clasts

Lithic clasts fragments of rock (e.G., Basalt) that are included in the sand-sized grains in sandstone, or in the larger grains in conglomerate

Lithification the conversion of unconsolidated sediments into rock by compaction and cementation

Lithosphere the rigid outer part of the earth, including the crust and the mantle down to a depth of about 100 km

Lodgement till sediment that accumulates at the base of a glacier and typically has a wide range of grain sizes (including clay) and is well compacted

Longshore current the movement of water along a shoreline produced by the approach of waves at an angle to the shore

Longshore drift the movement of sediment along a shoreline resulting from a longshore current and also from the swash and backwash on a beach face

Love wave a surface seismic wave, with horizontal motion, that develops in relatively weak (e.G., Unconsolidated) materials at surface

Luvisol a cold climate forest soil formed in which clay has been removed from the a horizon and relocated into the b horizon

Ma (megaannus) millions of years before the present

Mafic silica poor (<45% sio2) in the context of magma or igneous rock

Magma molten rock typically dominated by silica

Magnetic chronology the study of the timing of reversals of the earth's magnetic field, and the application of that understanding to dating geological materials

Magnitude a measure of the amount of energy released by an earthquake

Mantle the middle layer of the earth, dominated by iron and magnesium rich silicate minerals and extending for about 2900 km from the base of the crust to the top of the core

Mantle plume a plume of hot rock (not magma) that rises through the mantle (either from the base or from part way up) and reaches the surface where it spreads out and also leads to hot-spot volcanism

Marble metamorphosed limestone (or dolostone) in which the calcite or dolomite has been recrystallized into larger crystals

Mass wasting the mass failure, by gravity, of rock or unconsolidated material on a slope

Meander cutoff the formation of a shorter stream channel across the narrow boundary between two meanders on a stream

Meandering the sinuous path taken by a stream within a wide flat flood plain

Medial moraine a lateral moraine that has been shifted towards the centre of a valley glacier at a point where two glaciers meet

Member a subdivision of a formation

Mesopelagic zone the upper middle zone of the open ocean extending from 200 to 1000 m depth

Metallic lustre the lustre of a mineral into which light does not penetrate but only reflects off of the surface

Metallic bond a type of bond in which abundant electrons are easily shared amongst cations

Metamorphism the transformation of a parent rock into a new rock as a result of heat and pressure that leads to the formation of new minerals, or recrystallization of existing minerals, without melting

Metasomatism metamorphism facilitated by ion transfer through water

Meteoroid a fragment of either stony or metallic debris in space

Methane hydrate a combination of water ice and methane in which the methane is trapped inside "cages" in the ice

Mica a sheet silicate mineral (e.G., Biotite)

Migmatite a rock that is a mixture of metamorphic and igneous rock, formed at very high grades of metamorphism when a part of the parent rock starts to melt

Milankovitch cycles millennial-scale variations in the orbital and rotational parameters of the earth that have subtle effects on the earth's climate

Mohorovičić discontinuity the boundary between the crust and the mantle

Moment magnitude a way of estimating earthquake magnitude based on the area of the rupture surface and the amount of displacement

Monogenetic a volcano that forms in a single eruptive event

Moraine lake a finger lake that forms within a glacial valley and is dammed by an end moraine

Mud crack a dessication crack formed in mud that has accumulated in a small body of water that later dries up or drains

Mudflow a mass-wasting event involving the flow of mud (sand, silt and clay) within a channel

Mudrock an inclusive term for mudstone, shale and claystone

Muscovite a potassium-bearing non-ferromagnesian mica

Native element (also native element mineral) a mineral that consists of only one element (e.G., Native gold)

Nebula a cloud of interstellar dust and gases

Negative feedback a process that results in a decrease in that process (in the context of climate change it is a process that reduces the change in climate, such as the enhanced growth of vegetation in response to an increase in atmospheric carbon dioxide)

Neutron a sub-atomic particle with a mass of 1 and a charge of 0

Nonconformity a geological boundary where non-sedimentary rock is overlain by sedimentary rock

Non-ferromagnesian mineral a silicate mineral that does not contain iron or magnesium (e.G., Feldsspar)

Non-metallic lustre the lustre of a mineral into which light does penetrate

Normal fault a non-vertical fault along which the hanging wall (upper surface) has moved down relative to the footwall

Normal force the component of the gravitational force that acts directly into the slope

North atlantic deep water deep atlantic ocean water that has descended in the far north of the basin in the area between scandinavia and greenland

Nunatuk a rocky peak that extends above the ice level of a continental glacier

Obliquity in the context of milankovitch cycles, the angle of the tilt of the earth's rotational axis with respect to the plane of its orbit around the sun

Ocean plain the extremely flat surface of the deep ocean floor in areas unaffected by plate tectonic processes and volcanism

Oil window the depth range, which is approximately 2000 to 4000 m, within which the temperature is appropriate for the formation of oil from organic matter in sedimentary rock

Ooid a small (approximately 1 mm) sphere of calcite formed in areas of tropical shallow marine water with strong currents

Olivine a silicate mineral made up of isolated silica tetrahedra and with either iron or magnesium (or both) as the cations

Oort cloud a spherical cloud of icy objects extending from between about 5,000 and 500,000 astronomical units (sun-earth distances) from the sun (thought to be the source area of comets)

Open-pit mine a mine that is open to the surface

Outcrop a surface exposure of rock that is part of the crust (bedrock)

Outwash plain an extensive region of sand and gravel deposited by streams flowing out of a glacier (same as sandur)

Overturned a geological feature that has been tilted to the point where it is upside down

Oxbow a part of a stream meander that has become isolated from the rest of the stream as the result of a meander cutoff

Oxidation the reaction between a mineral and oxygen

Oxide a mineral in which the anion is oxygen (e.G., Hematite fe2o3)

Pahoehoe a lava flow with a ropy surface texture formed when the surface cools and hardens while the lava beneath is still flowing

Paleomagnetic past variations in the intensity and polarity of the earth's magnetic field

Pangea (10) the supercontinent that existed between approximately 300 and 180 ma

Paraconformity an interruption representing a period of non-deposition, without tilting or erosion, in a sequence of sedimentary rocks

Parasitic fold a fold within a fold

Parent rock the rock that was already in existence when a process of metamorphism started

Partial melting the process during which a only specific mineral components of a rock melt in response to changing conditions

Parting a narrow gap between individual sedimentary layers

Passive margin a boundary between a continent and an ocean at which there is no tectonic activity (e.G., The eastern edge of north america)

Paternoster lake one of a series of rock basin lakes

Pebble a sedimentary particle ranging in size from 2 to 64 mm (includes granule)

Pelagic the part of a lake or the ocean that is not close to shore

Permafrost ground that remains frozen for two or more years

Permanentism the now discredited theory that the features on the earth have not changed significantly over geological time

Permeability an expression of the ease with which liquid will flow through a porous medium

Phaneritic a rock texture in which the individual crystals or grains are visible to the naked eye

Phanerozoic the most resent eon of geological time, encompassing the paleozoic, mesozoic and cenozoic eras

Phenocryst a relatively large crystal within an igneous rock

Phyllosilicate a silicate mineral in which the silica tetrahedra are made up of sheets

Phosphate a mineral in which the anion is po43-

Photic zone the upper 200 m of the ocean or a lake, where, depending on the turbidity of the water, light can penetrate

Phreatic eruption a steam-drive volcanic eruption that takes place when surface or near-surface water is heated by volcanic activity

Phyllite a metamorphic rock with slaty cleavage and a sheen on the surface produced by aligned micas

Pillow a pillow-shaped mass of volcanic rock (typically basalt) formed when magma erupts beneath the surface

Pillow lava a volcanic rock (typically basalt) that is made up primarily of pillows

Pipe a cylindrical body of igneous rock, typically resulting from a feeder conduit to a volcano

Plate a region of the lithosphere that is considered to be moving across the surface of the earth as a single unit

Plate tectonics the concept that the earth's crust and upper mantle (lithosphere) is divided into a number of plates that move independently on the surface and interact with each other at their boundaries

Plinian eruption a large volcanic eruption in which a column of hot tephra and gases rises many kilometres into the atmosphere

Pluton a body of intrusive igneous rock

Podsol a soil with well-developed horizons formed in temperate forested regions

Podsolization the process of the formation of podsol

Polar wandering path a path of varying magnetic pole positions defined by paleomagnetic data (in fact it is now understood that the continents have wandered, not the poles, so a more appropriate terms is "apparent polar wandering path")

Polymerize the formation of molecular chains within a fluid (e.G., A magma) that lead to an increase in the fluid's viscosity

Polymorphs two or more minerals with the same chemical formula but different crystal structures

Porosity the percentage of open pore space within a body of rock or sediment

Porphyritic an igneous texture in which some of the crystals are distinctively larger than the rest

Porphyry deposit a mineral deposit (of copper or molybdenum especially) in which part of the host rock is a porphyritic stock

Positive feedback a process that results in an increase in that process (in the context of climate change it is a process that enhances the change in climate, such as the reduced reflectivity of the earth's surface when ice melts)

Potassium feldspar feldspar with the formula kalsi3o8

Potentiometric surface the imaginary surface defined by the levels to which water would rise in a series of wells drilled into a confined aquifer

Precession in the context of milankovitch cycles, the variation in the direction at which the earth's rotational axis is pointing

Principle of cross-cutting relationships the principle that a body of rock that cuts across or through another body of rock is younger than that other body

Principle of faunal succession the principle that life on earth has evolved in an orderly way, and that we can expect to always find fossils of a specific type in rocks of a specific age

Principle of inclusions the principle that inclusions within a body of rock must be older than the rock

Principle of original horizontality the principle that sedimentary beds are originally deposited in horizontal layers

Principle of superposition the principle that in a sequence of layered rocks that is not overturned or interrupted by faulting, the oldest will be at the bottom and the youngest at the top

Proglacial referring to the area in front of a glacier

Proton a sub-atomic particle with a mass of 1 and a charge of 1

Protoplanetary disk a rotating cloud of gas and dust surrounding a young star

Pumice a highly vesicular felsic volcanic rock (typically composed mostly of glass)

P-wave a seismic body wave that is characterized by deformation of the rock in the same direction that the wave is propagating (compressional vibration)

Pyroclastic volcanic material formed during an explosive eruption

Pyroclastic density current a body of hot pyroclastic rock and gases that is flowing rapidly down the flank of a volcano

Pyroxene a single chain silicate mineral

Quartz a silicate mineral with the formula sio2

Quartz sandstone a sandstone in which more than 90% of the grains are quartz

Quartzite a metamorphic rock formed from the contact or regional metamorphism of sandstone

Radial a pattern of streams radiating out from a central point, typically an isolated mountain

Radioactivity the natural transformation of unstable isotopes into new elements

Radiolaria microscopic (0.1 To 0.2 Mm) marine protozoa that produce silica shells

Rayleigh wave a surface seismic wave, with vertical motion

Recharge the transfer of surface water into the ground to become groundwater

Recharge area an area of an aquifer where recharge is predominant over discharge

Rectangular drainage a drainage pattern in which tributaries typically flow at right angles to each other and meet at right angles

Recumbent fold a fold that is overturned such that its limbs are close to horizontal

Redshift the increase in wavelength of light resulting from the fact that the source of the light is moving away from the observer

Reef a mound of carbonate formed in shallow tropical marine environments by corals, algae and a wide range of other organisms

Regional metamorphism metamorphism caused by burial of the parent rock to depths greater than 5 km (typically takes place beneath mountain ranges, and extends over areas of hundreds of km2)

Remnant magnetism magnetism of a body of rock that formed at the time the rock formed and is consistent with the magnetic field orientation that existed at that time and place

Reservoir rock rock into which petroleum has migrated and is now trapped

Residual soil soil formed by weathering of the underlying rock or sediment

Retrograde metamorphism metamorphism that takes place at a lower temperature than that at which the rock originally formed or was previously metamorphosed

Reverse fault a non-vertical fault along which the hanging wall (upper surface) has moved up relative to the footwall

Rhyolite a felsic volcanic rock

Ridge push the concept that at least part of the mechanism of plate motion is the push of oceanic lithosphere down from a ridge area

Rip current a strong flow of water outward from a beach

Ripple on a series of small parallel ridges formed within sediment that has accumulated in moving water or wind

Rip-rap angular rock fragments, typically boulder sized, used to armour slopes and shorelines against erosion

Roche moutonée a product of glaciation in which a bedrock protrusion is eroded into a streamlined shape that has a broken or jagged leading (down-ice) edge

Rock avalanche a rapid turbulent flow of broken bedrock fragments down a steep slope

Rock basin lake a lake situated in a rock basin carved at the upper end of an alpine glacier

Rock cycle the series of processes through which rocks are transformed from one type to another

Rock fall the near-vertical fall or bouncing of rock released from a steep slope

Rock slide the translational motion of an essentially intact body of rock down a slope (rock slides are typically slow, because once they start to move fast the rock body becomes fragmented and then flows as a rock avalanche)

Runoff flow of water down a slope, either across the ground surface, or within a series of channels

Rupture breaking of rock subject to stress, typically resulting in an earthquake

Rupture surface the area over which rock rupture takes place during an earthquake

Sackung an escarpment or trough at the top of a slow-moving rock slide (sackungen)

Salatation the bouncing of particles along a stream bottom or desert floor

Sand a mineral or rock fragment ranging in size from 1/16th to 2 mm

Sandstone a rock that is primarily comprised of sand-sized particles

Sandur an extensive region of sand and gravel deposited by streams flowing out of a glacier (same as outwash plain)

Saturated zone the part of an aquifer, or any body of rock, that is saturated with water

Schist a metamorphic rock with visible aligned mica crystals

Sea cave a shallow cave formed on a rocky shore by wave erosion

Sea cliff a coastal escarpment that is typically eroding inland as a result of wave action

Sea-floor spreading the formation of new oceanic crust by volcanism at a divergent plate boundary

Sector collapse the sudden collapse of a significant part of the flank of a volcano

Sedimentary rock rock that has formed by the lithification of sediments

Sediments unconsolidated particles of mineral or rock

Seismic pertaining to earthquakes

Seismic moment a measurement of an earthquake's energy based on longwave vibrations, or on the product of the fault area and displacement

Seismic reflection sounding measurement of the properties of sediments based on detection of sounds generated at surface and reflected from layers beneath the surface

Septae calcareous partitions between the successive living chambers in a cephalopod

Septic system a system constructed to facilitate the dispersion and detoxification of sewage (typically includes a septic tank and a drainage field)

Shaft a vertical opening at a mine

Shale a silt- and clay-rich rock that has evidence of layering

Shear force the component of the gravitational force in the direction parallel to a slope

Shear strength the strength of a body of rock or sediment that counteracts the shear force

Shear stress the stress placed on a body of rock or sediment adjacent to a fault

Sheeted dykes a series of near-vertical dykes formed in the vicinity of a spreading ridge when magma from depth flows into fractures formed by extensional forces

Sheet silicate a silicate mineral in which the silica tetrahedra are combined within sheets

Sheetwash overland flow of water, typically related to a heavy precipitation event

Shield a region of ancient (typically precambrian) crystalline rock (equivalent to a craton)

Shield volcano a low-profile volcano formed primarily from eruptions of low-viscosity mafic magma

Sial (sialic) referring to rock or magma in which silica and aluminum are the predominant components (generally equivalent to felsic)

Silica a form of the mineral quartz (sio2)

Silica tetrahedron a combination of 1 silicon atom and 4 oxygen atoms that form a tetrahedron

Silicate a mineral that includes silica tetrahedra

Silicon the 14th element

Silicone resin or caulking made from silicon-oxygen chains and various organic molecules

Sill an igneous intrusion that is parallel to existing layering in the country rock

Silt sedimentary particles ranging is size from 1/256th to 1/16th of a mm

Sima (simatic) referring to rock or magma in which silica, magnesium and iron are the predominant components (generally equivalent to mafic)

Skarn the contact metamorphism (and metasomatism) of limestone

Slab pull the concept that at least part of the mechanism of plate motion is the pull of oceanic lithosphere down into the mantle

Slate a fine-grained metamorphic rock that splits easily into sheets

Slaty cleavage the tendency for slate or phyllite to split into sheets (note that this is the only situation in this textbook where the term "cleavage" is applied to a rock as opposed to a mineral)

Slide the downward movement of rock or sediment on a slope as an intact mass

Slump a slide in which the nature of the motion is rotational (typically only develops in unconsolidated sediments)

Smectite a fine-grained sheet silicate mineral that can accept water molecules into interlayer spaces, resulting is swelling

Smelter a refinery at which minerals are processed to produce pure metals

Snowline in astronomy the radius around a star at which represents the boundary between gases (or liquids) and solids

Soil horizon a layer, within a well-developed soil, that is physically or chemically different from layers above or below

Solar system a star and the planets surrounding it

Solar wind a stream of ionized (charged) particles away from the sun

Solid solution the substitution of one element for another in a mineral (e.G., Iron can be substituted for magnesium in the mineral olivine)

Solifluction the flow of water saturated sediment or soil over a stronger and less permeable substrate

Source rock the sedimentary rock from which petroleum originates prior to its migration into a reservoir rock

Speleothem a solutionally-formed feature within a limestone cave (e.G., A stalactite)

Spit a sand or coarser deposit extending from shore out into open water

Spring the flow of groundwater onto the surface

Stack a prominent rocky island that is a remnant of the erosion of a headland

Stage the level of water in a stream

Stalactite a cone-shaped speleothem that is suspended from the roof of a cave

Stalagmite a cone-shaped speleothem that forms on the floor of a cave

Step-pool a characteristic of stream flow in which water flows from one pool to another, typically on a stream with a steep gradient

Stock an irregular pluton with n exposed area less than 100 km2

Stoping the fracturing and incorporation of fragments of country rock as a magma body moves upward through the crust

Strain the deformation of rock that is subjected to stress

Streak the mark left on a porcelain plate when a mineral sample is ground to a powder by being rubbed across the plate (typically considered to provide a more reliable depiction of the colour than the whole sample)

Stream any body of flowing water

Stress a force applied to a rock

Stress transfer the change in the pattern of stress on a region of rock as a result of an earthquake (typically stress is reduced in the area of a rupture zone, but is increased elsewhere in the vicinity)

Strike the compass direction of a horizontal line on a sloped surface (e.G., Bedding plane, fracture etc.)

Strike-slip fault a fault that is characterized by motion that is close to horizontal and parallel to the strike direction of the fault

Subaerial eruption a volcanic eruption that takes place on land

Subaqueous eruption a volcanic eruption that takes place under water

Subducted when part of a plate is forced beneath another plate along a subduction zone

Subduction zone the sloping region along which a tectonic plate descends into the mantle beneath another plate

Subglacial beneath a glacier

Sulphate a mineral in which the anion is so42-

Sulphide a mineral in which the anion is s2-

Supergroup a stratigraphically-continuous series of related groups

Superterrane a number of terranes that are contiguous

Supraglacial on the surface of a glacier

Surf zone the near-shore zone where waves are breaking into surf

Suture the line on the surface of a cephalopod that marks the boundary between a septum and the outer shell

Swash the upward motion of a wave on a beach (typically takes place at the same angle that the waves are approaching the shore)

S-wave a seismic body wave that is characterized by deformation of the rock transverse to the direction that the wave is propagating

Symmetrical a fold in which the limbs are at the same angle to the hinge

Syncline a downward fold where the beds are known not to be overturned

Synform a downward fold where it is not known if the beds are overturned

Tailings the fine-grained waste rock from a plant used to concentrate ore minerals

Talus slope a sloped deposit of angular rock fragments at the base of a rocky escarpment

Tarn a lake within a rock basin

Tectonic plate a region of the lithosphere that is considered to be moving across the surface of the earth as a single unit

Tectonic sea level change relative sea level change related to the vertical motion of a crustal block caused by tectonic processes

Tephra fragments of volcanic rock (including volcanic ash) ejected during an explosive eruption

Terminal moraine and end moraine that marks the farthest forward advance of a glacier

Terrane a block of crust that has geological features that are distinctive from neighbouring regions, and is assumed to have been moved from elsewhere by tectonic processes

Terrestrial planet a planet with a rocky mantle and crust and metallic core (e.G., Earth)

Terrigenous referring to sedimentary particles that originated on a continent

Test the shell-like hard parts (either silica or carbonate) of small organisms such as radiolarian and foraminifera

Thrust fault a low angle reverse fault

Till unsorted sediment transported and deposited by glacial ice

Tiltmeter a sensitive instrument used to monitor subtle changes in the tilt of the land, particularly in studies of active volcanoes

Tombolo a sand or coarser deposit connecting an island or rocky prominence to a larger body of land

Traction a force that contributes to the movement of particles situated on a stream bed or desert floor

Transform fault a boundary between two plates that are moving horizontally with respect to each other

Travertine a deposit of calcium carbonate that forms at springs, hot springs or within limestone caves

Trellis a drainage pattern in which tributaries typically flow parallel to one other but meet at right angles

Trigger an event, such as an earthquake or a heavy rainfall, that triggers the onset of a mass wasting event

Trough the lowest point of a wave

Truncated spur the steep end of a ridge or arête that has been eroded by a main-valley glacier

Tsunami a long-wavelength wave produced by the vertical motion of the floor of the ocean or a large lake, typically related either to an earthquake or a sub-marine mass wasting event

Tufa a form of travertine that is especially porous as it forms around existing vegetative material.

Tuya a flat-topped volcanic hill or mountain that formed when an eruption took place beneath a glacier and the melting led to the formation of a lake that then resulted in the wave-erosion of the top of the volcano

Unconfined aquifer an aquifer that is not overlain by a confining layer

Unconformity a geological boundary at the base of a sedimentary layer

Unconformity-type uranium deposit a uranium deposit that has formed at a nonconformity between sandstone and older rock

Uncompressed density the density of planetary material that it would have it was not compressed by the planets gravitational force

Underground storage tank (ust) an underground tank for storing liquids, most commonly for liquid fuel

Unsaturated zone the rock or sediment above the water table

U-shaped valley a relatively straight valley with a flat bottom and steep sides that has been carved by a valley glacier

Valley glacier a glacier formed in a mountainous region and confined to a valley (same as alpine glacier)

Varve a recognizable layer within sediments that represents a single year of deposition

Vesicular an igneous texture characterized by holes left by gas bubbles

Volcanic glass magma that has cooled within minutes, not allowing time for the formation of crystals

Volcanic-hosted massive sulphide a mineral deposit hosted by volcanic rocks and including zones where most of the rock is made up of sulphide minerals (including ore minerals and pyrite)

Wacke a sandstone with more than 15% clay and silt

Water table the upper surface of the saturated zone in an unconfined aquifer

Wave base the depth of water that is affected by the sub-surface orbital motion of wave action (approximately one-half of the wavelength)

Wave-cut platform a nearly-horizontal bench of rock eroded by waves within the surf zone (equivalent to wave-cut terrace)

Wavelength the distance between the crests of two waves

Weathering a range of processes taking place in the surface environment, through which solid rock is transformed into sediment and ions in solution

Western canada sedimentary basin a large basin in the western interior of canada, east of the rocky mountains, extending from the northern united states to the northwest territories

Wisconsin glaciation the most recent advance of the pleistocene glaciations, extending from 85 to 11 ka

Xenolith a fragment of country incorporated into igneous rock, commonly as a result of stoping

Youthful stream a stream that is actively down-cutting its valley in an area that has recently been uplifted

Zone of ablation the part of a glacier, below the equilibrium line, where there is net loss of ice mass due to melting and calving

Zone of accumulation the part of a glacier, above the equilibrium line, where there is net gain of ice mass because not all of the snow that falls each winter is able to melt during the following summer

INDEX

www.ingramcontent.com/pod-product-compliance
Lightning Source LLC
LaVergne TN
LVHW062213200926
848484LV00006B/1381